Springer Advanced Texts in Chemistry

Springer
New York
Berlin
Heidelberg
Barcelona
Budapest
Hong Kong
London
Milan
Paris
Santa Clara
Singapore
Tokyo

Springer Advanced Texts in Chemistry

Series Editor: Charles R. Cantor

Principles of Protein Structure
G.E. Schulz and R.H. Schirmer

Bioorganic Chemistry: A Chemical Approach to Enzyme Action
(Third Edition)
H. Dugas

Protein Purification: Principles and Practice (Third Edition)
R.K. Scopes

Principles of Nucleic Acid Structure
W. Saenger

Biomembranes: Molecular Structure and Function
R.B. Gennis

Basic Principles and Techniques of Molecular Quantum Mechanics
R.E. Christoffersen

Energy Transduction in Biological Membranes: A Textbook of Bioenergetics
W.A. Cramer and D.B. Knaff

Principles of Protein X-ray Crystallography
J. Drenth

Robert K. Scopes

Protein Purification

Principles and Practice

Third Edition

With 165 Figures

Springer

Robert K. Scopes
Department of Biochemistry and the Centre for Protein and Enzyme Technology
La Trobe University
Bundoora, Victoria 3083
Australia

Series Editor:
Charles R. Cantor
Boston University
Center for Advanced Biotechnology
Boston, MA 02215, USA

Cover Illustration: schematic representation of protein molecules undergoing purification. Art provided by Dr. David Goodsell, University of California at Los Angeles.

Library of Congress Cataloging-in-Publication Data
Scopes, Robert K.
   Protein purification : principles and practice / Robert K. Scopes.
 — 3rd ed.
     p.   cm. — (Springer advanced texts in chemistry)
   Includes bibliographical references and index.
   ISBN 0-387-94072-3 (alk. paper). — ISBN 3-540-94072-3 (alk.
paper)
     1. Proteins—Purification. I. Title. II. Series.
   [DNLM: 1. Proteins—isolation & purification. QU 25 S422p 1993]
QP551.S4257   1993
547.7'58—dc20
DNLM/DLC
for Library of Congress   93-31923

Printed on acid-free paper.

Acquiring editor: Robert C. Garber

Production coordinated by Chernow Editorial Services, Inc., and managed by Theresa Kornak; manufacturing supervised by Jacqui Ashri.
Typeset by Best-set Typesetter Ltd., Hong Kong.
Printed and bound by R.R. Donnelley & Sons, Inc., Harrisonburg, VA.
Printed in the United States of America.

9 8 7 6 5 4 3

ISBN 0-387-94072-3 Springer-Verlag New York Berlin Heidelberg
ISBN 3-540-94072-3 Springer-Verlag Berlin Heidelberg New York   SPIN 10533259

Series Preface

New textbooks at all levels of chemistry appear with great regularity. Some fields such as basic biochemistry, organic reaction mechanisms, and chemical thermodynamics are well represented by many excellent texts, and new or revised editions are published sufficiently often to keep up with progress in research. However, some areas of chemistry, especially many of those taught at the graduate level, suffer from a real lack of up-to-date textbooks. The most serious needs occur in fields that are rapidly changing. Textbooks in these subjects usually have to be written by scientists actually involved in the research that is advancing the field. It is not often easy to persuade such individuals to set time aside to help spread the knowledge they have accumulated. Our goal, in this series, is to pinpoint areas of chemistry where recent progress has outpaced what is covered in any available textbooks, and then seek out and persuade experts in these fields to produce relatively concise but instructive introductions to their fields. These should serve the needs of one semester or one quarter graduate courses in chemistry and biochemistry. In some cases the availability of texts in active research areas should help stimulate the creation of new courses.

Charles R. Cantor

Preface to the Third Edition

During the 12 years since the first edition of this book was prepared, there have been rapid developments in protein purification. One of the largest changes has been in the nature of the proteins being purified, and the requirements put on the end product. In 1980 most protein purification was laboratory based, usually concerned with enzymes, and aimed at getting at least a few milligrams of pure protein for characterization of bioactivity and structure. A decade later we can say that there is far more interest in nonenzymic proteins. There is the rapid development of commercial production of proteins on a large scale; on the other hand, much laboratory isolation is done with the aim of just getting sufficient pure product to do some amino acid sequencing, which in turn leads to the isolation of the gene. In the former case the end product may be measured in grams or kilograms; in the latter case micrograms can be sufficient. For both purposes, the development of new equipment and materials has been the main advance. Although there are some new methods, for the most part the principles of protein purification remain the same whether at microgram, milligram, gram, or kilogram scales.

In preparing this new edition, decisions had to be made on whether to completely rewrite sections, or simply to update where required. In some cases there have been few changes; for instance, there has been little to update in the "precipitation methods" chapter. In other cases, I have reorganized chapters as well as substantially rewriting them to reflect the increased or decreased importance of particular methodologies, while keeping the overall length about the same.

Probably the greatest change, still occurring rapidly, is the move toward isolation of recombinant proteins rather than isolation from the natural raw material. Whereas this usually requires an initial purification from the raw material, it is becoming increasingly common to isolate from an expressed gene a protein that has never been purified from its original source. With the exploding information on gene structure, assisted by the Human Genome Project and parallel projects on the genomes of yeasts, bacteria, and other organisms, there are many genes being described that can be expressed without going through a protein isolation stage first, including many which could be expressed without any knowledge of what the gene product is. Often, purification from recombinant expression is relatively easy, since in many cases the expression level can be so high that the foreign product dominates the host proteins, sometimes making up 50% or more of the total protein present. In such circumstances, purification to the "95%+" stage can be simple, and does not necessarily require sophisticated techniques or equipment. But if the product is to be used therapeutically, much higher degrees of purity are required, together with strict overseeing of each step in the procedure. In general, we can say that protein purification from natural sources involves, typically, a several hundred-fold purification to reach 95%+ purity, whereas a recombinant protein may need less than twenty-fold purification, but there may be a requirement for 99.9%+ purity. So, techniques for protein purification on a large scale or even as a routine requirement are becoming matters of developing a suitable recombinant expression system to maximize the level of the desired protein in the crude extract, and "polishing" steps, taking the almost-pure protein to a highly homogeneous product meeting tight specifications.

I have dealt, somewhat briefly, with the specifics of purification of recombinant proteins, and with techniques used for both large and very small scale isolations. But the book remains principally about the purification of proteins from natural sources, on a laboratory scale, and stressing the many methods that can be used without necessarily having to resort to complicated or expensive equipment. It should be remembered that there are many alternative ways for purifying a particular protein, and if a procedure you are trying to follow is not working well, do not be afraid to try something completely different.

I acknowledge once again the help that many colleagues have given me over the years, not only in suggesting improvements to the book, but also in carrying out experiments in protein purification that have led to new insights into protein behavior in real life situations.

Bundoora, Victoria   Robert K. Scopes
Australia

Preface to the Second Edition

The original plan for the first edition of this book was to title it *Enzyme Purification: Principles and Practice*. However, because all enzymes are proteins, and most methods applied also to nonenzymic proteins, I somewhat reluctantly agreed to the more general title. As a result, it has become increasingly clear during the past couple of years that the first edition was deficient in describing techniques that are in the main used for nonenzymic proteins. Indeed, there seems to be developing a separation between studies on enzymes and on nonenzymic proteins. With the advent of the new biotechnologies, a great amount of interest has been created in proteins such as hormones, hormone receptors, antigens, viral proteins, and others that for the most part are not enzymes and have structural properties that allow them to be treated quite differently. Most are extracellular, and as a result have a sturdier resistance to environmental factors. Many are very small proteins, they may have no subtle bioactivity to lose, and it is often not necessary to isolate them in a native, active state. Moreover, minute amounts, even a few micrograms, may be all that is needed to get a partial amino acid sequence to lead to gene isolation and expression in a host organism. Enzymes, on the other hand, are generally less sturdy, larger molecules, whose native environment is usually within the protection of the cell, and they cannot always be handled in the same way as small, extracellular proteins. Also, the enzymologist likes to have at least several milligrams of pure product to carry out kinetic and structural studies.

Techniques less suited for enzymes include reverse-phase high-performance liquid chromatography (RP-HPLC), sodium dodecyl sulfate-polyacrylamide gel electrophoresis, and extremes of pH during extraction. In this second edition I have included a more extensive discussion on

HPLC, but caution against automatic selection of this technique as being generally suitable for all enzyme isolations. With large-scale enzyme production for commercial application becoming more widespread, the costs of the methodology (and HPLC certainly has a high capital cost factor) become important in assessing the usefulness of any method.

I have also taken the opportunity to give a detailed, and therefore disproportionate, description of techniques developed in our laboratory and in others for operating dye-ligand column chromatography. This is a technique that is complementary to HPLC, being mainly suited to large-scale work, having low capital costs, and in the crude sense a low resolution. But careful choice of dyes and operating conditions can make this a very successful procedure, which can revolutionize large-scale protein (and especially enzyme) purification as much as HPLC has revolutionized small-scale operations.

Once again I acknowledge the advice and help of numerous colleagues throughout the world who have made suggestions for improvements on the first edition.

Bundoora, Victoria   Robert K. Scopes
Australia

Preface to the First Edition

This book marks twenty years of research involving enzyme purifications, initiated partly because at the time of my higher degree (I needed some creatine kinase for ATP regeneration) we could not afford to buy commercial purified enzymes; there were hardly any available at that time anyway. The fascination of isolating a reasonably pure enzyme from a complex natural soup of proteins remains with me; as a challenge and an academic exercise I still spend much time on enzyme purification even when I have no real use for the final product! In those days everything was empirical: try an ammonium sulfate cut; try organic solvents; will it adsorb on calcium phosphate gel? There was not much else available. Now there is such a vast range of methods, of materials, and of approaches to the problem of enzyme purification that the difficulties lie not in the subtleties of manipulating a few available techniques, but in trying to decide which of the plethora of possibilities are most suitable.

It is the purpose of this book to guide the newcomer through the range of protein fractionation methods, while pointing out the advantages and disadvantages of each, so that a choice can be made to suit the problem at hand. Thus, traditional procedures such as salt fractionation are presented, along with the many modern developments in a affinity chromatography and related techniques, which have been so successful in many cases, but have not proved to be the answer to all problems. During the two years of preparation of the manuscript, many new techniques and, equally important, new commercial products have been reported; preliminary drafts were updated as a new product came on the market, but by the time this is published there will undoubtedly be still more. In some cases the new product will be so superior that it will quickly supersede the older variety. Increasingly often, however, the question of cost-effectiveness is raised: this new system may be the best, but will it ruin our budget? For this reason I feel that the older, cheaper products (and more classical techniques) will be with us for many years to come.

A few sentences on what this book does, and does not attempt to do: It is not a comprehensive presentation of all the methods ever used in enzyme purification, nor does it give detailed examples or precise instructions; if it did, it would occupy far more shelf space and would only duplicate excellent multi-volume treatises already available—in particular the series *Methods in Enzymology and Laboratory Techniques in Biochemistry and Molecular Biology*, and *The Proteins* (3rd ed., Vol. 1, eds. Newarth and Hill). Instead, it gives a brief account of the main procedures available, with some simple (even simplistic) theoretical and thermodynamic explanations of the events occurring. A basic background in biochemistry and protein chemistry is assumed: I expect the reader already to have on his shelves textbooks describing protein structure, simple enzyme kinetics, and thermodynamics. It is aimed to assist all students and researchers involved in the process of isolating an enzyme, from whatever source, whether it be a new project or simply following a published procedure.

In most places in this book the words "enzyme" and "protein" can be interchanged. All enzymes are proteins, but the reverse is not so. For the most part, I have adopted the system of using "enzyme" when referring to the particular protein being purified (even though it may not be an enzyme), and "protein" when referring either to the complete mixture at hand, or else to the proteins *other than* the particular one being purified.

There are many audiences to satisfy in a book such as this. Three, not necessarily exclusive, categories are: (i) those purifying a protein that has never successfully been purified before from any source; (ii) those purifying a protein from a new source, there being an adequate method available using some other source; and (iii) those who are simply following a recipe in an attempt to obtain a pure protein equivalent to that reported previously. As to the first category: it is becoming less common now for anyone to be purifying an enzyme for the first time, simply because there are fewer enzymes being discovered and, one presumes, few to be discovered. But in the second category, it is more common than it used to be for people to be purifying enzymes from new sources: no longer do biochemists restrict themselves to *E. coli*, yeast, rat liver, rabbit muscle, pig heart and spinach! Each a small shift in evolutionary terms (e.g., from pig to bovine heart) results in different behaviors not only of the enzyme being isolated but, equally importantly, of the other proteins present. Many standard methods are so critically dependent on precise conditions that a shift in source material may result in complete failure because of a minor variation in protein properties. Nevertheless, it is rare for the one enzyme to vary much in molecular weight even over large evolutionary distances; within phyla other properties such as ion exchange behaviors (dependent on isoelectric points) are not often widely at variance, and solubility behavior may well be similar.

The third category contains those who wish only to duplicate a preparation that someone else has reported. If sufficient important details have been published, and the raw material and laboratory operating conditions are essentially identical, then there is no need for further help; unfortunately these criteria are not always met. Without experience and general knowledge of the principles, most beginners are unable to reproduce a method the first time—or even at all. Also, if it does not work they are afraid to depart one iota from the written word in case things get worse. For these people my advice will be to make use of the information given but, if things are not going well, not to be afraid to change the conditions (they may have been reported incorrectly anyway, and a misprint not spotted in proof-reading could waste months); also, not to worry if they do not have a PX28 rotor for an SS26 model D centrifuge, if all that is needed is the pellet. If things are still going wrong, then it is time to introduce greater variations, using new materials or methods that have been developed since the original publication. Because enzyme purification is essentially a methodology-oriented practice, it is usually worth manipulating conditions repeatedly until the ideal combinations are found. This can be time-consuming, but is worth the effort in the end, as you can be sure of reproducibility if you are confident of the limitations on each step.

Finally I wish to acknowledge the contributions over the years of many students and the staff at La Trobe University. I should also like to thank Prof. K. Mosbach, Dr. C.R. Lowe, Dr. I.P. Trayer, Dr. C.-Y. Lee and Dr. F. von der Haar for making available reprints and preprints of their work on enzyme purification. Financial support from the Australian Research Grants Committee is gratefully acknowledged.

Bundoora, Victoria   Robert K. Scopes
Australia

Contents

Chapter 6
Separation by Adsorption II: Ion Exchangers and Nonspecific Adsorbents

Chapter 7
Separation by Adsorption—Affinity Techniques

Chapter 1

The Protein Purification Laboratory

1.1 Apparatus, Special Materials, and Reagents

There are many complex, sophisticated pieces of equipment designed for protein separations and, as in every other walk of life, these are becoming more automated for convenience and simplicity of operation. Yet, as the apparatus becomes more enclosed in black boxes, controlled by microprocessors, we become further removed from the realities of what is happening, and in some ways have less control over what we want to do. Some of the more complex equipment is designed for repeating routine operations reliably, and as such may not be always the most appropriate for developing new methods. But apart from these considerations, there is the one of cost effectiveness; duplication of cheap equipment is often more useful than getting one of the most expensive on the market; money spent on the simplest things, such as plenty of pipettes, test tubes, and beakers, can be a better investment than purchasing an esoteric apparatus for carrying out one particular type of protein separation process that may not be used often (and may break down).

Having said that, there remains a baseline of minimal apparatus that any protein purification laboratory must have available. Measures of enzyme activity will usually be based on either spectrophotometry or radiochemistry. Although one can do without a scintillation counter if all assays can be done spectrophotometrically, a spectrophotometer is indispensable since it is required for protein concentration measurements (Section 3.1). A spectrophotometer with UV lamp, preferably with an attached recorder for timed reactions, is regarded as a prime requirement.

1

Centrifuging can rarely be avoided. On an industrial scale, filter systems are commonly used for separating precipitates, but in the laboratory a centrifuge is more convenient. The principles and practices of centrifuging are described below (Section 1.2). For most purposes the standard workhorse centrifuge is a refrigerated instrument capable of maintaining temperatures at or somewhat below 0°C, while centrifuging ability, expressed as relative g force × capacity in liters, should be of the order of 10,000–15,000. Many other types of centrifuge can be useful, from the benchtop small-scale machine to preparative ultracentrifuges. But these are not generally necessary and would rarely get as much use in enzyme purification work as the basic machine described above.

Column chromatography is so generally employed that equipment for this is essential. Although column work can be carried out with a home-made glass tube or syringe (Section 1.3), with manual fraction-collecting and subsequent spectrophotometric measurements on each tube, much more work can be done more easily with the basic automated setup, which costs about the same as the spectrophotometer and less than the centrifuge. This would consist of a variety of sizes of columns, a fraction collector with UV monitor, a peristaltic pump, and a magnetic stirrer (for gradient formation). A wide variety of column sizes and shapes should be available so that the optimum amount of column packing material can be used for the sample available; columns with adjustable plungers are advisable. A range of sizes suitable for "desalting" by gel filtration is described in Section 1.4. For adsorption chromatography still other sizes may be required. A given amount of money may be better spent on two or more columns of simple design rather than one made to optimum specifications.

Homogenizers for disrupting cells are needed and are described in Section 2.2. Routine laboratory equipment such as balance, pH meter, stirrers, and ice machines are as necessary in protein purification as in other biochemical methodology. Volumetric measurements can be made with graduated cylinders (rarely is the accuracy of a volumetric flask necessary), pipettes, and, for small amounts, microliter syringes. Enzyme assays frequently demand the addition of a very small sample of the test solution, perhaps less than $1 \mu l$. Modern syringes are capable of delivering such volumes with sufficient accuracy. A range of syringes from 0.5 up to $250 \mu l$ is highly desirable. In addition, continuously adjustable "Eppendorf"-type pipettes, using disposable tips, are useful for volume ranges from $1 \mu l$ up to $5 \, ml$.

Analysis of protein mixtures by electrophoresis is carried out by gel electrophoresis or isoelectric focusing, most often on polyacrylamide gel which may or may not contain dodecyl sulfate as denaturant. Methodology is described in Chapter 11. Many excellent equipment designs are now available commercially; starting from scratch, a newcomer would be advised to obtain a thin-slab apparatus. There can be few biochemical laboratories in the world that do not have any equipment for analytical gel electrophoresis.

There are a number of special reagents that the protein purifier requires. These are mainly packing materials for column chromatography and chemicals that are frequently used. Ion exchangers (Section 6.1), gel filtration materials (Section 8.1), and affinity adsorbents (Section 7.1) should be on hand. More mundane, but vitally important, are such things as dialysis tubing (a range of sizes), a range of buffers (Section 12.3), bulk ammonium sulfate of adequate purity (Section 4.3), thiol compounds such as 2-mercaptoethanol and dithiothreitol, EDTA (ethylenediaminetetraacetic acid), sodium dodecyl sulfate, protamine sulfate, phenylmethylsulfonyl fluoride, Folin-Ciocalteau reagent, Coomassie blue, etc., the uses for which will be described in various places throughout this book.

1.2 Separation of Precipitates and Particulate Material

Filtration

The development of efficient refrigerated centrifuges provided a cleaner, more efficient method than filtration for the separation of precipitated proteins and other matter, especially on a small scale. However, there are still occasions when filtration is preferable. On a very large scale, handling of large volumes by centrifugation will require specialized equipment, and the process will take a long time, perhaps comparable with large-scale filtration (see Section 10.2). On the other hand, the size and softness of most biological precipitates lead to rapid clogging of filtration materials so that after an initial burst of clear filtrate, the flow reduces to a trickle, even under suction, and the filtration material must be replaced. On an industrial scale it is possible to use special rigid filters with automatic scraping equipment for removing the slime-like precipitate as it collects, but such systems are not normally available in a laboratory.

Filtration can be greatly improved using filter aids such as Celite, a diatomaceous earth consisting mainly of SiO_2 (see Figure 1.1), as used for swimming pools. The best results are obtained by mixing the Celite with the sample to be filtered, then pouring the suspension onto a Büchner funnel under slight suction. By creating a very large surface to trap the gelatinous precipitate, the filter aid allows much more filtrate to be obtained before eventually clogging. The process is mainly used for removing small amounts of unwanted particulate material; it cannot deal successfully with large quantities of precipitate.

For removing obviously lumpy precipitates, hair, dirt, insoluble salt residues, etc., filtration through coarse filter paper can be carried out;

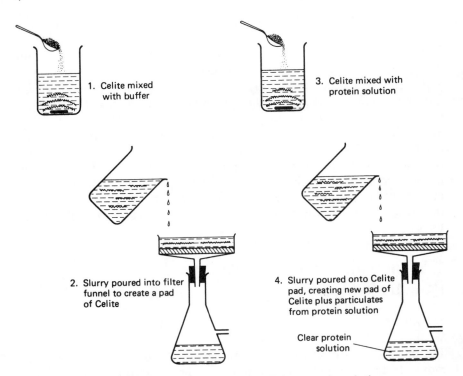

Figure 1.1. Use of Celite as a filter aid for clarifying protein solutions.

this is particularly appropriate immediately prior to a column procedure where the precipitate would otherwise be "filtered" on the packing material. To remove small amounts of fine particulate material, forcing small samples through a disposable filter disc with a syringe is beneficial. Centrifugation is usually equally satisfactory.

Removal of particulates on a large scale can be achieved by ultrafiltration (applying pressure to force liquid through a membrane while retaining particles too large to pass through it). A number of different systems are available, but these are only really suited to circumstances when centrifugation is not appropriate, e.g., when dealing with very large volumes (Section 10.2).

Centrifugation

Centrifugation relies on the sedimentation of particles in an increased gravitational field. The forces acting on a particle are illustrated in Figure 1.2. For rigid spherical particles, the time required for a particle to sediment in a given medium from the meniscus to the bottom of the tube is given by:

$$t = \frac{9}{2} \cdot \frac{\eta}{\omega^2 \cdot r^2 (\rho - \rho_0)} \cdot \ln \frac{x_b}{x_t} \tag{1.1}$$

where x_t = radial distance of meniscus
x_b = radial distance of bottom of tube

and the other symbols are defined in the caption to Figure 1.2.

Thus, the time required is proportional to the viscosity, but inversely proportional to the density difference and to the square of both the particle radius and angular velocity. The other factors are geometrical considerations for a particular rotor. From this we see that, other things being equal, a particle half the radius requires twice the angular velocity to sediment in the same time. More critical is the density difference; this is large for aqueous solutions containing little solute, but when there is a salt concentration such as in salting out (Section 4.3), this factor can be quite small. The density of anhydrous protein is about 1.34, and an aggregated protein particle contains about 50% protein and 50% trapped solution. Thus, ρ for an aggregate in water would be about $\frac{1}{2}(1.34 + 1.00)$ = 1.17, so $\rho - \rho_0 = 0.17$, whereas in 80% saturated ammonium sulfate it would be close to 1.27, the solution ρ_0 being 1.20 and $\rho - \rho_0 = 0.07$. On the other hand, 33% acetone has a density of 0.93, for which $\rho - \rho_0$ works out as 0.20, even greater than for pure water. This illustrates one advantage of using organic solvent precipitation (Section 4.4), especially

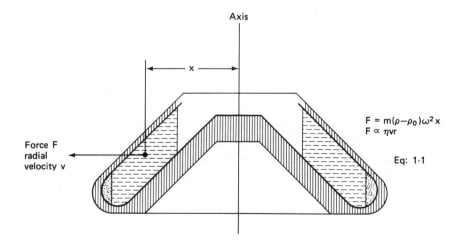

Figure 1.2. Forces acting on a particle in a centrifuge rotor. Mass of particle = m, density of particle = ρ, density of solution = ρ_0, angular velocity of rotor = ω, viscosity of solution = η, and Stokes's radius of particle = r.

for initially dense, viscous solutions, i.e., the precipitates sediment more rapidly.

The optimum objective of centrifugation in enzyme purification is to obtain a tightly packed precipitate and a clear supernatant; exceeding the minimum centrifugation time for this to occur is no disadvantage. With precipitates formed by salting out, isoelectric precipitation, organic solvents, and most other precipitants described in Chapter 4, centrifugation at about 15,000 g for 10 min, or 5,000 g for 30 min is usually sufficient. Note that on a small scale, 5,000 g for 30 min might be adequate in a centrifuge tube with sedimentation distance of 5 cm, but on a large scale where the distance might be as much as 20 cm, 30 min may not be enough. The g values above are average figures, since at a given rotation speed the g value increases as the particle sediments. However, if one quotes not g values but rotation speed ω, Eq. (1.1) indicates that it is not the absolute distance of sedimentation that determines the time taken, but the logarithm of the ratio between x_b and x_t; this dimensional factor does not vary much between different rotors. Thus, 5,000 rpm for 30 min would produce a similar result in most centrifuges.

Particle size is clearly critical in determining the amount of centrifuging required. Any precipitation process will result in a range of particle sizes, but the numbers quoted above will generally be sufficient to sediment all aggregated protein that forms an observable cloudiness; smaller protein aggregates that are not sedimented are also not likely to be visible in the solution. But many cellular organelles, and fractions derived from them, are not by any means pure protein. They can contain much lipid material, which drastically reduces the term $(\rho - \rho_0)$ in Eq. 1.1. In fact, this term may even become negative, in which case the particles move to the top of the tube as a floating layer.

Crude homogenates containing organelles or fragments can be fractionated by differential centrifugation processes in which the value of $\omega^2 t$ is increased at each step, causing particles with successively smaller $r^2(\rho - \rho_0)$ values to be sedimented. Differential centrifugation carried out in this way is a first step in purifying organelles from a variety of tissues if the protein required is present not in the soluble fraction but in a particulate form.

Modern centrifuges are sturdy pieces of equipment, designed to survive a certain amount of mishandling, and to switch off automatically if things go very wrong. The main thing that can be wrong is an imbalance in the rotor. This may be caused by a tube cracking during the run, or by a misbalance of the tubes in the first place. Obviously, tubes must be placed opposite each other or, in the case of heads with 6 spaces (or a greater multiple of three), three equally balanced tubes must be arranged evenly (see Figure 1.3).

Although balancing volume by eye is adequate for smaller tubes, volumes larger than 200 ml should be weighed to avoid imbalance. How-

Figure 1.3. Arrangements of 2, 3, 4, and 6 tubes in a 6-place centrifuge rotor. In the cases of 4 and 6, only the tubes opposite each other need to be identically balanced (see shading).

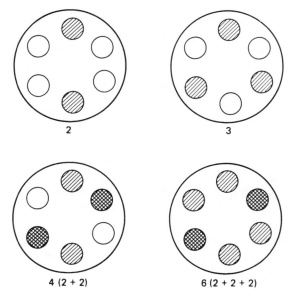

ever, this refers to the situation when all tubes contain the same liquid. To balance a single tube containing a sample in, say, 80% saturated ammonium sulfate (density about 1.20) with water, one could allow for the extra weight by increasing the water volume (Figure 1.4). But this is not strictly correct since (1) it is the inertia, not the mass of the liquid that should be equal, and (2) as particles sediment in the sample tube, the inertia increases. It is better to use two tubes less than half full (provided that the tubes are rigid enough to resist the g forces) or to use smaller tubes in adaptors or in a different rotor. Centrifuge tubes made

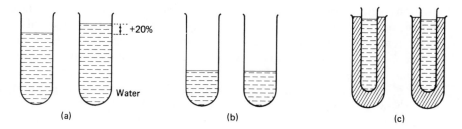

Figure 1.4. Alternative ways of balancing a dense liquid in pairs of centrifuge tubes: (a) On a small scale a rough balance is obtained with extra water to equalize masses. (b) Partial filling of both tubes with the dense liquid is preferable to (a), provided that the tubes are sufficiently rigid to avoid collapse when not filled. (c) The ideal situation is to have two tubes filled with the dense liquid, using adaptors to fit the rotor head, if necessary. However, note that centrifugation will be faster in (b) than in (c), as particles have less far to travel.

from polypropylene or polycarbonate are strong enough for most purposes. Glass centrifuge tubes may be desirable on occasion, but conventional glass (Pyrex) centrifuge tubes withstand only $3,000-4,000\,g$; special toughened glass (Corex) tubes are available for forces up $15,000\,g$.

Finally, here are a few words on the care of centrifuges. Although biochemists rarely use strongly corrosive acids or alkalis, it is not always appreciated how corrosive mere salt can be. Centrifuge rotors cast from aluminum alloys corrode badly if salts, such as ammonium sulfate solution, are in contact with the metal for a period. Any spillage should be immediately rinsed away; it is best to rinse and clean a centrifuge rotor after every use, or at least every day. Contamination of the outside of centrifuge tubes with salt solutions (they may have been immersed in an ice–salt bath for cooling) can result in corroded pits inside the rotor if it is not washed frequently. It is also worth noting that constant use of the machine at top speed will wear out both rotor and motor more quickly; for many manipulations in protein purification, half-speed is quite adequate.

1.3   Principles of Column Chromatography

Much of the detailed discussion of operating ion exchange columns, gel filtration, etc. will be described in Chapters 5–7. This short introduction gives the fundamental principles and describes the apparatus used in column chromatographic procedures.

Chromatography involves a retardation of solutes with respect to the solvent front progressing through the material. The name literally means "color drawing" and was used to describe the separation of natural pigments on filter papers by differential retardation (Figure 1.5). This same principle is still widely used. It can be called one-dimensional chromatography since there is no substantial thickness to the paper. A second dimension is introduced by including a column of matrix material—equivalent to layering together large numbers of strips of paper as in Figure 1.5c. The starting sample is now a disc or cylinder, rather than a thin line. As the solvent moves through the column, solutes present in the original sample can do one of three things. Completely partitioned into the solvent they will run with the solvent front ($R_f = 1$) and be washed out quickly. Totally adsorbed to the matrix ($R_f = 0$) they will remain in their starting position. The usefulness of chromatography lies in the third possibility ($0 < R_f < 1$), where partial adsorption retards the solute, but eventually it is eluted. Two solutes with closely similar retardation constants (R_f) would be eluted with midpoints at different times; if the column is long enough and longitudinal diffusion is not too great, a complete separation can be achieved. The separation ability depends partly on the number of effective adsorptive steps in the process. Because the sample

Figure 1.5. Principles of chromatography. In (a) the sample applied at the center of a filter paper disc spreads radially as more buffer is added, and the components separate in rings according to their R_f value. In (b) one-dimensional paper chromatography results in separation of components in bands. (c) Illustration of the extra dimension acquired in column chromatography.

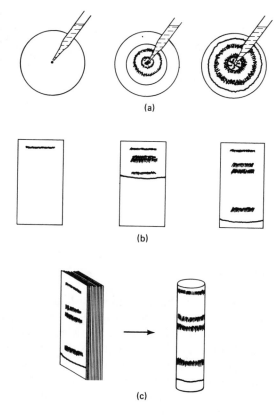

must have a finite depth, the number of processing steps, called the effective plate number, depends on the length of the column relative to the depth of the sample, as well as on the fineness of the adsorbent, plus consideration of diffusion. These factors are described in detail in Chapter 5. Since diffusion will spread any sample below a certain size to much the same amount, the width of the eluted fraction is the important feature, being dependent on both the depth of sample originally and the diffusion during passage down the column. As we shall see later, column chromatography usually involves changing buffer conditions, either by a stepwise instantaneous change or by an applied gradient moving from one condition to another; in these cases the concept of plate number is less relevant.

The wide variety of adsorbents suitable for proteins has made column chromatography the most important basic procedure for protein separation. Some adsorbents have very high capacities for certain proteins ($>100\,\mathrm{mg\,cm^{-3}}$), whereas others may have capacities 200-fold less than this; consequently, the *size* of column required may vary substantially. For an ideal system the *shape* should not matter greatly, since the ratio of

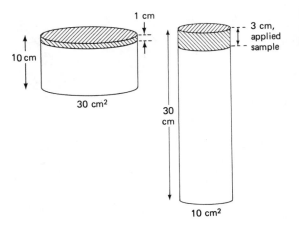

Figure 1.6. Two columns of identical volume but differing proportions.

length to sample depth remains the same if the volumes of adsorbent and sample are the same (Figure 1.6). But particle size is also relevant; sharper separations are obtained if more adsorbent particles are "passed" by the protein bands—giving a higher plate number—so the longer column has an advantage for a given sized particle. On the other hand a fatter column can use finer particles, since the slower flow rate per square centimeter that is usually necessary with fine, closely packed adsorbent is compensated for by the large cross section. In practice, uneven flow and non-ideal operation mean that a longer column is likely to give a better separation (see Figure 8.9). But it will take proportionately longer to run, which could be disadvantageous. Columns with capacities from 1 cm^2 to approximately 2 liters should be available in the laboratory, with the possibility of obtaining any given capacity ranging from a squat form where diameter ≈ height, to a long form whose diameter ≈ 1/30 height; there are uses for all shapes and sizes.

The best columns have carefully designed surfaces for the top and bottom of the packing material so that even flow down the cylinder of material is not disturbed at either end. An efficient porous net or disc at the bottom allows liquid to flow through to the collecting space—which should be as small a volume as possible to minimize mixing. Similarly, on the application end the sample and subsequent eluting buffers must be distributed evenly over the surface of the material (Figure 1.7).

In a teaching laboratory with dozens of students it is not economically possible to give each student such a column. Simple cheap systems include (for small scale) a Pasteur pipette with a plug of glass wool in its constriction (Figure 1.8a); a syringe either with or without the plunger, with a sintered plastic disc placed at the bottom (Figure 1.8b); or a simple glass tube with a rubber stopper, and some filter material such as a sintered disc or layers of filter cloth to assist in the smooth collection of eluant (Figure 1.8c). Uneven flow onto the surface of the packing

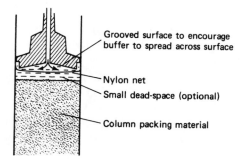

Grooved surface to encourage
buffer to spread across surface

Nylon net
Small dead-space (optional)

Column packing material

Figure 1.7. Typical design of applicator for a chromatography column.

material can create very uneven patterns of flow through the column; a small layer of liquid above the material can help to even out irregularities, provided that the adsorbent is a material that packs firmly and is not disturbed by buffer turbulence above it. Some examples of causes of irregular flow are illustrated in Figure 1.9. As many protein solutions are colorless, it is often not appreciated how irregularly a column might be working.

A second cause of irregularity is due to uneven packing of the column. The column material is virtually always of a range of particle sizes; this means that undisturbed settling will result in the largest particles falling to the bottom, and the top part of the column consisting of the smallest. This in itself may not matter too much except for reproducibility considerations. On the other hand, disturbed flow during settling may set up irregularities that result in a column with patchy particle distribution and patchy packing, causing uneven flow characteristics. Ideally, a column

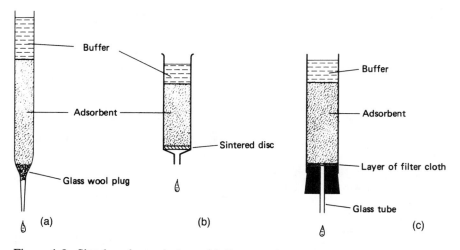

Figure 1.8. Simple column designs: (a) Pasteur pipette, (b) a syringe, (c) a glass tube, all adapted as columns for chromatography.

should be poured in one step with continuous stirring of the slurry, as illustrated in Figure 1.10. Short columns can be conveniently poured using a thick slurry, not more than 1 vol of buffer to 1 vol settled packing material, allowing it to settle rapidly with the exit tube open. After pouring the column, several volumes of buffer should be allowed to pass through to ensure equilibration; this is particularly important with ion exchangers, where the slurry pH should be adjusted to the correct value before pouring. It is best if the column can be packed at a flow rate *greater* than that at which it is to be operated. With small particles this may be possible only if the packing is done in a closed system under pressure, which is not easy to organize in a routine laboratory. Consequently, most high performance columns (see Section 5.4) come ready-packed, and repacking is not advised without appropriate equipment.

The flow through a column (for large particles) can be under gravity; if a meter or so of water pressure does not result in much flow, then the system is probably partially blocked or the tubing used is too fine and causes a resistance. Peristaltic pumps that deliver a constant rate of buffer are used widely, and enable the sample, buffer, and gradient former to be placed at a convenient height.

Separation of proteins is achieved in the column, and the results are monitored as soon as the liquid exits from the column. The remaining

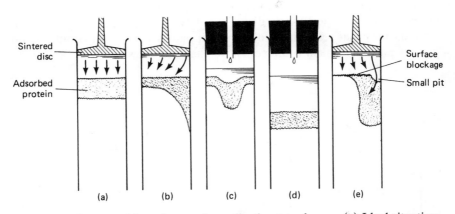

Figure 1.9. Some problems in sample application to columns. (a) Ideal situation; the flow is evenly distributed over the surface. (b) Due to a small gap in the side of the applicator sinter disc, most liquid flows through that gap and uneven flow onto the adsorbent occurs. (c) Drops from tube application create pit in adsorbent resulting in excessive flow in the center of the column. (d) Situation in (c) is remedied by having a large reservoir of buffer to absorb local disturbances. (e) Particulate material in sample results in clogging of adsorbent surface. If a small pit occurs in one side of the surface, all the flow may pass through it, causing highly irregular result. Remedy—centrifuge or filter sample before running on to column.

Figure 1.10. Ideal method for pouring column; the slurry is continuously stirred in the reservoir above the column to ensure that particles of all sizes are settling throughout the column.

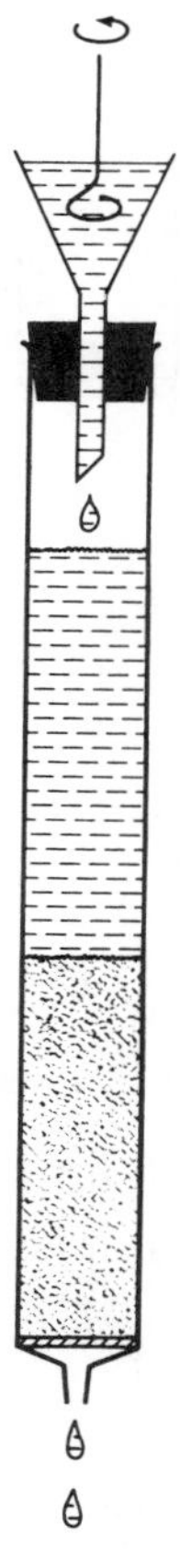

apparatus, shown in Figure 1.11, is for convenience—continuous monitoring of UV adsorption at 280 nm for protein (Trp + Tyr band), or more sensitivity at lower wavelengths, e.g., 220–205 nm (peptide band, see Section 3.1), and finally, collecting fractions of equal volume using a timed control.

If a peristaltic pump is not used, flow rates could change, and drop counting may prove more satisfactory for obtaining equal fractions. On the other hand, changes in surface tension as proteins emerge can result in different-sized drops. The flow from the column may be split, with a small amount being used for a continuous enzyme assay, which if synchronized properly could record the enzyme activity superimposed on the protein curve. Gradient elution can be accomplished using a gradient former which may be simply two beakers connected by suitable tubing for siphoning. Somewhat more reliable and reproducible gradients are formed using two cylinders connected at the bottom, with an efficient paddle stirrer (see Figure 6.7). Finally, the "black box" type of gradient

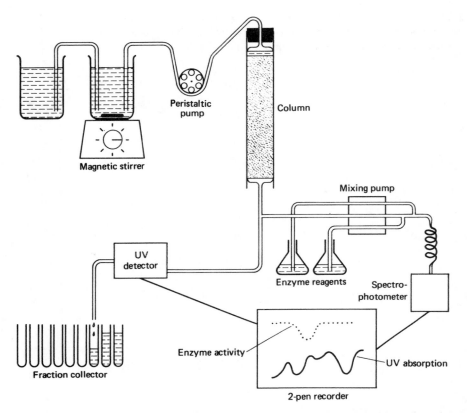

Figure 1.11. Complete arrangement for column chromatography, with automatic collection of eluate in fraction collector and continuous monitoring of enzyme activity. The simple mixing device illustrated using the magnetic stirrer would be replaced with an automated mixer and pump with preprogrammable gradient formation in modern apparatus.

former can be used to generate automatically any sort of gradient required.

1.4 Manipulation of Protein Solutions

As a final section in this introductory chapter on general methods, some notes on the handling of protein solutions are presented. Special techniques for maintenance of enzyme activity and stability are described in Chapter 12. Very often you have a solution of protein that is not in the right state for the next fractionation procedure. It might be in the wrong buffer, at the wrong pH, have too much salt, be too dilute, or too turbid. Turbidity is indicative of particulate material, which can be centrifuged or filtered as described in Section 1.2. The pH can be changed readily using a pH meter and not too strong an acid or base (Section 12.3). Concentration of solutions and ways of changing buffers are described now in more detail.

Concentration

Concentrated solutions can be diluted easily, but the second law of thermodynamics interferes with the converse. Concentration of protein solutions can be done in a number of ways, and the best way will depend on the particular application and requirement. Concentration can be carried out by precipitation, followed by redissolving in a small volume. Typically, this would use ammonium sulfate, though acetone is sometimes useful (Sections 4.3 and 4.4). The redissolved precipitate now contains precipitant, which will probably need removing. Precipitation methods are suitable only from a protein concentration of above $1\,mg\,ml^{-1}$; lower concentrations either fail to precipitate if the solubility is not low enough, or denature during the aggregation process. However there are many exceptions to this statement; some proteins in dilute solution are much more stable than others.

Adsorption, e.g., to an ion exchanger, may be appropriate and can achieve successful concentration of very dilute protein solutions. (As long as the required protein is totally adsorbed, it does not matter much whether other proteins are). For example, a very dilute protein solution eluted from DEAE-cellulose at low ionic strength, pH 6, would adsorb to a fresh, small column of DEAE-cellulose if its pH is raised to 8, and then could be eluted with a small pulse of salt.

The most important methods for concentrating dilute protein solutions involve simply removing water (plus low-molecular-weight solutes). This can be done by a variety of procedures based on the semipermeable membrane (dialysis) or gel filtration principle, according to which proteins cannot pass through a membrane or surface that water and small molecules can. The most commonly employed system is that known as ultrafiltration, in which water is forced through the membrane, leaving the more concentrated protein solution behind. This is done either by reducing the pressure to "suck" the water out (Figure 1.12a) or, more commonly, by applying compressed air or nitrogen at up to 5 atm in a pressure cell, which is constantly stirred to prevent clogging of the membrane (Figure 1.12b). The latter, when operating efficiently, can reduce 50 ml to 5 ml, or 200 ml to 10 ml within an hour or less, especially if the final concentration is still fairly low; ultrafiltration speed drops off rapidly as protein concentration rises.

For concentration of small volumes, centrifugation through a membrane is now widely used. Various designs of tubes containing an ultra-

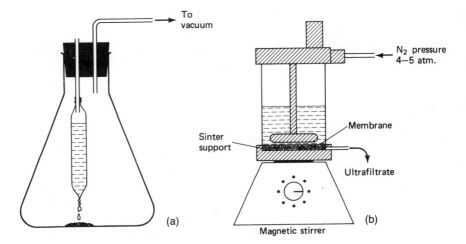

Figure 1.12. Ultrafiltration of protein solutions. In (a) the suspended dialysis bag is open to atmospheric pressure, while outside it is evacuated—maximum pressure difference less than 1 atm. In (b) the ultrafiltration cell can be operated at up to 5 atm positive pressure, forcing filtrate through membrane.

filtration membrane are available to fit into standard centrifuge rotors. The water and low-molecular-weight components are forced through the membrane by centrifugal pressure, and 1 to 3 ml of solution can be concentrated to μl volumes in less than an hour.

Dry gel filtration particles that exclude protein when swollen can be employed for rapid concentration. Since it is not possible to remove all protein solution from the outside of the particles without washing (causing a re-dilution), this method is really suitable only if concentration of the solution is more important than overall recovery; at best a 10–15% loss of protein can be expected. For instance, 100 ml of solution could be treated with 20 g of *dry* Sephadex G-25 (Pharmacia). The powder rapidly swells as it takes up water, eventually occupying almost the whole volume. The slurry can then be filtered under suction, obtaining about 30 ml of solution containing some 80% of the original protein—approximately a twofold concentration. This is very useful if rapid concentration of an already concentrated protein solution is desired, but is not generally used in protein purification procedures because of the losses involved. For small volume concentration using Sephadex G-25, centrifugation to recover the protein has been employed [1].

The final method of concentration by water removal involves osmotic forces. The sample is placed in dialysis tubing, which is then immersed in a solution, or surrounded by a powder, which attracts water from the bag. Commonly used substances are polyethylene glycol (MW 20,000+) and highly substituted carboxymethyl cellulose. Both of these polymers attract water very strongly, maintaining a low water activity outside the bag and high activity inside, forcing the water out. Calbiochem has adopted the trade name of Aquacide for these products. On a small scale it is very convenient: 10 ml can be concentrated to practically nothing in an hour by dabbling the dialysis bag in powder, and occasionally stripping off the hydrated gel forming on the outside. For large amounts, a concentrated solution (say 20%) of polyethylene glycol can be used, and the dialysis bag left, with stirring, for hours without attention. Provided that the polymers do not contain low-molecular-weight impurities damaging to an enzyme, the method is quite satisfactory, but be careful not to contaminate the protein solution with the polymer solution.

Removal of Salts; Changing Buffers

Concentration of protein solutions and dialysis are closely related. Conventional use of dialysis bags involves the removal of unwanted low-molecular-weight solute from the sample and replacement with buffer present in the "dialysate." Osmotic forces are usually the opposite to that described above: high concentrations of salt or organic solvent in the sample cause water to enter the bag before the salt leaves; consequently, there is an increase in volume during the early stages if the solute concentration is high. Dialysis is used for removing excess low-molecular-weight solute and simultaneously introducing a new buffer solution (it may be just water) to the sample. Dialysis tubing is available in a variety of sizes, and, conveniently, does not allow molecules larger than 15,000–20,000 Da to pass through. All low-molecular-weight molecules diffuse through the tubing and, eventually, the buffer composition on each side equalizes. Complete removal of endogenous salts from a sample cannot occur in one dialysis simply because at equilibrium, what was originally inside the dialysis bag now is distributed throughout the buffer *and* the dialysis bag. If the buffer volume is 50 times that of the bag, then the best that can be achieved is a 51-fold dilution of the original salt/low-molecular-weight material. Consequently, changing the buffer at least once is needed, and to speed up the process, stirring of the buffer *and* movement of the dialysis bag should occur—a well-mixed system can reach more than 90% equilibration in 2–3 h. A large volume of buffer is preferable since fewer buffer changes are needed (Table 1.1). Dialysis is a convenient step to be carried out overnight, as it needs no attention and equilibrium is reached by morning. However, as will be discussed later (Section 12.2), the possibility of proteolytic degradation during dialysis may make its use undesirable.

There are methods for pretreating dialysis tubing to remove a variety of chemicals introduced during manufacture. Boiling with an EDTA-NaHCO$_3$ solution has been recommended, and long-term storage im-

Table 1.1. Dialysis Protocols for Decreasing Salt Concentration from $1\,M$ to $<1\,\text{m}M^a$

Procedure 1

Dialysis against 1 liter of water	Complete equilibrium
→ swells to 100 ml, at equilibrium = 95 mM	would take 3–5 hr at
Change dialysate, further 1 liter of water	least; with shorter
→ swells to 110 ml, at equilibrium = 9.3 mM	times, a fourth change
Change dialysate, further 1 liter of water	would be needed.
→ no further swelling, at equilibrium = 0.9 mM	

Procedure 2

Dialysis against 5 liters of water	A single change would be
→ swells to 110 ml, at equilibrium = 20 mM	sufficient even without
Change dialysate, further 5 liters of water	complete equilibrium.
→ no further swelling, at equilibrium = 0.4 mM	

---

^{a} Initial sample is protein in 50 ml of buffer containing $1\,M$ ammonium sulfate (redissolved precipitate from an ammonium sulfate fractionation).

mersed in such a solution gives a good result. Chemical contamination problems are more likely to arise when dialyzing a very dilute protein solution, since then the membrane/protein ratio is high. Protein solutions of the order $10\,\text{mg}\,\text{ml}^{-1}$ and more are unlikely to be affected; in this situation it is usually safe to use the tubing direct from the dry roll. After wetting, a knot (preferably two) is tied in one end (clamps are also available for closing the ends) and a filter funnel placed in the other for pouring in the sample. With the end of the tubing resting on a convenient surface, the sample is poured in (Figure 1.13). It is best to have the tubing flat to start with so that air does not have to get out through the funnel. The tubing is then closed with a further knot and placed in the dialysis buffer. Remember to leave space for expansion if there is a high solute concentration in the sample. Otherwise, expansion will cause swelling and a high pressure inside the bag. This might result in bursting, or else leakage through the knots.

Before gel filtration techniques became widely used, dialysis was a routine standard procedure, involved in almost all protein purifications. Although it is simple and requires little equipment, it has two major disadvantages: the need to change dialysate, perhaps at awkward hours, and the relative slowness of the method. The importance of speed is discussed later. Except when it would be a case of carrying out dialysis overnight compared with not doing anything overnight, the rapidity and complete separation achieved by gel filtration makes this a much preferable method—especially on a small scale.

The principles of gel filtration are described in detail in Section 8.1. For removing salts and buffer changing, the "all or none" principle of gel filtration is involved; the medium chosen totally excludes proteins, but

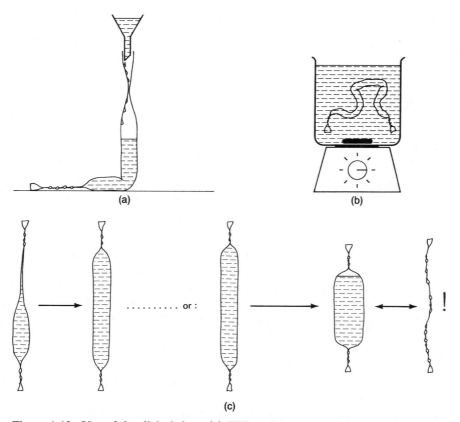

Figure 1.13. Use of the dialysis bag. (a) Filling with sample. (b) Agitation during dialysis. (c) Results of swelling due to osmotic forces.

includes small-size solutes. Probably the most widely used material is Sephadex G-25 (Pharmacia); Biogel P-30 Bio-Rad has approximately the same properties. Packed into a column preequilibrated with the buffer required in the sample, rapid "desalting" can be achieved in one pass, provided that the sample volume is no greater than one-fifth that of the column. Satisfactory separation is achieved at flow rates of 50 to 100 cm h^{-1} (*note*: 50 cm h^{-1} equals 50 ml h^{-1} per cm^2 cross section). A range of prepacked columns should be on hand so that changing the buffer or desalting a solution can be a routine and immediate operation. Table 1.2 lists the range that we routinely use. Small columns should be packed with finer grade beads to retain optimum resolution. Rapid desalting of small samples with virtually no dilution can be achieved by a centrifugal method [2] in which a small column of Sephadex G-25 is used inside a bench-top centrifuge.

The sample should not have too high a concentration of either salts or protein if optimum separation is needed. After dissolving an ammonium

Table 1.2. Dimensions of Columns for Desalting Using Sephadex G-25

Column dimensions, cross-section area (cm^2) × height (cm)	Resin grade	Column volume (cm^3)	Sample size (cm^3)	Normal flow rate (ml hr^{-1})	Typical time for completion (protein emergence) (min)
1 × 8	Fine	8	1–1.5	40	7
6 × 10	Medium	60	2–10	200	15
8 × 30	Medium	240	10–30	250	30
8 × 60	Medium	480	30–80	250	45
16 × 90	Coarse	1,440	80–300	500	90
50 × 100	Coarse	5,000	300–1,000	1,500	90

sulfate precipitate with a *minimum* amount of buffer, it should be diluted with at least a further equal volume of buffer before desalting on a column. Protein concentration should not be more than 30 mg ml^{-1}. Protein emerging from the column, if it is not colored and not being monitored continuously in a UV monitor, can be detected by spot tests using trichloroacetic or perchloric acid precipitation. Ammonium sulfate can be detected by spot tests using barium chloride. (If the column buffer contains phosphate, use barium chloride in HCl to avoid confusion with barium phosphate precipitates.) Remember that a gel filtration column should be preequilibrated with the buffer that the protein is to be exchanged into, using at least one column volume before applying the sample; but it does not matter what the column is washed with, since the protein travels ahead of the solvent front. When using expensive buffers, rather than equilibrate the whole column with buffer, one can use a "desalting buffer" such as a very low concentration of Tris-EDTA. About 0.2 mM of the acid form of EDTA is neutralized to pH 7.5–8 with Tris; this has no effect on the proteins unless (1) EDTA inactivates by pulling out an essential metal ion, or (2) the proteins "salt in" (Section 4.1) at very low ionic strength and so precipitate in the column. Provided that these circumstances are not occurring, the emerging protein can be made up to the correct buffer composition by, for instance, adding 0.11 vol of 10 × concentrated stock solution.

As with dialysis, the final volume is likely to be considerably larger than the initial volume before "desalting"; nevertheless, further dilution might be required before application to an adsorption column (Chapters 5–7). Changing of buffers, rather than removal of excess salt, is carried out in exactly the same way; however, if the protein concentration is already low and the volume large, it may be better to "salt out" first, dissolve in a small volume of buffer, and then use a relatively small desalting column.

This chapter has described some of the basic operations in the protein purification laboratory. Chapters 4–9 go into details of fractionation procedures used for isolating proteins. But first, one needs an extract of the raw material to work with; obtaining this extract is described in the next chapter.

Chapter 2
Making an Extract

2.1 The Raw Material

To many people embarking on a protein purification project, much of this section will not be relevant, for they have no choice of raw material. A wide variety of different circumstances could prevail; the laboratory may be working solely on a particular organ or species, and the only considerations about the raw material may be questions of availability in sufficient quantities at times when needed, and the possibility of frozen storage before use.

At the next level, the researcher may have some choice of tissue to use for purifying the protein, or, alternatively, the tissue/organ may be defined but the species source allowed some variation. Species used for protein purification have been chosen principally on the ease of raising or growing. With animals, the principal species used are the rat (especially for liver studies), rabbit (especially for skeletal muscle), and meat animals, mainly cow and pig, for organs which are rather small in size in the laboratory animals (e.g., heart, brain, kidney, thymus). In addition, one must not forget the human animal, on behalf of whom so many laboratory animals are sacrificed, on the assumption that what is true for the rat is likely to be true for the human. Studies on proteins from human tissues have of necessity been limited to those materials relatively easily obtained, such as blood and placenta. But studies on a wider variety of species have become commoner, partly through the realization that many important features of metabolic control, and enzyme localization and characteristics do differ markedly even within the vertebrate phylum, and humans are not always the same as rats!

Invertebrates' proteins have been rather neglected, mainly due to the fact that most invertebrates are very small, and a large number are required to get a reasonable amount of starting material. If the organ of interest is not dissected from each individual, an arduous task with insects, whole-body homogenates cause great problems because of the large quantity of digestive juices, including proteolytic enzymes, that are released into the extract. Most well-known invertebrate proteins and enzymes have been purified from the larger crustacea such as crayfish and crabs. Many a feast of lobster claws has been enjoyed in laboratories isolating enzymes from the tail muscle!

Plant biochemistry has long been the poor relation to animal work, largely because of the medical implications of the latter, but also because plants pose particular difficulties to the protein chemist. The variation of quality with season can be overcome only by growing the plants in special growth chambers; otherwise, one has to put up with this variation in material grown in the open. Of the vast variety of species, those studied have been, for obvious reasons, mainly those of economic importance, and these are not necessarily the best choices biochemically. Nevertheless, one or two plants have tended to predominate because of the ease of extraction of their leaves, especially when it is easy to isolate chloroplasts from these extracts. Most commonly used is spinach, *Spinacia oleracea*, or alternatively the (not closely) related silver beet *Beta vulgaris*, which is easier to grow in a range of climatic conditions. Plant cells are highly compartmented; in most cases the bulk of the volume is vacuolar space, which can be filled with quite acid solutions, proteases, and a variety of other detrimental compounds. There is also a large amount of cell wall (cellulose), and the chloroplasts, starch granules, and other organelles occupy much of the cytoplasmic space. Indeed, the cytoplasm (which, together with the internal space of the chloroplasts, contains most of the enzymes) may often be no more than 1–2% of the total cell volume in plants. Consequently, plant extracts may be very low in protein even if very little fluid is added when making the extract.

Microorganisms again present a different range of problems. Whereas the animal source may be available at the local abattoir or at least in the institution's animal house, and a plant source may be at the supermarket, microorganisms (with the notable exception of yeast) will have to be specially grown in controlled conditions on a fairly large scale. Algae, fungi, yeasts, and bacteria each have special requirements for growth conditions, harvesting, and extracting. The particular problem of marked changes in enzyme composition during different growth phases must be carefully studied. Collection of bacteria and other unicellular organisms during the log phase of growth is usually desirable (Figure 2.1), though an enzyme being studied might not be at a maximum activity at this time; preliminary investigations on the organism to determine what physiological state contains the most of the enzyme required should be carried out. *Saccharomyces cerevisiae*, baker's yeast, is so readily available in large quantities that, if this can be used, the whole problem of raw

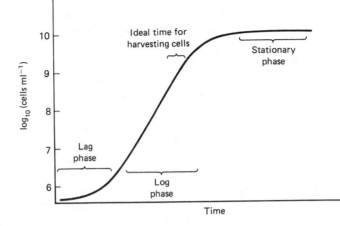

Figure 2.1. Growth of microorganisms in nutrient-rich medium. The ideal time for harvesting is toward the end of the log phase before growth rate slows, giving a high yield of cells. But specific enzymes may be maximum at an earlier or later stage, so some trials at different times are desirable.

material is solved. Compressed yeast cakes are sold that represent nearly 100% pure yeast cells at the end of their growth phase. Grown on a variety of nutrients, the main carbon source is usually molasses or ethanol. Yeast cakes remain viable (and so the enzymes remain active) for some weeks at 0°C, and if frozen will be usable months later. Also very popular with bakers and biochemists are the packs of dried yeast which reconstitute on adding water. These are convenient sources for many proteins and enzymes, though the drying process may tend to partly inactivate a few of the more sensitive ones. Sealed under vacuum or in nitrogen and stored at 0–4°C, these can be used years later, and represent perhaps the most convenient raw material source there is—provided that you can break open the cells (see below).

Molecular biology techniques have so much dominated biochemical and genetic research during the past decade and more, that many people commencing a protein isolation project will be starting with a raw material that contains the protein expressed as a recombinant form in a foreign host. Moreover, the protein may not be in its native form; it may be insoluble; it may have been genetically fused to another protein or peptide fragment, or it may differ in some other way from the protein as found naturally. In almost all cases, it will be present in the host cells at a higher concentration than found naturally (as that is one of the principal reasons for using a recombinant form), and so a purification procedure already established for the natural source may be quite inappropriate for a recombinant source—in addition, the unwanted proteins are different. Further discussion of recombinant protein purification is reserved for Chapter 9.

Freshness and Storage

Usually, the sooner the raw material is used, the better and more physiologically relevant the preparation will be, for if the tissue has been "dead" for some time, natural degradative processes will have commenced. Yet there are occasions when absolute freshness can be a nuisance. For example, skeletal muscle contains a high level of ATP, with phosphocreatine and glycogen present which maintain that level for some hours after death. The structural proteins myosin and actin are normally insoluble at ionic strengths below about 0.25, but the presence of ATP causes the disrupted myofibrils to swell, and a portion of these proteins go into solution. Even at physiological ionic strength (~0.16) some myosin can be solubilized. This can subsequently upset enzyme preparations by precipitating and mixing into fractions that should not contain it. A particular example is the preparation of AMP deaminase [3]. Extraction with a buffer of ionic strength 0.25 is recommended—with really fresh muscle much actin and myosin go into quasi-solution and upset the next step, adsorption on phosphocellulose. But if the muscle is allowed to go into *rigor mortis* first, indicating a loss of ATP, a much better preparation is obtained. This is a rare example; freshness should always be the prime aim unless there is some good reason for delay.

The availability of fresh material does not always coincide with one's ability to use it. Frozen storage, either of the material as received or of an extract from it, must be considered. During freezing a great many events occur. First, the free water freezes and ice crystals grow. The ice crystals are very destructive for membranous layers and organelles, but do not normally upset proteins directly. As the temperature decreases further, the remaining liquid becomes increasingly concentrated in salts as well as protein, until their solubility is exceeded. The less soluble of a pair of salts comprising a buffer will come out of solution first; consequently, the pH can change drastically before complete solidification takes place [4,5]. If the storage is at a temperature of between $-15°C$ and $-25°C$ (typical of the temperatures in domestic refrigerator freezing compartments and simple freezers), the remaining concentrated solution, perhaps at a quite different pH from the original solution, remains unfrozen. Proteases liberated from lysosomes during the freezing can go to work (albeit slowly at such low temperature) in the concentrated protein soup, and a few weeks of storage may do a lot of harm. Thus, freezing should aim at reaching a temperature below $-25°C$ quickly, and storage should be at even lower temperatures if possible. Commercial freezers operating at temperatures down to $-80°C$ are available.

Freezing of extracts is sometimes preferable, because the composition of the medium can be manipulated to optimize storage conditions. Proteolytic enzyme inhibitors can be added after making the extract, and the pH adjusted with a suitable buffer to the best value for stability of the enzymes. Also, large amounts of raw material can all be processed into one extract, so that subsequent samples taken from frozen storage are identical except for their storage time; variation in behavior cannot then be blamed on biological variation of the frozen material.

Remember that thawing speed can be important—the faster, the better, provided that local overheating does not occur. The best way is to immerse the container in warm (40–50°C) water and agitate frequently. It is unlikely that in these conditions the melted solution would rise above 10°C, despite the outside temperature, as long as ice remains inside.

Now that we have the raw material, either fresh or thawed, it is time to homogenize it and obtain an extract containing the enzyme in solution, the first step in the majority of enzyme purification procedures. Cases where preparation of subcellular particles or other insoluble material precedes the release of the enzymes into solutions are described briefly later (Section 2.4).

2.2   Cell Disintegration and Extraction

Most of the proteins and enzymes studied in the early days of protein chemistry were isolated from extracellular fluids. The reason for this is not just because it was easy to obtain the raw material, but because extracellular proteins are for the most part more stable, often as a result of disulfide cross-links, and they tend to be small molecules; early studies on protein structure naturally concentrated on such small proteins. Thus lysozyme, ribonuclease, and chymotrypsin were among the earliest proteins studied in detail, and all are from extracellular sources. But most enzymes are found inside cells, and are very often much less stable; disulfides are generally absent because of the more reducing intracellular environment. The purpose of the present section is to describe methods of disrupting the cells and releasing the enzyme into an aqueous "extract" which is the first stage of enzyme purification techniques.

There are many methods of cellular disintegration, for there are many types of cell. Most cells have particular characteristics which need special attention during disintegration. Animal tissues vary from the very easily broken erythrocytes to tough collagenous material such as found in blood vessels and other smooth-muscle-containing tissue. Plant cells are generally more difficult to disrupt than animal cells because of the cellulosic cell walls. Bacteria vary from fairly fragile organisms that can be broken up by digestive enzymes or osmotic shock, to more resilient species with thick cell walls, needing vigorous mechanical treatment for disintegration.

Table 2.1. Cell Disintegration Techniques

Technique	Example	Principle
Gentle		
Cell lysis	Erythrocytes	Osmotic disruption of cell membrane
Enzyme digestion	Lysozyme treatment of bacteria	Cell wall digested, leading to osmotic disruption of cell membrane
Chemical solubilization/ autolysis	Toluene extraction of yeast	Cell wall (membrane) partially solubilized chemically; lytic enzymes released complete the process
Hand homogenizer	Liver tissue	Cells forced through narrow gap, rips off cell membrane
Mincing (grinding)	Muscle etc.	Cells disrupted during mincing process by shear force
Moderate		
Blade homogenizer (Waring type)	Muscle tissue, most animal tissues, plant tissues	Chopping action breaks up large cells, shears apart smaller ones
Grinding with abrasive (e.g., sand, alumina)	Plant tissues, bacteria	Microroughness rips off cell walls
Vigorous		
French press	Bacteria, plant cells	Cells forced through small orifice at very high pressure; shear forces disrupt cells
Ultrasonication	Cell suspensions	Microscale high-pressure sound waves cause disruption by shear forces and cavitation
Bead mill	Cell suspensions	Rapid vibration with glass beads rips cell walls off
Manton-Gaulin homogenizer	Cell suspensions	As for French press above, but on a larger scale

It is generally not advisable to use a disruption treatment more vigorous than necessary, since labile enzymes may be inactivated once liberated into solution. Table 2.1 gives a list of techniques that can be used, with illustrations in Figure 2.2.

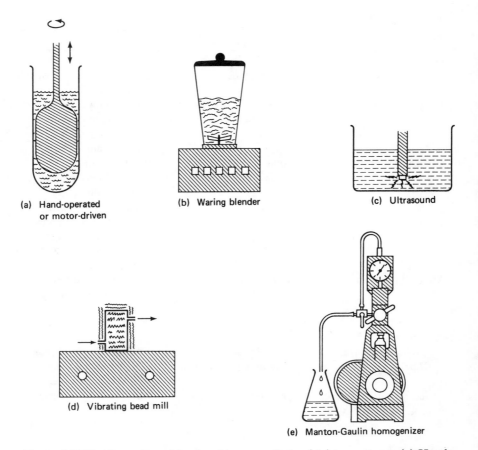

(a) Hand-operated   (b) Waring blender   (c) Ultrasound
    or motor-driven

(d) Vibrating bead mill

(e) Manton-Gaulin homogenizer

Figure 2.2. Equipment used for breaking up cells to obtain an extract. (a) Hand-operated or motor-driven glass homogenizer. (b) Waring blade-blender (food processor). (c) Ultrasonic probe. (d) Vibrating glass bead mill. (e) Manton-Gaulin cell disintegrator.

These methods are for disrupting cells and releasing proteins into solution. On occasions when an enzyme is present in an organelle, the methods may still be suitable in that they also disrupt the organelles. On the other hand, it may be desirable to isolate the organelles themselves first, so disposing of contaminating cytoplasmic proteins, before extracting the enzyme from the organelle. Less vigorous techniques are needed; preliminary digestion of collagenous or cellulosic extracellular structures with digestive enzyme preparations enables the cells to be broken subsequently by a gentler treatment while maintaining the integrity of the organelles. Preparation of mitochondria from tissues such as skeletal or cardiac muscle using proteinase treatment is an example of such a procedure [6]. However, this is generally only applicable on a small scale, more suited for metabolic studies on the organelles than enzyme purification. Yields of purified organelles may be very low, and it is often a better procedure to do a complete tissue disruption and then approach the problem of isolating the protein required from the complex mixture of proteins in the extract. Thus, mitochondrial and chloroplast enzymes would often be prepared from a complete tissue homogenate rather than from the isolated organelles.

Finally, the protein may not be soluble in the extraction buffer. In this case special techniques are required and are discussed briefly in Section 2.4.

The "extract" is prepared, after cell disintegration, by centrifuging off insoluble material. Before centrifuging, the mixture is usually described as a homogenate; after centrifugation as much as possible of the desired protein should be present in the liquid layer. Liquid is trapped within the precipitated residue, and the total loss will be related to the proportion of residue to liquid. Most animal tissues have large amounts of insoluble cell material which bind a lot of water. They give a volume of residue about as much as the original tissue (dependent somewhat on the amount of centrifuging). Thus, when making a homogenate of, say, liver, *at least* 2 vol of a suitable extractant buffer should be added when making the homogenate. The more that is added, the larger proportion of the soluble fraction will be extracted, but the extract will be more dilute; its greater volume may create difficulties when working on a large scale. Two-and-one-half volumes of extractant is typical for liver, heart, or skeletal muscle homogenates. Sometimes, with valuable raw material, a reextraction of the residue can be carried out.

Plant tissues are quite different; as mentioned in the previous section, only a small fraction of the volume of plant tissue is truly intracellular; large vacuoles (being regarded here as extracellular) and intercellular spaces mean that on disruption much liquid is released, making additional extractant liquid almost unnecessary. The residue after centrifugation may occupy only 20–40% of the volume of the original plant tissue. Nevertheless, it may be important to use some added extractant liquid so as to control undesirable processes during homogenization. These include acidification and oxidation of susceptible compounds. A particular problem with many plants is their content of phenolic compounds, which oxidize—mainly under the influence of endogenous phenol oxidases—to form dark pigments. These pigments attach themselves to proteins and react covalently to inactivate many enzymes. Two approaches to this problem are useful. First, the inclusion of a thiol compound such as β-mercaptoethanol minimizes the action of phenol oxidases. Second, addition of powdered polyvinylpyrrolidone is often beneficial as it adsorbs the phenolic compounds.

Microorganisms are more like animal tissue with respect to the volume of residue after centrifuging; yeasts and similar fungi with thick cell walls give a residue which nearly equals the volume of cells initially. With microorganisms, the first step is to harvest the cells by centrifugation, flocculation, or ultrafiltration; On a large scale this will need special equipment (Chapter 10). Even on a laboratory scale, with organisms which grow poorly, it may be necessary to grow batches of tens of liters, which must be harvested. Centrifuges capable of dealing with up to 6 liters at a time are useable for this purpose; alternatively, cross-flow or hollow fiber filtration with $0.2-0.4\,\mu m$ membranes (Amicon Corp Inc., Beverly MA; Millipore, Millford, MA) can concentrate the culture. It may be possible to either develop a flocculant strain, which allows decantation of the bulk of the spent medium, or to cause flocculation by addition of polyelectrolytes (cationic) such as chitosan [7]. Breakage of the cells can vary from very easy to nearly impossible; a summary of methods is given below. They include mechanical breakage of cells through to gentler lysis techniques, and combinations of these may be most suitable for particular purposes [8].

Cell debris may centrifuge down easily, particularly the coarse residues from animal and plant tissues, although fine intracellular particulates will remain in suspension after, say, $10,000\,g$ for 15 min. This cloudiness need not be of any concern, as fractionation steps will remove it. But use of the more vigorous disruption techniques such as bead-milling can break down cell wall material into fine particles which may have a relatively low density, since lipids are usually present. This can be difficult to sediment at the available g force for large volumes, and the extract may be very turbid indeed. Methods for clarification of the extract are described in Section 2.3.

The following outlines some procedures for extracting various tissues.

Mammalian Tissues

Dice tissues and cut away connective tissue and fat as far as possible. Place in Waring blender with 2–3 vol cold extraction buffer per gram of tissue. Blend for 30 sec, then repeat if still lumpy. Stir the homogenate for 10–15 min, checking that its pH is suitable, then transfer to centrifuge buckets. Centrifuge at $5,000-10,000\,g$ for up to 60 min ($2-3 \times 10^5\,g$ min). Decant extract through Miracloth (Calbiochem., San Diego, CA), cheese-cloth, or glass wool to trap fat particles.

Very fatty tissues may not be manageable this way. A traditional, but still useful way of preparing such tissues is to make what is known as an acetone powder. The homogenization takes place in the presence of large quantities of acetone that have been chilled to a temperature well below 0°C (hazard warning: do *not* place acetone in deep freezers—use an ice–salt bath or dry ice). The acetone solubilizes most of the fat, whereas the proteins are precipitated, but remain in the native conformations provided the temperature remains below zero. The powder precipitate is washed with cold acetone, then dried, and can be stored before use. The extract is made by stirring the acetone powder with an appropriate buffer in the warm, solubilizing the required proteins. Membrane proteins that would otherwise be insoluble may now go into solution, as the fatty membrane has been dissolved.

Erythrocytes

Red blood cells are easy to extract after collection by centrifugation, and a rinsing step in isotonic NaCl (0.9%, 0.15 M) followed by centrifugation again. The cells are osmotically lysed with water, e.g., 2 vol water to 1 vol packed cells. However, approximately 90% of the protein going into solution is hemoglobin, and unless this is what you are purifying, a method for selectively removing it is more useful. Ethanol-chloroform has been used for over 60 years to denature hemoglobin [9].

Soft Plant Tissues

Using only 0.5–1 vol of cold extraction buffer containing 20–30 mM mercaptoethanol, homogenize in a blender for 30 sec. Alternatively, pass material through domestic juicer, washing through with the buffer. Centrifuge as soon as possible to minimize oxidative browning (2–3 × $10^5 g$ min). Decant carefully from soft material on surface of precipitate. Addition of powdered polyvinylpyrrolidone can be beneficial in adsorbing phenols.

Yeasts

1. *Mechanical disruption.* Complete disruption can be obtained with a Manton-Gaulin homogenizer (Gaulin Corp., MA, USA) or a Vibrogen Cell Mill (Bühler, Tübingen, Germany), using 2–5 vols buffer per gram wet weight.
2. *Toluene autolysate.* A variety of toluene methods have been used [10,11,12]; certain of them are not fully successful with all qualities of commercial cake yeast. The principle is to treat the yeasts with toluene, usually at a temperature of 35–40°C, when after 20–30 min the yeast "liquefies" due to extraction of cell wall components. Buffer is then added and the slurry stirred in the warm for a few hours or left overnight in the cold. As the method is autolytic, in which cell wall structure is degraded by enzymic action, some cellular enzymes become degraded during the treatment. Ethyl acetate has been used in lieu of toluene [13,14] but does not work with tougher yeast strains.

3. *Ammonia cytolysis* [15]. This method is best suited to dried yeast; it is simple, but some enzymes unstable at pH 10 are lost completely. "Active dried" yeast is stirred with $0.5\,M$ NH$_4$OH, 2 vol per g dry weight, at room temperature, and stirring is continued for 16–20 h. A little toluene, 1–2% of the total volume, sometimes improves the overall yield of soluble proteins. A further 1–2 vol of water plus enough acetic acid to bring the pH down to an appropriate value is stirred in before centrifuging off the cell debris.

4. *Bead mill.* Equipment for shaking cell suspensions with glass beads is available. A slurry of yeast cells (ca. 1 g wet weight to 3 ml buffer) is mixed with an approximately equal volume of glass beads of 0.5–1.0 mm diameter, placed in the container, which is usually of stainless steel, and this is then vigorously shaken while cooling water circulates around it. Within a few minutes, the cells have been sheared apart and the contents released. Continuous flow adaptations of the equipment enable larger volumes to be processed.

5. *Manual beading* [16]. This method is surprisingly effective, rapid, and requires no special apparatus other than glass beads of diameter 0.5 or, preferably, 1.0 mm. A suspension of yeast cells is put into a suitable bottle, and beads are added so that almost the whole suspension in full of beads. The bottle is sealed and shaken manually for 5–10 min. The beads are filtered under suction and washed, and the homogenate centrifuged.

6. *Enzymic lysis* [17,18]. A number of enzymes that act on yeast cell walls have been described; in fact, a mixture containing mannanases, glucanases, and chitinases is needed. Such enzymes are produced by organisms that feed on yeasts. Preparations from *Arthrobacter, Oerskovia*, and *Cytophaga* (Lyticase) have been used; these are useful for small-scale work, but may be uneconomical on scale-up.

Bacteria

Vigorous treatments with sonication, bead-milling, or the French press successfully disrupt bacteria, though not all are convenient for large-scale extraction. Grinding with alumina is often successful on a small to medium scale. Gram-positive species are mostly susceptible to lysozyme; stirring a suspension of cells in buffer plus $0.2\,\text{mg ml}^{-1}$ egg white lysozyme at 37°C for 15 min may be sufficient to release the cytoplasmic components. Inclusion of deoxyribonuclease I $(10\,\mu\text{g ml}^{-1})$ improves the quality of the extract obtained by reducing its viscosity. A combination of enzymatic lysis with mechanical disruption may be the optimal process in many cases [19].

Gram-negative species are less susceptible to lysozyme without prior treatment. A combined nonionic detergent–osmotic shock–lysozyme treatment has been described [20]. Inclusion of nonionic detergent to solubilize cell membranes greatly improves the cell lysis. We routinely use a lysozyme–nonionic detergent method to extract the gram-negative bacterium *Zymomonas mobilis*; this also works quite well with *Escherichia coli* and many other gram-negative bacteria [21], as well as gram-positives. The method extracts most of the soluble cytoplasmic proteins, but avoids total disruption and release of cell wall material, so a relatively clear extract can be obtained without high-speed centrifugation.

To 10 g wet weight of cells, 40–60 ml of extraction buffer, preferably at a pH above 7, containing 0.1% Triton X-100 or Nonidet P-40, 0.2 mg ml^{-1} lysozyme, and 10 μg ml^{-1} DNase, is added. The mixture is stirred for 1–2 h and the effectiveness of extraction tested on a small sample. We often leave the extraction to continue overnight. Proteolytic inhibitors may be added (see Chapter 12). A clear extract is obtained after centrifugation at 20,000 g for 20 min. The effectiveness of this procedure depends greatly on the growth conditions of the cells. Better extraction may be achieved by the inclusion of mercaptoethanol up to 0.1%, or of certain partially miscible organic solvents such as toluene, *n*-butanol, or ethyl acetate at about 1% v/v.

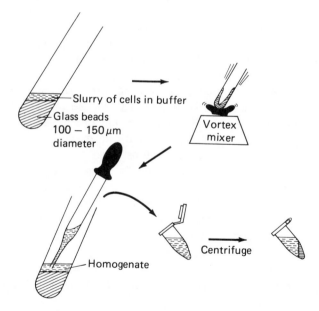

Figure 2.3. Small-scale disintegration of bacteria. Glass beads are added to a small test tube (e.g., 15 mm diameter) to a depth of about 10 mm. A slurry of cells in buffer (e.g., 1 g wet weight in 5 ml of buffer) is added until up to 5 mm above the glass beads. The mixture is vortex-mixed for 2–5 min. The extract is sampled from the surface of the beads and centrifuged in a microcentrifuge tube.

There are other lytic enzymes than the commonly used hen egg-white lysozyme. Culture supernatant fractions from several microorganisms such as *Cytophaga* and *Streptomyces* produce enzyme mixtures that successfully lyse certain cells that are otherwise difficult to break open [17].

Small-scale extraction of bacterial cells can be achieved with all but the hardiest organisms using glass beads of a smaller size than described above for yeast extraction, preferably around 0.1 mm (100 μm). The method is illustrated in Figure 2.3.

Fatty Tissues

Many animal and some plant materials contain a lot of fat which can make homogenization difficult, and the extract may contain a lot of finely dispersed lipid which complicates the subsequent purification steps. There are two ways of overcoming this. First is the use of quite large amounts of detergent to solubilize the fats and at the same time release proteins associated with them (see also under extraction of membrane proteins). The major problems with this are the need for large amounts of detergent, and the difficulty of removing it from the aqueous extract so obtained. Second is the classical technique which still finds use both in the laboratory and commercially, the preparation of an "acetone powder." After washing with acetone and perhaps ether as well, the powder is dried in air, and can be stored in the cold for long periods [22]. The proteins are solubilized with an aqueous buffer; in some cases, detergent may be needed also, but the mass of fat has been removed.

These are just a few examples of procedures that can be used for obtaining extracts from various sources. In practice there will often be a reported method to follow, but remember that raw materials can be very variable, as can the equipment used, and one may need to modify and develop the method for particular circumstances.

Thus, the basic methods of obtaining an extract are simply disruption of the raw material in the presence of a suitable buffer, followed by centrifugation to remove insoluble residues. The next section describes how the quality of the extract as a starting material for protein purification may be optimized.

2.3   Optimization and Clarification of the Extract

At this stage we will assume that a suitable method for cell disruption has been found and an extract made. It is important for reproducibility that each time an extract is made, it is as close to identical with previous extracts as possible. Often the properties of proteins in fractionation procedures depend not only on their own particular characteristics, but also on the composition of the solution, including other proteins present. Things that can go wrong include suboptimum operation of the homogenization apparatus—low pressure in a pressure cell, worn beads in a bead mill, or blunt blades on a Waring blender, resulting in less than complete cell disintegration. As a result, the extract would be more dilute and may have a different proportional composition since some components may be selectively extracted by suboptimum disruption. It is important to know exactly how much of an enzyme is present in a gram of raw material to evaluate a particular extract. Probably the best method is to make a small-scale extract using a large volume (say $10 \, \text{ml g}^{-1}$) of extractant, and measure the amount of activity after different times of treatment. Extended times may eventually result in less activity because the treatment is denaturing proteins, either by heating or by the vigorous shearing nature of the disintegration method. It is very often necessary to take measures to cool the system during cell disruption; ice-cold buffer and short bursts of treatment followed by a cooling period may be needed. As mentioned in the preceding section, the volume of extractant per gram of material is an important consideration; on a large scale the amount used will tend to be a compromise between maximum extraction and minimum volume of extract.

Up to this stage no mention has been made of the nature of the extractant buffer, other than to suggest inclusion of things like β-mercaptoethanol for specific purposes. Plain water can be used, but it will not extract all proteins—this may be a good thing if the protein one is interested in is extracted by water. Cells contain many salts and much insoluble material that is charged, e.g., proteins, phospholipids, nucleic acids. The ionic strength inside the cytoplasm of a typical cell is in the range $0.15–0.2 \, M$; under these conditions the cytoplasmic proteins are "soluble," in that they can move around the cell. If homogenized with 2–3 vol water, the ionic strength of the homogenate may be 0.05 or less. Under these conditions the charged particulates can act as ion exchangers and adsorb proteins, especially basic ones.

There have been many reports in the literature of enzymes being distributed between soluble and particulate fractions. When homogenates were made with dilute buffers, the more buffer used, the more protein appeared in the particulate fraction [23]. The ion exchange adsorbent properties (as opposed to physiologically significant, specific interactions) of cellular constituents have not always been appreciated. Thus, to ensure that all "soluble" cell constituents are extracted, one should use a buffer of ionic strength similar to the physiological one, and, of course, at the appropriate pH. If this leads to too much extraction of unwanted compounds, a compromise may be better. Typical buffers would be 20–50 mM phosphate, pH 7–7.5; 0.1 M Tris-Cl, pH 7.5; 0.1 M KCl with a little buffer in it; and for isolating organelles, isoosmotic buffers containing sucrose, mannitol, or sorbitol, as well as salts and buffer ions would

be appropriate. Also included may be EDTA (1–5 mM), β-mercapto-
ethanol, or cysteine (5–20 mM), and specific stabilizing agents for particular proteins, e.g., Zn^{2+} for zinc-containing protein, or pyridoxal
phosphate for enzymes using this as cofactor.

After the cells have been disrupted in the appropriate buffer, it may be
desirable to check the pH of the homogenate. Although really fresh
material, homogenized in the appropriate buffer, will give a homogenate
of consistent pH, that value may well change downward due to metabolic
processes leading to acidification. The best example is that of skeletal
muscle, in which glycogen can be rapidly converted to lactic acid, and the
pH of a homogenate could drop from 7.0 to 6.0 in 30–60 min (including
during centrifugation). In such circumstances the pH can be raised *above*
the desired value (e.g., with M Tris) before centrifuging. Alternatively,
inclusion of a glycolytic inhibitor can arrest the process; fluoride (10–
30 mM) is sometimes used.

These general comments refer to maximizing extraction conditions of
all soluble components. For a particular protein it may be better to use
other conditions, purposely preventing a complete solubilization of all
components provided that the protein required is totally extracted.

Often, the extract obtained is turbid, perhaps with fat particles floating
on the surface after centrifugation. The fat can be removed by coarse filtration through a plug of glass wool or a fine mesh cloth. The particulate
material in suspension will include organelles and membrane fragments
that may sediment fully only at about 100,000 g, an impractical procedure
for liters of extract. Filtration will usually be useless, since if the filter is
fine enough to trap the particles, it will rapidly clog. However, if the
cloudiness is relatively slight and a clear extract is needed—for instance,
for passing straight on to an adsorption column—filtration can be carried
out with a filter aid such as Celite (Section 1.2).

Clarification of cell extracts on a large scale can be carried out by
membrane ultrafiltration, using a membrane with a pore-size of about
0.2 μm.

Let us consider the different types of extract. Sometimes the particulates will not constitute a large proportion of the extract and can be
ignored; they will aggregate in, say, the first cut of an ammonium sulfate
fractionation and be discarded then. On the other hand, many animal tissues contain fats and membranous structures which end up in suspension
in the extract. Acidification to a pH between 6.0 and 5.0 will usually
cause aggregation so that some of the particulate material can be centrifuged off at relatively low speed. Ribosomal and other nucleoprotein
material can be removed by this acidification, which can be regarded as
a form of isoelectric precipitation (Section 4.2); the phosphate groups
protonate to neutralize, or at least lessen, the charge on the particulate
suspension. Provided that the protein wanted (1) does not also isoelec-
trically precipitate at the pH used, (2) does not adsorb to the precipitate

forming, and (3) remains stable at the subphysiological pH, this treatment can be most beneficial. The extract should be kept cold, and the pH lowered with a suitable acid (e.g., $1M$ acetic acid). After stirring for 10–20 min, the precipitate is centrifuged off and the supernatant pH readjusted, if required, before commencing the first fractionation step proper. Plant tissues are more acidic in the first place, and particulate material such as chloroplast fragments, which are more fatty, will not aggregate so readily. Otherwise, there is a relatively small amount of particulate material in plant extract (apart from coarse particles like starch, which sediment down easily after making the extract).

Microorganisms cause a variety of problems with their extracts. First, there is the large amount of nucleic acid material extracted from rapidly proliferating cells. Second, during cell disruption the cell wall and extracellular material may either be finely dispersed to give a turbid extract, or partially solubilized so that, as well as protein and nucleic acids, there is a large amount of gumlike polysaccharide in solution. This can cause considerable problems in the early steps of fractionation.

A common procedure is to treat with substances that cause precipitation of the nucleic acids and associated compounds. These include:

Streptomycin. An antibiotic which acts by interacting with ribosomes. Addition of a solution of streptomycin sulfate to the extract precipitates ribonuclear proteins and clarifies the extract.

Protamine. A natural DNA-binding protein from sperm. Addition of a neutralized solution of protamine sulphate to the extract precipitates most DNA and RNA complexes. A number of proteins bind biospecifically (e.g., nucleic-acid binding proteins) or by chance, and so may be lost from the extract, but can be reextracted from the precipitate (Section 4.5).

Polyethyleneimine. A synthetic, positively charged polymer that interacts with nucleic acids and causes precipitation of aggregated nucleoproteins; similar to protamine in its effect, and only a very small amount (e.g., 0.1% w/v) is needed (see also Section 6.6).

Yeast extraction presents several problems, the main one being that the cell walls of fungi are mostly very thick and difficult to break open. Consequently, the more vigorous methods are needed to obtain the extract; if this is to be achieved in a short time, use of a glass bead vibrating mill or a Manton-Gaulin homogenizer is the only reliable method that can be used on a large scale. These methods cause a substantial amount of low-density particulate material to be liberated, and centrifugation does not remove it all. The turbidity can upset the efficiency of precipitation by ammonium sulfate, and may cause clogging of chromatographic columns. Toluene autolysis of yeasts generally gives clearer extracts, but the conditions needed can result in proteolysis.

Strong ammonia was originally reported as a cytolysis method for preparing the very stable enzyme phosphoglycerate mutase [15]. This procedure has been adapted and used extensively on "active dried" yeast, to give a clear extract containing most, but not all, glycolytic enzymes [24]; but some are totally lost because of the high pH. Proteolysis does not seem to be a big problem with this method, as the proteases have little activity at pH 9.5–10, the pH of the mixture. The efficiency of extraction has tended to vary from one batch of yeast to another.

At this stage, one should be ready to go ahead with the purification method; an extract containing as much of the enzyme as possible has been made, and, where appropriate, particulate and most nonprotein materials have been removed. But before concluding this chapter, there is one important class of enzymes that the preceding sections have not dealt with, that is, the insoluble, membrane-associated enzymes which by definition would not be in the clarified extract.

2.4   Extraction of Membrane Proteins

Many proteins and enzymes are not naturally present in an aqueous phase either inside or outside of cells. In prokaryotes, there are cell membrane proteins and, in gram-negative organisms, periplasmic proteins which are partly immobilized between the outer and inner membranes of the cell envelope. In multicellular eukaryotes, as well as individual cell membranes, there are many organelles within cells that are membranous and have proteins associated with them. These include mitochondria, nuclei, endoplasmic reticulum, Golgi, vacuolar, and lysosomal membranes. When attempting to purify a protein from a membranous structure, the first thing to consider is whether the trouble involved in isolating the organelle/membrane is worth the substantial degree of purification it achieves. Usually it is, for although membrane isolation can often involve considerable losses, this need be no great concern at the first step if the raw material is readily available. On the other hand, the time, equipment, and labor costs involved in first isolating the membranous structure may sometimes be better employed on dealing with the more impure "total extract" made on the whole material. There is no general advice here; each circumstance must be taken individually. But if the ultimate objective is a scale-up process, steps to isolate an organelle, such as differential centrifugation, easily carried out on a laboratory scale, may not be possible on a production line.

If the decision is to isolate the organelle, then there are standard ways of going about this, depending on the nature of the organelle and the tissue being used. Some form of homogenization that does not disrupt the organelle (or if it does, keeps the desired proteins on the membrane fragments) must be used. This is followed by a centrifugation, resuspension of the particulate matter in a suitable buffer, and recentrifugation for washing off remaining soluble proteins. Differential centrifugation is commonly carried out, the heaviest fragments such as nuclei being removed at low g force, large but low-density organelles such as mitochondria sedimenting next, and the smallest particles such as ribosomes and other microsomal fractions sedimenting only at $100,000\,g$ or more. The supernatant fraction after $100,000\,g$ contains soluble proteins only. But some proteins which were membrane-associated may have been released during the processing of the tissue and so be present in this fraction; it is important to analyze each fraction carefully for the desired protein.

Membrane proteins are categorized as either peripheral or integral. Peripheral membrane proteins are only loosely associated on the surface of the membrane, and can normally be released by mild treatments that do not involve solubilization of the membrane itself. These are the type which might be partially released during the membrane isolation, in which case a whole extraction process may be more suitable. Having obtained the organelle or membrane fragment preparation enriched in the desired protein, release as a soluble form can be achieved by one or more of a combination of the following treatments [25]:

1. Sonication;
2. Metal chelators such as EDTA, EGTA at $1-10\,mM$;
3. Mild alkaline conditions (pH 8–11) at low ionic strength;
4. Dilute non-ionic detergent;
5. Low concentrations of partially miscible organic solvents such as n-butanol;
6. High ionic strength, e.g., $1\,M$ NaCl;
7. Phopholipase treatment.

The protein, once released into solution, should remain soluble during further processing even without the extractant present, since the process has been to disrupt the weak salt or hydrophobic interactions with the membrane. Removing the extracted membrane by centrifugation is the first step after extraction.

Integral membrane proteins present a much greater problem. Since they are embedded within and often right across the membrane, it is normally necessary to solubilize the membrane itself in order to release the protein. And once in solution, it will be necessary to retain a lipophilic component (usually a detergent) together with the solubilized protein to prevent it from aggregating. For the surfaces of an integral membrane protein are hydrophobic in so far as they interact with the lipids themselves, and in isolation, such hydrophobic patches will mutually attract, resulting in aggregation. Solubilization requires detergent, and this will need to be retained throughout the purification. The presence of detergents causes further problems as described below.

Table 2.2. Some detergents commonly used for extraction of membrane proteins (many of the names are trade names). In general, the ionic detergents are more solubilizing, but also are more likely to denature the solubilized proteins

Nonionic:
    Tween 80
    Triton X-100
    Triton X-114
    Emulgens
    Lubrol
    Digitonin
    Octyl glucoside
Zwitterionic:
    Lysolecithin
    CHAPS
    CHAPSO
    Zwittergents
Ionic:
    Cholate
    Deoxycholate
    Cetyl trimethylammonium bromide
    Dodecyl sulfate

Whereas virtually all proteins can be solubilized with ionic detergents such as sodium dodecyl sulfate (SDS) or cetyl trimethylammonium bromide (CTAB), they will also mostly be denatured, often irreversibly. Even if a denatured product is acceptable, normal separation procedures are not suitable for use with denatured proteins. So the initial extraction of the membranes must be with a gentler detergent that retains the native conformation of the protein, even if bioactivity is lost as a result of its extraction. Many and varied detergents are available and described in greater detail than here in several sources [26,27,28]. Some of the more commonly used detergents for solubilizing membranes are listed in Table 2.2. The structure of one of the most frequently used nonionic detergents, Triton X-100, is shown in Figure 2.4.

By definition, the "gentler" detergents are less efficient at solubilizing, and higher concentrations are required than with the more "vigorous" types. It is unfortunate that the gentlest and most refined detergents not only need to be used at higher concentrations, but are also the most expensive. Solubilization is not as clear-cut as simply centrifuging at $30,000\,g$ and seeing if the desired protein is in the supernatant. Detergent treatment of membranes may cause them to fragment into vesicles, with the protein still in an insoluble form, but not sedimenting except at high g force. If the desired protein is still in the supernatant after a $100,000\,g$ spin for 60 min, then it can be considered solubilized. Nevertheless, it could still be associated with detergent micelles (see below), or partially aggregated with other membrane proteins.

When considering what concentration of detergent to use, some account must be taken of the *amount* of detergent relative to the membrane fraction. If, for example, the membrane suspension is $5\,mg\,ml^{-1}$, then 0.1% detergent is only one-fifth of the amount of membrane in milligrams, and will certainly not be enough to solubilize it. As a rule-of-thumb, 1 mg of membrane requires a minimum of 2 mg of detergent. This is approximately 1 mg of detergent for each mg of protein and 1 mg for each mg of lipid. So, a $5\,mg\,ml^{-1}$ suspension requires at least 1% $(10\,mg\,ml^{-1})$ of detergent. On the other hand, at a later stage in purification, when there is no lipid present, and the protein concentration is less than $1\,mg\,ml^{-1}$, 0.1% detergent may be more than adequate to retain solubility and bioactivity.

All detergents form micelles. These are enclosed clusters of detergent molecules with the hydrophilic portion on the outside, and the lipophilic portion inside (Figure 2.5). (In organic solvents, reversed micelles form, with the lipophilic portion outside.) But at very low concentration, there is not enough detergent for the micelles to form. Above a certain value, called the critical micelle concentration, the detergent molecules cluster together into micelles, and can cause complications in protein purification protocols. The micelle clusters have a size comparable with proteins; typically, micellar molecular weights are in the range 30,000 to 100,000 [27]. Consequently, they behave in a similar fashion in size-fractionation procedures, and by associating with the proteins can cause very poor resolution in other classical methods.

Extraction of membranous fractions with detergent is likely to result in a mixture of solubilized lipid, free protein, protein tightly associated with detergent, and free detergent, both as single molecules and in micellar form. The first step in fractionation (see Chapter 9) must be able to deal with this, remove excess lipid, and remove some unwanted protein components.

Some membrane components are so intractable that in order to get them into a soluble form, they must be denatured using a "vigorous" ionic detergent. Purification of proteins in a denatured state is not an easy process, but one bonus is that preextraction in milder conditions solubilizes and so removes most of the other proteins. It may be sufficient to extract with sodium dodecyl sulphate and run an electrophoretic sepa-

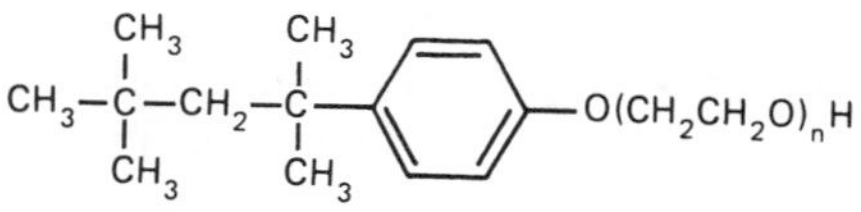

Figure 2.4. Structure of Triton X-100.

$n = 9 - 10$

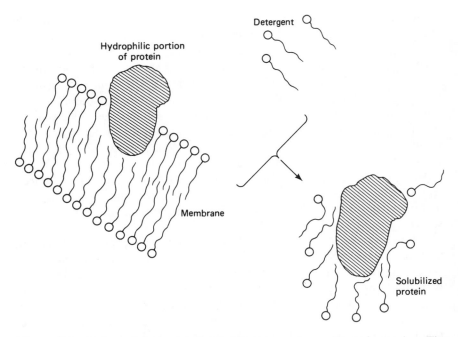

Figure 2.5. Action of detergent in solubilizing membrane-located proteins. The solubilized protein may be as illustrated, or bound in a more complex detergent-micelle, depending on the relative proportions of proteins and detergents.

ration to purify the desired component—provided that the latter can be identified.

A useful method of extraction of membrane proteins which should be included is the use of Triton X-114 [29]. Many detergents have limited water-solubility at elevated temperatures, and on warming a solution, separation into an aqueous phase containing little detergent, and a heavier, detergent-rich phase will occur. This can be a quite abrupt phenomenon with a critical temperature characteristic of the detergent. With Triton X-114 this occurs as low as 20°C. Extraction close to 0°C using 1–3% Triton X-114 should solubilize all peripheral and most integral membrane proteins. After centrifuging off insoluble residue, the extract is warmed to 25°C, at which point phase separation takes place; a brief, gentle centrifugation results in an oily, detergent layer containing most of the integral membrane proteins, and an upper, aqueous layer, which has only a small concentration of detergent in it, containing most of the peripheral proteins. It is also possible to induce phase separation at low temperatures in other detergents by increasing salt concentrations. Because the integral membrane proteins are contained in a phase consisting of a very high detergent concentration, they may be inactivated in the process.

There have been a number of attempts at purifying proteins in non-aqueous solvents, especially membrane proteins which are particularly apolar. Included in these is the use of 100% aqueous w/v choral hydrate [30]. (This is not totally nonaqueous; its use may become restricted due to the stratospheric ozone-destroying nature of chloral hydrate.) Purification of proteins with detergent into reverse micelles has been reported [31,32], and manipulation of some proteins which are soluble in non-aqueous solvents has been successfully achieved [33,34,35].

Chapter 3
Analysis—Measurement of Protein and Enzyme Activity

3.1 Methods for Measuring Protein Concentration

During a protein purification procedure, the most important thing to follow is the recovery of the protein you are isolating: by enzyme activity if it is an enzyme; by bioactivity of a non-enzymic protein; or simply by some convenient method of quantitating the desired component, e.g., electrophoretic analysis. The second important thing to know is the *total amount* of protein present in the fraction in which you have collected your enzyme/specific protein. From these two measurements, the percent recovery and the degree of purification can be calculated. Methods used for measuring enzymic activity will be described in Section 3.2; methods for measuring other sorts of bioactivity or quantitating specific proteins are, obviously, characteristic for each problem and so cannot be discussed in general.

In addition to enabling the recording of the degree of purification, the measurement of total protein at a given step may be important in the following circumstances:

1. When a fractionation step is critically dependent on the protein concentration;
2. When it is necessary to know whether a particular step has really removed much unwanted protein;
3. To test the specific activity of the final preparation to see if it has reached the maximum value expected for the pure protein.

At other times, it is useful to know how things are progressing as far as protein content is concerned, but often one gets sufficient information from experience—just how large a particular precipitate is, how big the peak on the recorder was, etc. There is often no real need to know accurately what the protein content is at all stages, although it is generally no trouble to find out. Indeed, it is possible to carry out routine enzyme preparations without *ever* measuring protein content, or perhaps just measuring it in the final preparation. One should not get held up during a fractionation procedure while measuring protein—small samples can be put aside and measurement done later.

Some of the available methods for measuring protein are now described. Several which require special apparatus (e.g., Kjeldahl nitrogen, refractometry) will be omitted. The most widely used methods are:

1. Biuret–alkaline copper reagent (now mostly in analytical laboratories);
2. Lowry-Folin-Ciocalteau reagent;
3. UV absorption at 280 nm (aromatic band) or 205–220 nm (peptide band);
4. Dye binding;
5. Bis-cinchonic acid (BCA) reagent.

Each of these has particular advantages and disadvantages, but it is worth noting that *no* method, with the exception of dry weight determination, gives an answer that is unambiguously correct to within a few percent unless it has been calibrated against a protein solution of identical composition. In practice, this means that really accurate measurements can be made only for pure proteins, after those pure proteins have been standardized for that method against a dry weight determination. If the protein contains nonproteinaceous prosthetic groups, carbohydrate, or nucleic acid, dry weight determination gives a value for the whole molecule rather than just the protein portion. The principles of the five methods are as follows.

Biuret Reaction

This involves a strongly alkaline copper reagent which produces a purple coloration with protein. The principal reason why it is not widely used in research is its low sensitivity; several milligrams of sample must be sacrificed for a reliable measurement. The method gives a fairly accurate measurement since there is little variation in color yield from protein to protein. This is because the copper reagent reacts with the peptide chain rather than side groups. Ammonia interferes by complexing with copper, so ammonium sulfate fractions do not give accurate results. A greater sensitivity of the biuret reaction is achieved by observing the copper–protein complex not at 540 nm, but at 310 nm in the near-ultraviolet [36]. Unfortunately, this method needs several blanks to subtract spurious absorption at this wavelength, so it has not been widely employed.

Lowry Method

The most cited reference in biochemistry is that of Lowry, Rosebrough, Farr, and Randall [37] on a method for protein estimation. A combination of the copper reaction used in the biuret method and the reaction of the Folin-Ciocalteau reagent [38] with phenols such as tyrosine in proteins, was found to give a strong, dark, blue color with proteins. The original paper surveyed ranges of concentrations of reagents to determine the ideal values. Since then many modifications have been reported, mainly to avoid interference by specific components. Unfortunately, many of the compounds used in enzyme purification interfere with the reaction, and different modifications are required to overcome each problem. But since it is a sensitive method giving a good color with $0.1\,\mathrm{mg\,ml^{-1}}$ of protein or less, interfering compounds are often diluted out to levels where their effect is insignificant. Extensive reviews of the method and the many variations on it have been published (e.g., [39]), the chemistry of the reaction reviewed [40], and a general formulation of the assay for specific purposes has been presented [41].

UV Absorption

This has been employed for measuring protein ever since Warburg and Christian reported a method based on 260 nm and 280 nm absorption, which would correct for nucleic acid and nucleotide content [42]. This correction is very important with crude extracts, but becomes less relevant as protein purification progresses and interfering compounds are removed; a simple 280 nm reading is then sufficient. Proteins absorb at 280 nm solely because of tyrosine and tryptophan residues (unless they also contain UV-absorbing prosthetic groups). Since the quantity of these two amino acids varies enormously from one protein to another, the extinction coefficient, usually expressed either as $E_{280}^{1\%}$ or $E_{280}^{1\,\mathrm{mg\ ml^{-1}}}$, varies considerably. Most proteins fall in the range (for $1\,\mathrm{mg\,ml^{-1}}$) 0.4–1.5, but extremes include some parvalbumins and related Ca^{2+}-binding proteins completely lacking either tyrosine or tryptophan ($E_{280} = 0.0!$), and the tryptophan-rich lysozyme ($E_{280} = 2.65$). So, absorption at 280 nm gives only a rough idea of the actual protein content, except with pure proteins of known extinction coefficient. The extinction coefficient for the pure protein should be standardized against dry weight (or at least against 205 nm absorption—see below). If this is done, then the reading at 280 nm provides the most accurate measure of a pure protein, since no manipulation other than appropriate dilution is required. Of course, the solvent must not absorb at 280 nm, or if it does, an accurate blank must be taken. The method is nondestructive, so, although up to a milligram may be required for accurate determination, this can be returned to the bulk sample.

Absorption in the far-UV around the peptide band is used as a far more sensitive method, much less affected by the protein amino acid composition [43,44]. Because of far-UV-light absorption by oxygen, it is not possible to measure at the peak of the peptide absorption at ca. 192 nm using routine spectrophotometers. But measurements on the side of the band give sufficiently accurate results provided that the absorbance is kept down to no more than 0.5. The extinction coefficients for proteins (1 mg per ml solution) are approximately:

220 nm:	11
215 nm:	15
210 nm:	20
205 nm:	31
200 nm:	45
Peak at ~ 192 nm:	60

These values vary from protein to protein because aromatic and some other residues have some absorbance in this range, and the secondary structure (α-helix, β-sheet, etc.) has some influence on the shape and exact position of the peptide absorption peak. By analyzing possible variations, it can be shown that the most consistent value is obtained around 205 nm [43]. The extinction coefficients of proteins nearly all fall in the range 28.5–33 at this wavelength. By making a correction for the tyrosine and tryptophan content, which are the predominant contributors to the side-chain absorbance at this wavelength, the extinction coefficient at 205 nm can be predicted with no more than 2% error [45]. The correction is simply made by measuring the absorbance (on a more concentrated sample) at 280 nm also. The formula used is:

$$E_{205}^{1 \text{ mg ml}^{-1}} = 27.0 + 120 \times \frac{A_{280}}{A_{205}}$$

Since 205 nm is close to the useful limits of routine spectrophotometers, measurements at 205 nm require very clean cuvettes, a relatively new deuterium lamp, and buffers with minimal absorption. Most salts absorb at this wavelength; about the only common anions that do not are sulfate and perchlorate. One can use a weak phosphate buffer (5 mM, pH 7.0) in the presence of 50 mM sodium sulfate—this salt helps to avoid adsorption of protein to the cuvette. The most convenient method for protein samples containing at least 2 mg ml^{-1} is to zero the spectrophotometer with 3 ml of buffer, then mix in 1–10 μl of the protein sample, and note the increase in absorption at 205 nm. If the buffer composition of the protein sample is known, then the same amount of that buffer can be used to determine a blank. Virtually all commonly used buffers absorb quite strongly at 205 nm, but because the dilution is of the order of 1,000-fold, the actual effect is usually very small. For more dilute protein solutions larger samples must be used, and the buffer blank is relatively larger, so use of the method may be impractical. If less accurate estimates ($\pm 10\%$) are sufficient, then use of $E_{205}^{1\,\mathrm{mg\,ml^{-1}}} = 31$ is an adequate compromise. For use of the 280/205 absorption method, the 280 nm determination must be done on a more concentrated sample and the appropriate factor used in the equation. A slight variant on this procedure has been reported, which gives effectively the same answers [46].

Dye Binding

This method [47,48] has become the most popular because it is very sensitive, fast, and at least as accurate as the Lowry method. The only problems are technical; the dye adsorbs to glassware and to cuvettes (and skin!), and may build up in cuvettes with multiple use. The dye used is Coomassie Blue G-250. This, when dissolved in a strong acid, turns a red-brown color due to protonation, but when it binds to a positively charged protein, the blue color is restored due to a shift in the pK_a of the bound Coomassie Blue. The procedure is simply to add a sample of protein to the reagent and measure the blue color at 595 nm. As the method is an equilibrium-binding process, the response is not linear, but approximates to a hyperbola. The response to a given amount of protein depends on the exact pH of the mixture, so it is important that the sample added does not include a significant amount of buffer, alkali, or strong acid. At lower pH the sensitivity is less, but the zero blank is low. A higher pH results in more color for a given amount of protein, but the blank also absorbs more strongly. The standard conditions give an absorbance of about 0.3 at 595 nm for $10\,\mu\mathrm{g\,ml^{-1}}$ of protein, but it is essential that a standard curve be carried out frequently to check the response of the reagent. Single use of disposable cuvettes is recommended to avoid buildup of color, but adsorbed blue dye can be quickly removed with a rinse in dilute sodium dodecyl sulfate or similar strong detergent. Be sure to remove all traces of detergent before using the cuvette again. The color that develops depends on the composition of the protein to a limited extent, so accurate determinations require standardization with an appropriate protein mixture. In our experience, using perchloric acid [49] instead of the more commonly used phosphoric acid [48] gives a more consistent response with different proteins, and a stabler reagent. A number of other refinements have been made (e.g., [50]).

Bicinchonic Acid

The bicinchonic acid (BCA) reagent is patented by Pierce Chemicals (Rockford, IL), and is provided by them in a ready-to-use form. The principle is somewhat similar to the Lowry method, in that it involves reaction of the protein with copper, which is reduced and then combines

Table 3.1. Relative Advantages and Disadvantages of Some Commonly Employed Methods of Protein Determination

Method	Amount of protein needed (mg)	Destructive?	Variation of response with amino acid composition	Comments
Biuret	0.05–5	yes	Low	Caustic reagent; NH_4^+ interference; rapid color
Lowry	0.05–0.5	yes	Moderate	Slow color development, many interfering compounds
Absorbance at 280 nm	0.05–2	no	Large	Interference by UV-absorbing materials; instantaneous
Absorbance at 205 nm	0.01–0.05	no	Low	Interference by UV-absorbing materials; instantaneous
Dye binding	0.01–0.05	yes	Moderate	Acid reagent: color adsorbs to glassware; rapid color formation
BCA reagent	0.005–0.05	yes	Moderate	Long incubation, warming required

with bis-cinchonic acid to produce a purple coloration. There is only one addition, the mixed reagent containing both the copper and the BCA. Color development for 30 min is recommended. The sensitivity is somewhat higher than the Coomassie dye-binding method, and the reagent is compatible with many detergents, but strong reducing agents, e.g., thiols, interfere. The principles of the method are described in [51].

There are several methods designed for extreme sensitivity; the detection of nanogram and even lesser amounts is possible using gold, iron, and Indian ink-based stains [52]. But these are not appropriate for general use.

This brief summary of some of the more popular methods of protein measurement gives an idea of what is possible. Each has its advantages and disadvantages, summarized in Table 3.1. The objective of an enzyme purification is to get rid of as much unwanted protein as possible while retaining enzyme activity, and the success of each step depends both on the retention of activity and the extent of improvement in specific activity, expressed in activity units per milligram protein.

Detailed recipes for some of the reagents used for protein estimation are given in Appendix B.

3.2   Measurement of Enzyme Activity— Basic Principles

Most people approaching the problem of enzyme purification will be following a standard procedure for measuring enzyme activity. But in some cases, it may be useful to develop an alternative assay, for instance, a rapid spectrophotometric method to avoid queuing for the use of a scintillation counter. On rare occasions a newly discovered enzyme will need a new assay method. Ultimately, the biochemist wants to know the relationship between an enzyme's activity *in vitro* and its physiological situation. But the pH, ionic strength, and, particularly, the substrate concentration *in vivo* are likely to be quite different from the conditions used to measure the enzyme activity during a purification process. What is required here is a reliable, reproducible assay which will relate the amount of enzyme in one fraction with another, regardless of what the "activity" may mean in a physiological context.

Enzyme activity is affected by the concentrations of its substrate(s), activators, and inhibitors specific for the enzyme, nonspecific effects of compounds such as salts and buffers, pH, ionic strength, temperature, and in some cases interactions with other proteins or membranous material that might be present. Usually one aims at having all conditions optimal, so that the enzyme exhibits maximum activity (V_{max}). But this is not always possible, for reasons of substrate cost or solubility, or perhaps compatibility with conditions required by coupling enzymes for activity detection. These conditions will be discussed below.

Substrate Concentration, Activators, and Inhibitors

Activity is nearly always greatest at the highest feasible substrate concentration. If the enzyme obeys Michaelis-Menten kinetics, then a substrate concentration of at least 10 times the K_m should be used. At $10 \times K_m$, the rate is 91% of the theoretical rate at infinite concentration. Remember that the concentration of one substrate will usually affect the K_m of another in a way that depends on the type of reaction mechanism. For instance, for many sequential mechanisms, the higher the concentration of substrate B, the smaller is the K_m for A, and vice versa. So a high concentration of one of the substrates (e.g., the cheapest) would mean that less of the other is required to achieve a concentration of $10 \times K_m$. On the other hand, nonsequential (Ping-Pong) mechanisms have the peculiar property that the higher the concentration of one substrate, the

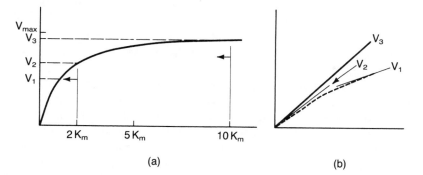

Figure 3.1. Effect of substrate consumption on enzyme activity: (a) relationship between rate and substrate concentration; (b) observed rates as substrate is consumed, starting with [substrate] $= 10K_m$ (*solid line*) and [substrate] $= 2K_m$ (*dashed line*).

higher the K_m (lower the apparent affinity) for the other, so to approach V_{max}, both substrates may be needed at quite high concentration.

There are two reasons for operating well above the K_m value. First, slight variation of substrate concentration from one set of assays to another makes very little difference. A 10% error may result in only 1% change in rate, barely detectable. Thus the concentration of the substrate, which may be difficult to determine, need not be known accurately. Second, consumption of substrate during the assay will make little difference to the rate (Figure 3.1), and, if products are accumulating, any inhibitory effect they may have is less if the substrate concentration is high. In the example shown in Table 3.2, the product binds to the enzyme as tightly as the substrate; if such a product is not removed during the assay, the rate will decline steadily and the true *initial* activity will be underestimated. Readily reversible enzymes such as isomerases fall into this category. It is sometimes technically convenient to measure

Table 3.2. Rate of Enzyme Under Different Situations of Substrate Concentrations

	Concentration of substrate	Concentration of product	Rate (% V_{max})
Initial concentration	$10\,K_m$	0	91
Final, with product removed	$9\,K_m$	0	90
Final, with product accumulating	$9\,K_m$	$1\,K_{ip}$	82
Initial concentration	$2\,K_m$	0	67
Final, with product removed	$1\,K_m$	0	50
Final, with product accumulating	$1\,K_m$	$1\,K_{ip}$	33

Key: K_m = Michaelis constant for substrate; K_{ip} = inhibitor concentration for product. K_{ip} is assumed to be equal to K_m.

an enzyme's activity in the direction where ΔG^0 is positive, i.e., the equilibrium favors the substrates; in such cases removal of *all* products during the assay is desirable. Removal of products is a natural result of coupled assays (see Section 3.4), but inclusion of systems to remove products even in a stopped assay can be useful.

Any assay takes a finite time to complete; disregarding the attainment of the initial steady state, which takes only milliseconds, the rate should be linear from zero time up to the time that the product formation can be observed or measured accurately. To overcome problems of product inhibition and approach to equilibrium, various strategies can be used. These include using very high concentrations of substrate, unphysiological pH (if a proton is involved in the reaction), or use of a very sensitive method for detecting product, so that little conversion is necessary during the course of the assay. Take as an example the enzyme malate dehydrogenase, which operates in both directions physiologically:

$$\text{malate}^{2-} + \text{NAD}^+ \rightleftharpoons \text{oxaloacetate}^{2-} + \text{NADH} + \text{H}^+$$

The equilibrium is far to the left at pH 7; nevertheless, a high NAD^+/NADH ratio in cells, combined with rapid removal of oxaloacetate, can result in the net reaction progressing from left to right. Inside the cell, of course, it will simultaneously be going in the other direction, it is just that, overall, *more* catalysis in the direction of oxaloacetate formation might be occurring. A lot of the enzyme present would be taken up with catalyzing the reverse direction, so the next reaction (flux) would be far less than the V_{max} condition, and would bear little relationship to the amount of enzyme present. Since one wants to know the amount of enzyme present, conditions must be arranged such that reversal of reaction does not occur. Malate dehydrogenase activity can be detected directly by formation of NADH (or, in the reverse direction, loss of NADH), because NADH absorbs at 340 nm, whereas NAD^+ does not (Figure 3.2). The enzyme can be assayed in the thermodynamically favorable direction using oxaloacetate plus NADH at a pH of 7 or somewhat less, in which case a certain value of activity is obtained. Alternatively, it can be forced to go in the forward direction by (1) a high concentration of malate, typically 50–100 mM; (2) a high pH to remove protons as they are formed—pH 9–10; and/or (3) removal of oxaloacetate by trapping as a hydrazone:

$$
\begin{array}{lcl}
\text{COOH}^- & & \text{COO}^- \\
| & & | \\
\text{CO} & & \text{C}=\text{N.NH}_2 \\
| & + \text{NH}_2\,\text{NH}_2 \rightarrow & | \\
\text{CH}_2 & & \text{CH}_2 \\
| & & | \\
\text{COO}^- & & \text{COO}^-
\end{array}
$$

Figure 3.2. (a) The absorption spectrum of NADH, compared with that of NAD⁺; (b) the fluorescence emission spectrum of NADH, with excitation at 365 nm.

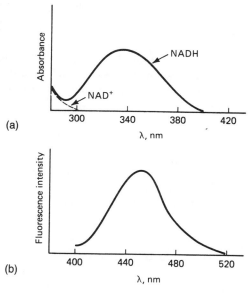

(a)

(b)

If all three of these conditions are used, then the enzyme's activity can be measured as NADH formation. But note that there is no reason why the activity in absolute units should be the same in this direction as in the other. With physiologically reversible enzymes these activities are often similar; in the case of malate dehydrogenase the maximum rate attainable in the forward direction is about half of that in the reverse direction.

The conventional picture of the effect of substrate concentration, hyperbolic kinetics with an asymptote approaching V_{max} (Figure 3.1a), may not apply if there are either allosteric effects or the phenomenon of "substrate inhibition" occurs. With allosteric enzymes, assay conditions are chosen so that any activators needed are present and substrate concentrations are high. Often under these conditions allosteric behavior is abolished. *Substrate inhibition* is the term used to describe the effect of reduced activity which sometimes occurs at high substrate concentration. For example, in the case of malate dehydrogenase assayed in the direction of malate formation, oxaloacetate concentrations even as low as $0.5 \, mM$ lead to a lowering of the rate, due to the formation of a tight oxaloacetate-NAD⁺-enzyme abortive complex. In other cases the reasons for substrate inhibition are not always clear, but with ionic substrates it can be due to increases in ionic strength inhibiting the enzyme.

It is also possible that a substrate preparation can include an inhibitor. Unfortunately, if it is a competitive inhibitor (most likely), this will not be detected except by comparison with another, better grade of substrate. Mathematically, one can describe these situations as follows:

For a competitive inhibitor:

$$\frac{V_{max}}{v} = 1 + \frac{K_m}{s}\left(1 + \frac{i}{K_i}\right)$$

where v = rate observed; s = substrate concentration; i = inhibitor concentration; K_i = enzyme-inhibitor dissociation constant.

If the inhibitor is present in the substrate, then i is proportional to s,

$$\frac{V_{max}}{v} = 1 + \frac{K_m}{s} + \beta$$

where β is a constant.

At infinite s, $V_{max}/v = 1 + \beta$; thus, both the determined V_{max} and K_m are less than the true values by a factor of $1/(1 + \beta)$, but hyperbolic kinetics apply (Figure 3.3).

If the inhibitor is noncompetitive, the general equation is:

$$\frac{V_{max}}{v} = 1 + \frac{K_m}{s} + \frac{i}{K_i} + \frac{K_m \cdot i}{s \cdot K_i'}$$

With i proportional to s:

$$\frac{V_{max}}{v} = 1 + \frac{K_m}{s} + \alpha s + \beta$$

where α and β are constants. Thus, a plot of $1/v$ against $1/s$ would be nonlinear, and apparent substrate inhibition would occur (Figure 3.3).

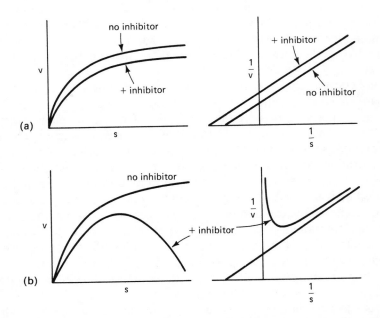

Figure 3.3. (a) The effect of a competitive inhibitor present in the substrate preparation; rate versus substrate, and reciprocal plots; (b) as (a), but with a noncompetitive inhibitor in the substrate.

In conclusion, there are a number of pitfalls when choosing a suitable substrate concentration. Questions of economy may override ideality; in any case, *routine* assays are not necessarily required to be absolutely correct and accurate. But for presentation of final values for publication, accurate figures are needed for reproducibility in other laboratories.

pH, Ionic Strength, and Temperature

Having stated that substrate concentrations should be, if possible, at least 10 times the K_m values, it should be made clear that K_m values are dependent on other influences, such as pH, ionic strength, and temperature. Similarly, the catalytic rate constant k_{cat} (where $V_{max} = k_{cat} \times$ total enzyme) is dependent on these variables. Thus, although the optimum pH is regarded as that pH value (or range of values) where a maximum activity is obtained, a lower activity on changing pH could be due to a sharply rising value of K_m, rather than a decline in V_{max}. Some enzymes have optimum pH values well away from the physiological value, when measured at high substrate concentration. For instance, alkaline phosphatases, as their name implies, may be optimally active at pH 10–11, yet naturally operate at a pH close to 7. This is because physiologically, K_m is much more important than V_{max}; the natural substrates occur only at very low concentrations. The K_m in this case is much lower at pH 7 than at pH 10.

To establish a reproducible assay for enzyme purification, these subtleties are relatively unimportant; using a suitably high substrate concentration, the rates at a range of pH values can be tested, and the optimal value of v (not necessarily V_{max}) is found. The nature of the buffer can be very important here, since many buffers can upset the behavior by directly or indirectly acting as inhibitors. Phosphate or other multiple-charged anionic buffers such as citrate often compete with a negatively charged substrate by binding weakly to the positively charged active site. Since buffer concentrations used are typically between 10 and 100 mM, even a weak association may inhibit significantly. On the other hand, salts may be essential for activity, especially with enzymes from extreme environments such as found in thermophilic or halophilic bacteria. Metal ions essential for the enzyme's activity may be complexed by the buffer; particularly effective metal-complexing buffers are citrate and histidine. Finally, the ionic strength of the buffer plus any other salts present could affect the enzyme's activity. Physiological ionic strengths are normally in the range 0.1–0.2 M. Enzymes are not often less active at lower than the physiological value (and sometimes are much more active), but as salt concentration increases above 0.2 M, activity is often depressed. (Remember when testing an ammonium sulfate supernatant not to use much sample, unless it has been established that this salt is not inhibitory.)

Finally, one must consider the temperature at which assays are carried out. The concept of "optimum temperature" of an enzyme is often misunderstood; it is not a fundamental property, but, rather, depends on the mode of assay, and, in particular, the time of incubation. Enzyme reaction rates, like almost every other chemical process, increase with temperature, typically by a factor of between 1.5 and 2.5 for every 10°C (2.0 is a good mean value). At high temperatures the counteracting force of protein denaturation means that, in the finite time needed to determine the activity, some enzyme is lost by denaturation, so increasing temperature eventually leads to decreasing product formation. The shorter the incubation time, the higher will be the *apparent* optimum temperature, since there is less time for enzyme to denature. Typical apparent optimums for enzymes from mesophilic organisms and animals, for assays taking 5–10 min to complete, are in the range 40–60°C. Much higher values necessarily occur with enzymes isolated from thermophilic organisms, with examples extending to over 100°C.

For a routine assay it would be inconvenient to use high temperatures. Moreover, some standardization of temperature would be a desirable principle. The standard temperature in chemistry has stood for a long time at 25°C (298 K), and this is often the temperature of choice. However, use of 37°C for animal enzymes is common, being more physiological. Thermophilic enzymes may have so little activity at 25°C that a more relevant temperature, such as 50°C, may be chosen. Sometimes it is not possible or convenient to maintain the temperature at exactly the value demanded, in which case a correction can be applied. The value of Q_{10} (activity at $T + 10$°C divided by the activity at T°C) is easily determined. Use a Q_{10} of 2.0 (7% per degree) if the exact value for a particular enzyme is unknown. Note that, as in the case of pH, K_m also changes with temperature, and Q_{10} refers strictly to V_{max} rather than just v. It is not always appreciated how much difference temperature makes; sometimes students try to squeeze activity to the last 1% of accuracy out of enzyme activity data, and then do not know to within several degrees what the temperature was! To quote an enzyme activity to within 5%, it is essential that you know the temperature of the assay solution to the nearest 0.5°C.

3.3   Measurement of Enzyme Activity Using Stopped Methods

To measure the activity of an enzyme requires a method for determining the amount of product formed (or sometimes the amount of substrate used). This may involve separating substrates and products after the reaction; alternatively, it may be possible to measure one without interference from the other. It is convenient to divide the techniques for measuring enzyme activity into two classes: stopped methods and continuous methods.

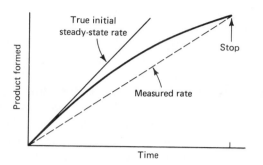

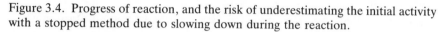

Figure 3.4. Progress of reaction, and the risk of underestimating the initial activity with a stopped method due to slowing down during the reaction.

The latter will be described in the next section; continuous methods allow the progress of reaction to be monitored continuously and instantaneously, which is a great advantage. However, continuous methods have some limitations, and with many enzymes there is no suitable method available. Stopped methods can be used for an almost limitless range of enzymes, but they are generally less convenient.

In a stopped method the enzyme is incubated with its substrates for a fixed period of time, and the reaction is then stopped by one of several techniques. The product formed is then measured, and the rate assumed to be linear from time zero. One of the first things to be checked when using a stopped method is this assumption of linearity. A single point may underestimate the initial steady state if, by the time the reaction is stopped, it is already slowing (Figure 3.4). Alternatively, a lag in the enzyme activity would result in an underestimate from a single point observation. By carrying out a series of incubations at different times, linearity can be checked; it is especially important with crude tissue extracts, since many side reactions are possible; the next enzyme in a metabolic sequence may be removing the product that one is trying to measure. Further discussion on measuring enzyme activity of crude tissue extracts can be found elsewhere [53].

Incubation Conditions

For stopped methods there are no restrictions on incubation conditions; the optimum pH can be chosen, but on occasion it may be preferable to work at a pH somewhat away from optimum to shift the equilibrium to a more favorable value. Temperature is easily controlled, and the reaction mixture can be a very small volume—50 to $100\,\mu l$ is common—which allows economization on reagents. The sample containing the enzyme is added at time zero and rapidly mixed in; the reaction occurs for a fixed time and is then stopped. To stabilize dilute enzyme solutions, bovine serum albumin $(0.2-1\,mg\,ml^{-1})$ is often included in the reaction mixture.

Stopping Methods

The enzyme reaction is stopped by a method which either causes instantaneous denaturation of the enzyme or shifts the conditions so that the enzyme is no longer active, though not necessarily denatured. The commonest methods involve acidification (usually with the protein-precipitation acids trichloroacetic acid or perchloric acid) or rapid heating. The latter method has some advantages, particularly if it is undesirable to dilute the sample; but the heating must be very rapid. Very small samples in thin-walled glass (not plastic) tubes placed in a boiling water bath can reach 90°C within a second, which should be sufficient to denature most enzymes. But the rate of denaturation must be high enough to completely inactivate the enzyme within that second. Otherwise, the portion of undenatured enzyme will be reacting very fast, and may yield a significant amount of additional product, especially if the activity incubation was short. Heat denaturation is best used with long incubations (>10 min), so that any additional product formed during the heating is proportionately small.

Trichloroacetic or perchloric acids at a final concentration of 3–5% stop the reaction instantly because of the very low pH (~0), and because the enzyme denatures and is precipitated. After removing the precipitate by centrifugation, the supernatant may need neutralization before product measurement. KOH containing some K_2CO_3 is used; most perchloric acid is removed because the salt $KClO_4$ is sparingly soluble, especially at 0°C. But K-trichloracetate remains in solution. Carbonate is useful in that it provides buffering at around neutrality, avoiding overshooting with possible destruction of any alkali-labile product. An internal indicator such as bromothymol blue can be used, provided it does not interfere with product measurement. Some very acid-stable enzymes may not totally precipitate with these acids, and, on neutralization, can reactivate; this has been known to occur with adenylate kinase and potato ATPase, among others. A zero-time stopping of reaction will detect such occurrences. The zero-time experiment should in any case be carried out as it represents the true "blank" for the assay. The product being measured could be present in the enzyme sample or in the reagents, causing a non-zero blank.

Another very useful stopping method, again provided it does not interfere with subsequent product measurement, is to use sodium dodecyl sulfate at about 1% concentration. Most enzymes are rapidly denatured by this detergent, but remain in solution.

The reaction may be stopped by nondenaturing methods, provided that subsequent manipulations do not restore the conditions to allow further reaction. For example, an enzyme active only below pH 8 may be stopped by raising the pH to 9 and keeping the pH at that value. An enzyme having an absolute requirement for divalent metal ions may be stopped with an excess of ethylenediaminetetraacetate (EDTA), provided that the product can be measured or be separated from the reactant in the presence of this compound. In radiochemical assays, the reaction can be stopped by swamping with cold substrate so that very little extra labeled product is formed. Similarly, adding a large dose of a known inhibitor can be employed. There are many ways of stopping enzyme reactions which apply only to particular enzymes.

Measurement of Product

There is no limitation on the methods that can be employed for measuring the product, provided that the stopping procedure has not introduced interfering substances. The three major methods are enzymatic, chemical, and radiochemical.

Enzymatic methods use purified enzymes to convert the product directly or indirectly to make an observable compound. Spectrophotometry or fluorometry are the commonly employed methods, though various other procedures (e.g., bioluminescence, gas production) are possible. If an enzymatic method is to be used, then it can often be adapted to a continuous rather than stopped assay method. But sometimes the stopped method is more convenient, more accurate, or less expensive. Because of the specificity of enzymes, separation of product from reactants should not be needed. The product is acted on by one or more enzymes, and the reaction should preferably go to completion. Since the added enzymes do not have to "keep up with" the enzyme being assayed, as they do in continuous methods (Section 3.4), much smaller amounts can be used, with the reaction progressing leisurely over a period of 10 min to an hour or so if necessary. Enzymatic analysis of metabolites requires amounts of enzyme that result in near-completion of reaction in an appropriate time. It can be shown that for 98–99% completion in 10 min, the minimum number of units of enzyme per milliliter (V_{max}) is given by $V_{max}/K_m = 0.5 \, \text{min}^{-1}$ [54], where K_m is the Michaelis constant of the coupling enzyme for the relevant substrate. The process is illustrated in Figure 3.5.

As an example, consider the assay of fructose 1,6-bisphosphatase in the presence of phosphate, in which case the alternative method of measuring phosphate production is not possible:

First incubation:
  Sample + fructose 1,6-bisphosphate + P_i + activators, inhibitors, etc.
  Stop reaction (e.g., perchloric acid)
  Measure fructose 6-phosphate: see Figure 3.5, in which

$$E_P = \text{phosphoglucose isomerase}$$
$$E_Q = \text{glucose 6–phosphate dehydrogenase} + \text{NAD(P)}^+$$

$$\text{Fructose 6-P} \xrightarrow{\quad E_P \quad} \text{Glucose 6-P} \xrightleftharpoons[\quad]{\quad E_Q \quad} \text{6-P-Gluconate}$$
$$\text{NAD}^+ \qquad \text{NADH}$$

Observe the total increase in absorbance at 340 nm (*not* the rate).

Using known values for K_m, we can show that the amount of E_P needed is about 0.5 unit ml^{-1} to give the desired 99% completion in 10 min, and somewhat less for E_Q as its K_m is lower. This assay for fructose bis-phosphatase can be carried out in a continuous assay, but *at least* 5 times more coupling enzymes are needed (see Section 3.4).

Chemical measurement of product P cannot be generalized because of the wide variety of techniques used. If it is not possible to measure P in the presence of remaining substrate S (both are likely to be similar structurally), a separation procedure will be needed. Separation is always needed for radiochemical measurement and is discussed below. If no separation is needed, the chemical reagents can be added directly to the stopped solution, often without prior neutralization (assuming acidification was the method of stopping). To simplify procedures and eliminate centrifugation of denatured protein, some stopping method may be desirable which does not cause precipitation of protein (either during stopping or after adding the chemical reagents). As mentioned above, dodecyl sulfate denaturation leaves the protein in solution. For example, it is convenient to measure phosphate by stopping the reaction with

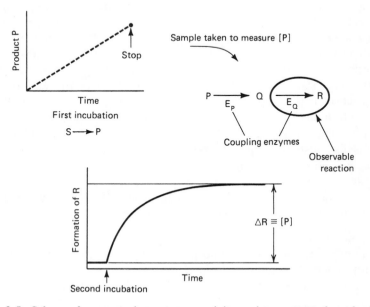

Figure 3.5. Scheme for measuring enzyme activity, using a stopped method followed by enzymatic detection of the products.

Figure 3.6. Problems in separation of substrate from product by chromatography.

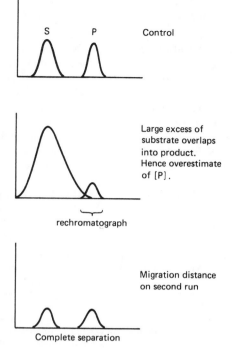

S P   Control

Large excess of substrate overlaps into product. Hence overestimate of [P].

rechromatograph

Migration distance on second run

Complete separation

sodium dodecyl sulfate, then adding nonprecipitating acid molybdate reagents (in sulfuric acid rather than the more commonly used perchloric acid) to measure phosphate. The dodecyl sulfate does not interfere with phosphate color formation, and no protein precipitate forms.

Radiochemical methods depend on the detection of labeled product, and this must be separated completely from the labeled substrate used. This separation step is likely to be the most time-consuming and introduce the largest error into the final answer. If the product and reactant can be separated on a phase basis (e.g., precipitation of a labeled protein from soluble reactants, removal of labeled CO_2 from mixture), the principle is simple, although there can be many practical problems. Otherwise, some chromatographic method is needed. Paper or thin-layer chromatography, and ion exchange or high-performance liquid chromatography are all possibilities. Sometimes they are quite rapid and simple, especially in an all-or-none situation where one of the product–reactant pair adsorbs totally and the other not at all. On other occasions the separation can be very tedious, especially if a very large excess of substrate is still present; the natural diffusion distribution may overlap the product position and give an erroneously high answer if the whole chromatogram is not checked (Figure 3.6).

Table 3.3. Examples of Uncoupled Continuous Assays

Enzyme	Reaction	Observation
Lactate dehydrogenase (1)	Lactate + NAD^+ → pyruvate + NADH + H^+ (high pH)	NADH formation at 340 nm
Lactate dehydrogenase (2)	NADH + pyruvate + H^+ → lactate + NAD^+	NADH loss at 340 nm
Nonspecific phosphatases	NO_2⟨⟩O –Ⓟ→ NO_2⟨⟩OH + P_i	Nitrophenol production at 400 nm
Glycosidases	Methylumbelliferyl-glucoside → sugar + methylumbelliferone	Fluorescence of methylumbelliferone
Xanthine oxidase	Xanthine + O_2 → uric acid + superoxide	Increase in absorbance at 295 nm due to uric acid
Endo-cellulase	CM-Cellulose → oligo β-glucosaccharides	Loss in viscosity
Hexokinase	Glucose + $MgATP^{2-}$ → glucose 6-P^{2-} + $MgADP^-$ + H^+ (high pH)	Acid production (pH-stat)
Chymotrypsin	N-t-BOC-L-phenylalanine p-nitrophenyl ester → N-t-BOC-L-phenylalanine + p-nitrophenol	Nitrophenol production at 400 nm

Note: In some of these examples stopped methods have greater sensitivity, since maximum absorption or fluorescence may be obtained at a pH outside the enzyme activity range.

High-performance liquid chromatography gives a level of resolution which largely avoids the problem described above. In many cases it is not even necessary to use labeled substrate, since very small amounts can be quantitated directly by refractometry or UV absorption, and the technique is becoming widely used for difficult assays requiring high-resolution separation.

In summary, stopped methods have both advantages and disadvantages compared with continuous methods (for comparison see Table 3.4). For routine purposes, the rapidity and ease of a continuous method should make it the method of choice *provided that* such a method is available and has sufficient sensitivity. Radiochemical methods must employ the stopping principle and are usually (though not always) the most sensitive.

3.4   Measurement of Enzyme Activity Using Continuous Methods

Continuous methods imply that the progress of reaction can be observed throughout, and this is generally an instantaneous observation. A fairly large group of enzymes fits into the simple category where the formation of a product of the reaction or the loss of a substrate can be observed directly because of difference in absorption spectrum or fluorescence between product and substrate. Also, other results of net reaction, such as acid production or uptake, viscometric changes, or even heat production, can be followed directly. The main enzymes to which these conditions apply are dehydrogenases using NAD or NADP, many hydrolases which can utilize synthetic substrates which liberate colored or fluorescent products, and polymer endohydrolases which cause viscosity decreases. In theory, any enzyme reaction which has an enthalpy change significantly far from zero can be followed using a sensitive calorimeter. Although not widely used, calorimetric enzyme assays are possible where no other assay is available. All of these techniques could be called "uncoupled continuous assays" to distinguish them from the coupled assays described below. Some examples of uncoupled continuous assays are given in Table 3.3.

Coupled Methods

The principle of coupled methods is that a product which cannot be observed directly itself is removed and reacted on, either chemically or enzymically, to form (directly or indirectly) a compound that can be observed. The added reagents must, of course, be capable of acting in the same conditions (pH, temperature, etc.) as the enzyme being measured, and must not interfere with the enzyme's activity. Most of the methods use the specificity of enzymes to select out the product and continue the mini-metabolic chain until something is formed that can be observed. But there are a few direct chemical methods in which a colored compound is formed. One of the commonest is the formation of p-nitrophenol from an adduct of that compound which is hydrolyzed, for example p-nitrophenol phosphate for phosphatase enzymes.

Above pH 7, p-nitrophenol is ionized to give the yellow phenolate ion with an absorbance maximum at 400 nm. A similar color is produced with the compound 5,5'-dithiobis(2-nitrobenzoic acid) (DTNB), which reacts at neutral pH with sulfhydryl groups to liberate the yellow 5-thio-2-nitrobenzoic acid (TNB), followed at 412 nm (Figure 3.7). Enzymes involving acyl-CoA, liberating free CoA, can be assayed by this method:

$$\text{Acyl-S-CoA} + \text{HX} \rightarrow \text{Acyl-X} + \text{HS-CoA}$$

$$\text{HS-CoA} + \text{DTNB} \rightarrow \text{TNB-S-CoA} + \text{TNB}$$

However, enzymes with a reactive sulfhydryl in the active center are likely to be inactivated. Another example involves electron transfer reactions by such enzymes as hydrogenases: hydrogen uptake and passage of electrons normally occurs to flavodoxins or ferredoxins; this can be followed using the synthetic compound methyl viologen which replaces the ferredoxin. The absorption of this compound in the ultraviolet is largely removed on reduction:

$$\tfrac{1}{2}H_2 \rightarrow [e^-] + H^+ \rightarrow \text{ferredoxin} \rightarrow \text{methyl viologen}$$

There are several other examples of chemical coupling, but enzymic coupling is the most widely used method in continuous assays. The principles are simple, but there are many pitfalls, especially when using the method on a crude extract. The product P is converted by a coupling enzyme E_p to another product Q:

$$S \rightarrow P \underset{E_P}{\rightarrow} Q \underset{E_Q}{\rightarrow} \text{etc.}$$

In the simplest case, the reaction of enzyme E_P can be observed directly. For instance, consider the measurement of the enzyme triose-P isomerase.

$$\text{glyceraldehyde 3-P} \rightarrow \text{dihydroxyacetone-P}$$

The formation of the product, dihydroxyacetone-P, cannot be detected directly. It is removed by the enzyme glycerol 1-P dehydrogenase, involving NADH:

$$\text{dihydroxyacetone-P} + \text{NADH} \rightarrow \text{glycerol 1-P} + \text{NAD}^+$$

Although the product glycerol 1-P cannot be detected, every molecule that is formed results in the loss of a molecule of NADH, so the absorption at 340 nm decreases. The coupling enzyme, glycerol 1-P dehydrogenase, is in this case the "indicator enzyme." It is important that the overall conversion from immediate product (P) to end products is thermodyamically favored, so that essentially all of P is converted.

A general enzymatic method for measuring ADP is illustrated in Figure 3.8. The ADP, produced typically from ATP by a kinase or ATPase enzyme which is being measured, is reconverted to ATP using pyruvate kinase and phosphoenol pyruvate. The pyruvate is then reduced to lactate, with oxidation of NADH by the indicator enzyme, lactate dehydrogenase. This pyruvate kinase–lactate dehydrogenase system is widely used; it

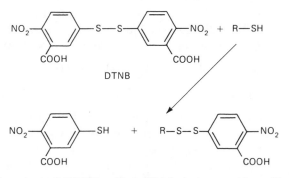

Figure 3.7. Reaction of DTNB with sulfhydryl compound to liberate colored product.

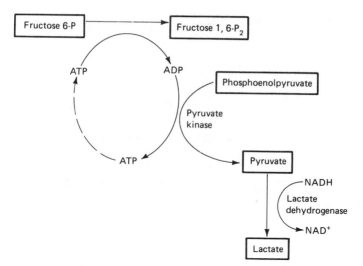

Figure 3.8. Principle of the coupled assay for phosphofructokinase using lactate dehydrogenase as the indicator enzyme.

illustrates an example where one of the primary substrates of the enzyme being measured, ATP, is regenerated, so avoiding problems of inhibition by product.

We can show that for efficient coupling, so that the indicator enzyme reaches a rate of at least 98% of the primary enzyme within 1 min, the minimum amounts of coupling and indicator enzymes needed are given by $V_{max}/K_m = 5\,min^{-1}$. The K_m values (in mM) are for the appropriate substrate in the conditions of the assay; V_{max} is in units of $\mu mol\,ml^{-1}\,min^{-1}$. In practice we try to use 2 to 5 times this amount, because there are many complications, such as unfavorable equilibria, constraining the coupling process. But if the coupling enzymes are expensive, or possibly not as pure as desirable, it may be necessary to try minimizing the total amount added. The progress of reaction for a two-step coupled enzyme assay is illustrated in Figure 3.9.

If the coupling enzymes available contain contaminating activities that affect the assay, then there can be problems. Sometimes it is not possible to get an accurate measure of the required enzyme in the crude extract. Sometimes the contaminant is the desired enzyme itself, in which case a rate will be observed even in supposedly blank samples. If this is not great, then a simple correction for it with each assay is all that is needed (Figure 3.10). Contamination of this sort is obvious and easy to correct for, although it will limit the sensitivity of the assay. Other contaminants may affect the result, especially when measuring relatively crude samples at an early stage of purification, by causing alternative reactions with substrates, removing products, or producing color that interferes with

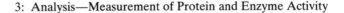

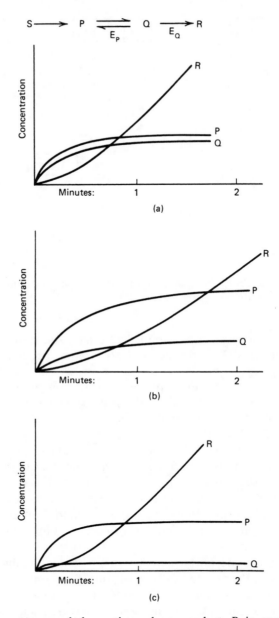

Figure 3.9. Two-step coupled reaction where product P is converted to Q and then to R by enzymes E_P, E_Q. Concentrations of P, Q, and R (a) when $V_{max}/K_m = 5\,min^{-1}$ for both E_P and E_Q, and $P \rightarrow Q$ is essentially irreversible. (b) $V_{max}/K_m = 5\,min^{-1}$ for both E_P and E_Q, and $P \rightarrow Q$ is reversible when equilibrium favoring P. Note that rate of formation of R is still increasing at 2 min. (c) as (b), but $V_{max}/K_m = 25\,min^{-1}$ for E_Q. Rate of formation of R reaches a steady state by 1 min, with $[Q]$ much lower than in (b).

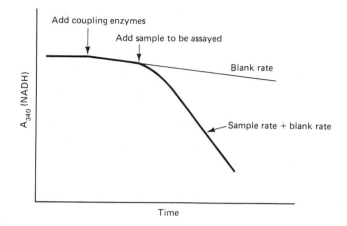

Figure 3.10. Progress of a coupled reaction involving oxidation of NADH, in which there is a blank rate due to contamination in the coupling enzymes.

the measurement. Most commercial enzymes are now manufactured to high standards, ensuring that potential problematical contaminants are minimized.

Assays of crude extracts commonly pose problems because, in addition to the desired enzyme, a whole host of others are present which might cause upsets. The problems divide into two categories: first, the crude extract may contain sufficient other substrates (as well as enzymes) to allow other detectable reactions to occur; second, other enzymes in the extract itself may be interfering with the coupling system being used. Problems occur mainly when the enzyme being measured is relatively weak in activity, since this means that larger amounts of sample must be added, increasing the probability of detectable interference. For instance, if a kinase is being measured as ADP formation by the pyruvate kinase–lactate dehydrogenase method described above, addition of glucose in the sample (likely in liver extracts) will activate hexokinase as well as the kinase being measured. NADH oxidation may occur due to the presence of an "NADH oxidase" system. NADPH formation can be decreased by the presence of glutathione reductase plus oxidized glutathione. Problems such as removal of substrate by another enzyme or removal of product by a side reaction may be overcome by various blank determinations or, possibly, introduction of inhibitors. This is discussed in more detail elsewhere [53], and further analysis of coupled methods can be found in [55,56].

Having described the principles of both stopped and continuous assays, a comparison of the relative advantages and disadvantages of continuous compared with stopped assays can be presented. These have already been pointed out earlier, and Table 3.4 lists them systematically.

Table 3.4. Comparison of Advantages and Disadvantages of Stopped and Continuous Enzyme Assays

Stopped assays	Continuous assays
No limitation on conditions	Conditions may be limited by requirement of coupling system
Very small volume saves on expensive reagents	Usually requires larger amounts of reagents, though microcells result in saving
High sensitivity for radiochemical methods	Sensitivity usually low compared with radiochemical methods
Single point determination may hide irregular activity	Continuous observation detects nonlinearity
Result not instantaneous	Instantaneous result
Multi-manipulation more time-consuming	Simple manipulation
No coupling enzymes required in assay (but may be needed to determine product)	Coupling enzymes needed may be expensive and/or contain contaminants
Product accumulation may cause inhibition[a]	Product may be removed by coupling system

[a] A nonobservable coupling system can be included to remove products.

3.5   Practical Points in Enzyme Activity Determination

The theory and basic practice of enzyme activity measurement have been broadly described above. This short section amplifies the practical laboratory aspect of actually carrying out the measurements.

A typical day during an enzyme preparation may involve only one or two assays, but more probably it will be 10 or 20; monitoring a column effluent may require even more. Most assay mixtures involve many components—buffers, salts, substrates, cofactors, etc.—and it is not only inconvenient to mix up each reaction mixture separately, it is also bad practice, since slight variations and errors will make each mixture somewhat different. The best method is to estimate how much will be needed in the day, and make up the whole mixture in bulk with the exception of one substrate, to be added at the last moment to each assay. Some reagents can be mixed and kept (refrigerated) for weeks—for instance, simple buffer and salt solutions. Thus "Reagent A" might include all of the buffers, salts, and perhaps cofactors, needing only one or two other components to complete the mixture. Many biochemical reagents are quite stable in solution unfrozen, provided that bacterial or fungal activity is prevented. One drop of 20% sodium azide for every 20 ml usually gives sufficient protection; many stock solutions can be kept in the refrigerator for months like this; others may need to be frozen. Others may be chemically labile and so need to be mixed fresh every day. Sodium azide rarely affects enzymes, and is in any case diluted out to low levels if the stock solutions of reagents are concentrated. As an example, below is a protocol for mixing up a hexokinase assay:

To make 20 ml of "Reagent A" for hexokinase assay:
Take 2 ml of stock buffer pH 8.0 (see below)
Add: 1–2 drops of 5% bovine serum albumin
      0.4 ml of stock 50 mM NAD$^+$
      0.4 ml of stock 50 mM ATP
      10 μl of glucose 6-P dehydrogenase (from *Leuconostoc mesen-teroides*), 1,000 units ml^{-1} suspension in ammonium sulfate
Make to 20 ml with water.
Stock buffer contains 0.3 M Tris, 0.5 M KCl, 30 mM MgSO$_4$, 1 mM EDTA (+ azide), pH to 8.0 with HCl.
Assay procedure:
    Take 1 ml of Reagent A in cuvette
    Add 20 μl of stock 1 M glucose ("Reagent B")
    Check A$_{340}$ for any blank rate
    Add sample (0.2–10 μl), measure increase in A$_{340}$.

$$\text{glucose} + \text{ATP} \xrightarrow{\text{hexokinase}} \text{glucose 6-P} + \text{ADP}$$

NADP$^+$ glucose 6-P
NADPH   dehydrogenase
6-P-gluconate

(*N.B.* Glucose 6-P dehydrogenase from either *Leuconostoc mesenteroides* or *Zymomonas mobilis* is used as it can utilize NAD; the yeast enzyme can only use NADP, which costs 10 times as much.)

Note that unless dealing with very labile compounds, Reagent A should *not* be kept on ice; the temperature should be close to that required in the assay to minimize temperature equilibration time in the cuvette. This is not so important with stopped assays, since a short preincubation is easily done. Temperature control in spectrophotometers can be arranged using a cuvette holder with circulating water; although this will control the temperature accurately, there are two points to remember. First, the actual temperature in an equilibrated cuvette may be somewhat different from the set temperature on the circulating water thermostat because of heat losses during transmission. Second, it could take up to 10 min for a cold sample in a cuvette to reach within 0.5°C of the steady-state temperature; this is obviously inconvenient, so preincubation is desirable. Because of these points, for routine work it is usually more convenient to carry out assays at room temperature, noting the actual temperature in the cuvette, and correcting to a standard temperature using an appropriate factor. A good average value for many enzymes is 7% per degree (Section 3.2). It does not take long to find the correct factor for one particular enzyme; simply assay the same sample over the range of temperatures likely to be used.

With a mixture containing all the ingredients, preincubated to a suitable temperature, the final step is to add the enzyme. With very low-activity samples the volume needed may be 20% or more of the final mixed volume. At the other extreme, concentrated samples of very active enzymes may need diluting as much as a millionfold into the assay mixture. Microliter syringes can deliver as little as $0.5\,\mu l$ with 1–2% accuracy, but smaller volumes than this inevitably involve larger errors. Micropipettes range from $1\,\mu l$ upwards, but one should not expect accuracies better than $\pm 0.1\,\mu l$ even with positive displacement pipettes. Microliter syringes and micropipettes from total capacity of $1\,\mu l$ up to $100\,\mu l$ should be on hand so that the volume desired can be added with optimum accuracy. If the smallest accurately deliverable volume still gives too fast a rate, then some dilution of the sample is necessary. This dilution should be done with maximum accuracy since it is an additional step which can introduce errors. One way to dilute samples is to use $10\,\mu l$ plus 90, 190, 490 or $990\,\mu l$ of dilution buffer from a graduated pipette to give $10\times$, $20\times$, $50\times$, or $100\times$, dilutions, respectively. The dilution buffer's composition is important. The protein concentration of the sample could already be quite low, and to avoid inactivation/adsorption losses due to low protein concentration, a neutral protein must be added. Bovine serum albumin is normally used, and a concentration of 1% in the dilution buffer is not unreasonable. The buffer itself should be optimal for stability of the enzyme; if these conditions are not known, then a pH 7 buffer, e.g., phosphate, containing magnesium ions ($1–2\,\text{m}M$) and a small amount of EDTA ($0.1\,\text{m}M$) is a good starting point. If the enzyme diluting buffer contains $5\,\text{m}M$ sodium azide as well, it also will keep for months.

It is not possible to detail methods for assaying nonenzymic proteins, since each will have a specific property to be followed. This could be an immunological assay such as ELISA, a bioassay, or even quantitation of a band on an electrophoretic gel. Let it be stressed that a protein purification procedure *cannot be developed* without having a reliable detection method for the protein being isolated, because you simply would not know where your protein was. But if you are following a standard protocol for purifying that particular protein, it is possible to manage without an assay, but it takes faith in the method.

Chapter 4
Separation by Precipitation

4.1 General Observations

In the early days of protein chemistry, the only practical way of separating different types of proteins was by causing part of a mixture to precipitate through alteration of some property of the solvent. Precipitates could be filtered off and redissolved in the original solvent. These procedures remain a vitally important method of protein purification, except that filtration has mostly been replaced by centrifugation. Protein precipitates are aggregates of protein molecules large enough to be visible and to be centrifuged at reasonably low g forces. In some cases, the aggregation continues and the precipitate flocculates, but usually the motions and collisions of the particles in suspension keep their size small. This results in clogging of filter papers and necessitates centrifugation at considerably more than $1g$. The various methods of obtaining a precipitate are described in the separate sections below.

Original classifications of proteins depended on solubility behavior, and although this classification has largely been forgotten today, the words are still used daily when talking of serum albumin or immunoglobulins. Globulins were defined as proteins insoluble in low-salt solution; serum globulins can be precipitated by simply diluting serum with water, while albumins remain in solution. This globulin behavior is an example of isoelectric precipitation in the salting-in range (see Section 4.2). The classification becomes confused since some proteins are insoluble at pH 6, but soluble in low salt at pH 8. Nevertheless there are many proteins that properly fit the globulin description. They are proteins of generally low solubility in aqueous media that tend to have a substantial hydrophobic amino acid content at their surface. Conversely there are many true albu-

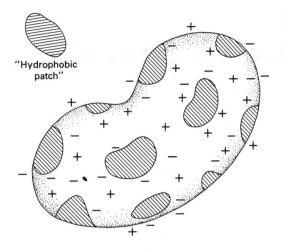

Figure 4.1. Distribution of charge and hydrophobic patches over surface of a typical protein.

"Hydrophobic patch"

mins, with high aqueous solubility and low hydrophobicity. Other classifications, such as prolamines, soluble at high ethanol concentration [57], have been used for particular purposes.

The distribution of hydrophilic and hydrophobic residues at the surface of the protein molecule is the feature that determines solubility in various solvents. Although hydrophobic groups tend to be concentrated around the interior of protein molecules, substantial numbers reside on the surface, in contact with the solvent (Figure 4.1).

Hydrophobic groups on the surface of protein molecules are as important in determining the behavior of the molecules as are charged and other polar groups. The solvent used for isolating proteins is almost always aqueous-based, though there have been some successful procedures using nonaqueous solvents [33,34,35]. The solvent properties of water can be manipulated to alter protein solubility by changes in ionic strength, pH, addition of miscible organic solvents, other inert solutes or polymers, or combinations of these together with temperature variation. The most widely used procedure is the use of neutral salts to cause selective precipitation. This is the subject of the next two sections.

4.2   The Solubility of Proteins at Low Salt Concentrations

Most enzymes exist in cell fluids as soluble proteins, often in a concentrated soup that may be as much as 40% protein [58]. Thus they are very soluble in physiological salt conditions, at ionic strength generally around $0.15-0.2 M$, and neutral pH. Once extracted, these conditions can be varied readily, though it is inconvenient to increase the protein concentration (by removing solvent), or to decrease the ionic strength substantially (by removing low-molecular-weight ions). The solubility is a result of polar interactions with the aqueous solvent, ionic interactions with the salts present, and, to some extent, repulsive electrostatic forces between like-charged molecules or small (soluble) aggregates of molecules. Small aggregates cancel out strong attraction charges (Figure 4.2). In the ionic strength range from zero to physiological, some proteins form precipitates because the repulsive forces are insufficient. For example, a high surface hydrophobicity means low interaction with the solvent and fewer charged groups to interact with salts. An overall charge near zero minimizes electrostatic repulsion, and may, close to the isoelectric point, cause hydrophobic forces to attract molecules to each other (Figure 4.3). This is called isoelectric precipitation; most proteins have a minimum solubility around their isoelectric point, which in the case of globulins is low enough to result in precipitation. In a protein mixture, the situation is greatly complicated by heterogeneous interactions; different proteins with similar properties aggregate to form the isoelectric precipitate. In many cases isoelectric precipitates can be formed in a tissue extract by lowering the pH to between 6.0 and 5.0. This is not just an aggregation of soluble protein; it includes aggregates of particulate material such as ribosomes and membranous fragments, as discussed previously (cf. Section 2.3).

The solubility of globulin-type proteins decreases as the salt concentration decreases. This can rarely be exploited with a crude extract since a

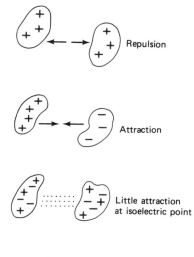

Figure 4.2. Electrostatic forces between protein molecules and small aggregates.

twofold decrease in salt can only be achieved easily by adding an equal
volume of water; if the solubility does not decrease by *more* than twofold,
nothing is gained! However, it can be a useful property at a later stage
when desalting by dialysis. On dialysis against a low-ionic-strength buffer,
precipitates can form in the dialysis sac. If the precipitate includes most of
the protein required, or alternatively none at all, a purification results.
The precipitate is simply centrifuged off. But if desalting is attempted by
gel filtration (see Section 1.4), development of an isoelectric precipitate
can be a nuisance. Aggregates may clog up the column, or, precipitated
and trapped, they remain behind until the salt removed from the original
protein mixture catches up and redissolves the precipitate. Gel filtration
desalting may simply not work under these conditions.

Isoelectric precipitation of low-solubility proteins can be improved by
inclusion of a solute that further decreases protein solubility. Thus organic solvents or polyethylene glycol (cf. Sections 4.4 and 4.5) together
with a pH adjustment can produce the desired precipitate. It is also possible to cause precipitation by addition of divalent metal ions, e.g., $50\,mM$
$MgCl_2$ [59]. The metal "salt" of the protein has a higher isoelectric point,
so aggregation can occur at a higher pH.

Within the ionic strength range of zero to about $0.5\,M$, the increase in
solubility (at a given pH and temperature) with *increasing* salt con-

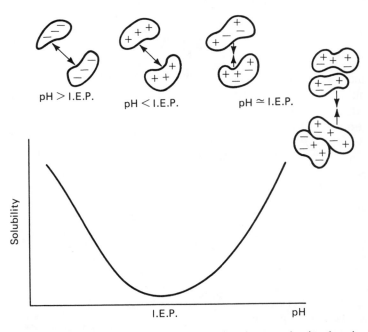

Figure 4.3. Solubility of a globulin-type protein close to its isoelectric point
(IEP).

Figure 4.4. Solubility of globulin-type protein at a fixed pH close to its isoelectric point (IEP), varying with salt concentration.

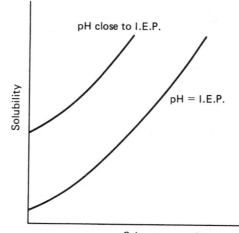

centration is known as "salting in." For this to be useful in a protein purification protocol, the protein must first be insolubilized at low ionic strength. A typical salting-in curve for a pure protein is shown in Figure 4.4; the curve's steepness in response to salt may be sufficient to induce reprecipitation on dilution with water.

Points to Note in Practice

Most isoelectric precipitates are aggregates of many different proteins and may include particulate fragments and protein–nucleic acid complexes. If the initial composition of the solution is changed, a desired protein may not exhibit the same apparent solubility behavior if its partners in precipitation are absent. Since isoelectric precipitates normally occur when the pH is lowered below physiological values, make sure the protein is stable at the pH needed. Keeping the temperature low will improve the chances of stability and decrease solubilities. If adjusting the pH at 0°C, do not forget to standardize the pH meter at this temperature, using ice-cold standard buffer (see Section 12.3).

In summary, the salting-in range can be exploited in purification in two distinct ways. First, aggregation may occur by dilution or dialysis, and if the required protein is present in the aggregate, a useful purification will have been achieved. Second, isoelectric precipitation can be used by exploiting the variation of solubility with pH, without changing ionic strength. In either situation, if the required protein remains in solution, the degree of purification achieved is usually minimal since a relatively small proportion of the total protein is likely to be in the precipitate. In this case, what one accomplishes is simply a cleaning up of the extract (cf.

Section 2.3). But if your protein is entirely in the precipitate *and remains fully active*, then this step is likely to be very effective.

4.3   Salting Out at High Salt Concentration

This section describes one of the most widely used techniques in enzyme purification, the salting out of proteins, using high concentrations of salts. It is quite distinct from the salting-in effect described in Section 4.2. Although there is a tendency for globulin-type proteins (those that have low solubility at low ionic strength near their isoelectric point) also to show relatively low solubility in high salt solutions, there are many exceptions. Thus serum γ-globulins, typical of proteins that are insoluble at low salt concentration but soluble in the physiological ionic strength range, precipitate with ammonium sulfate in the salting-out range at the relatively low concentration of around $1.5\,M$. On the other hand, some other serum globulins (α- and β-) have fairly average ammonium sulfate solubilities, precipitating around $2.5-3\,M$. Salting out is largely dependent on the hydrophobicity of the protein, whereas salting in depends as much on surface charge distribution and polar interactions with the solvent.

A typical protein molecule in solution has hydrophobic patches on its surface (side chains of Phe, Tyr, Trp, Leu, Ile, Met, Val). Forcing these into contact with the aqueous solvent causes an ordering of water molecules, effectively freezing them around the side chains (Figure 4.5) [60]. This ordering is a thermodynamically unstable situation since it represents a large decrease in entropy compared with (a theoretical) unsolvated protein plus free water molecules.

If the process of dissolving a dry protein in water is represented by:

$$P + nH_2O \rightarrow P \cdot nH_2O \qquad (4.1)$$

where the nH_2O represents ordered water around hydrophobic residues (we shall ignore the hydrogen-bonded and polar-bound water in this description), then ΔS^0 is large and negative because of the degree of order introduced. The process is known to occur; thus ΔG^0 is likely to be negative. Consequently ΔH^0 is also highly negative ($\Delta H^0 = \Delta G_0 + T\Delta S^0$). Consider two protein molecules coming together and interacting through the hydrophobic residues, releasing the water:

$$P \cdot nH_2O + P' \cdot mH_2O \rightarrow P{-}P' + (n + m)H_2O \qquad (4.2)$$

Clearly, ΔS^0 will be large and positive in this case; the enthalpy change is expected to be approximately the reverse of the process in Eq. (4.1), and so would also be positive. Now ΔG^0 may be small, and its sign would depend on the relative magnitudes of ΔH^0 and ΔS^0. In low salt concentration, for a typical water-soluble protein, ΔG^0 would be posi-

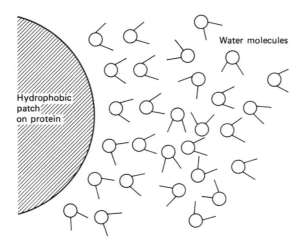

Figure 4.5. Ordering of water molecules around hydrophobic residues on the surface of a protein.

tive. Since the protein is soluble, aggregation (Eq. 4.2) is not occurring to any significant extent. However, if the water molecules liberated by the process described in Eq. (4.2) are trapped, the reaction can be made to progress and protein aggregation will occur. Salt ions become hydrated; thus, adding large amounts of salt traps the water:

$$\underset{\text{(salt)}}{AB} + (n_1 + n_2)H_2O \rightarrow \underset{\text{(ions)}}{A^+ \cdot n_1H_2O + B^- \cdot n_2H_2O} \qquad (4.3)$$

Once a sufficient amount of water has been attracted away from the protein, the aggregation described by Eq. (4.2) occurs. Note that ΔH^0 is positive for this reaction. As a result, the Arrhenius equation predicts that aggregation occurring due to hydrophobic forces increases with temperature. This is a recognized feature of hydrophobic interactions. Salt precipitation, in most cases, has a positive ΔH^0, since more precipitate forms *at a given salt level* at higher temperatures.

Another way of describing the aggregation of hydrophobic patches on protein surfaces due to high salt concentration is to say that, as the salt ions become solvated, there is a greater tendency, when freely available water molecules become scarce, to pull off the ordered "frozen" water molecules from the hydrophobic side chains, so exposing the bare, fatty areas that want to interact with one another. Those proteins with a larger number or bigger clusters of such residues on their surface, will aggregate sooner, whereas proteins with few nonpolar surface residues may remain in solution even at the highest salt concentration. The nature of the salt is extremely important here; those salts that actually bind and interact directly with the proteins have a destabilizing effect and are called chaotropic. The optimum salts are those that encourage hydration of

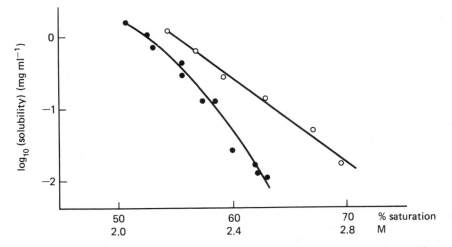

Figure 4.6. Solubility of rabbit muscle aldolase at pH 7.0 in ammonium sulfate solutions: (○), pure enzyme solubility curve; log (soly) = 6.30 − 2.84 (salt conc); (●), solubility in a mixture of rabbit muscle enzymes: aldolase, pyruvate kinase, glyceraldehyde phosphate dehydrogenase, and lactate dehydrogenase in proportions 3:3:3:1. (Author's unpublished results.)

polar regions (and dehydration of the hydrophobic regions) on the protein without direct interaction themselves [61]. In general, these can be correlated with the molal surface tension increments of the salts [62]. Salts that increase the surface tension of the solvent enhance interactions between solutes such as proteins, or indeed between proteins and ligands such as used in hydrophobic chromatography. Thus, multivalent anions, especially sulfate and phosphate with "innocuous" cations such as potassium, sodium, or ammonium, are preferred. Because high concentrations are often required to cause salting out, the solubility of the salt is an important consideration.

Starting with a crude mixture, co-aggregation is extensive; individual identical molecules are not going to search each other out when another sticky molecule presents itself alongside. Nevertheless, many proteins are precipitated from solution over a sufficiently narrow range of salt concentration to make the procedure a highly effective method of fractionation. Some results on the effects of the presence of other proteins on the solubility of some specific enzymes are shown in Figures 4.6 and 4.7. Pure protein solubility follows the empirical formula:

$$\log(\text{solubility}) = A - m(\text{salt conc})$$

where A = a constant dependent on temperature and pH
      m = a constant independent of temperature and pH

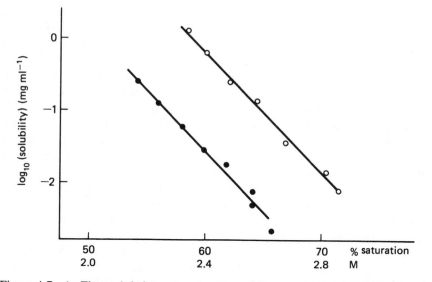

Figure 4.7. As Figure 4.6, for pyruvate kinase: (○), pure enzyme solubility curve; log(soly) = 10.2 − 4.34 (salt conc); (●), mixture of enzymes. (Author's unpublished results.)

A few examples of values for pure proteins are given in Table 4.1. Impure samples do not follow this rule, since coprecipitation occurs to an amount depending on the properties of the other protein present. In general, solubility of a particular component is decreased by the presence of other proteins.

Table 4.1. Some Values of the Constants A and m in Eq. (4.3) for Pure Enzymes at pH 7.0 (Solubility Expressed in $mg\,ml^{-1}$)

Protein	A	m	Ammonium sulfate concentration for solubility of $1\,mg\,ml^{-1}$ (M)
Rabbit muscle aldolase	6.30	2.84	2.20
Rabbit muscle pyruvate kinase	10.2	4.34	2.34
Rabbit muscle glyceraldehyde phosphate dehydrogenase	8.6	2.84	3.34
Rabbit muscle lactate dehydrogenase	8.0	3.97	2.02
Rabbit muscle phosphoglycerate mutase	9.85	4.26	2.32
Bovine serum albumin	9.2	3.26	2.84

Source: Author's unpublished observations.

Although salting out depends strongly on hydrophobic interactions, other features do affect solubilities. Thus pH may change the solubility; a neutralization of a charge makes the surface less polar. Solubility in salts is usually highest around pH 7, where proteins have the most charged groups. On the other hand, aggregation may occur more easily close to the protein's isoelectric point. A classic example of this is in the one-step purification of glyceraldehyde phosphate dehydrogenase (EC 1.2.1.12) from rabbit muscle [63]. At pH 6, this enzyme is soluble in $3.2\,M$ ammonium sulfate, but if the pH is adjusted to 7.5 or higher, it precipitates and may even crystallize from the crude extract. Its isoelectric point measured at low salt concentration is around 8.5.

Temperature affects the solubility of proteins at high salt concentration in an unusual way because of the effects of temperature on hydrophobic interaction as described above. In the salting out range, the solubility of proteins generally decreases with increasing temperature. This is an important consideration when crystallizing proteins from ammonium sulfate (cf. Section 12.5); a clear solution in the cold can rapidly become turbid with amorphous protein precipitate when warmed.

Salting out of proteins involves dissolving the salt into the solution containing the proteins. Several features of this process should be considered. The nature of the salt is all-important. One must consider physical characteristics of the salt, such as solubility, besides its effectiveness in causing precipitation. The most effective salts are those with multiple-charged anions such as sulfate, phosphate, and citrate. The cation is relatively less important, but should not complex or otherwise interact with the protein; this virtually eliminates all multivalent metal ions. Cost rules out many salts that might otherwise be considered. The salting out ability of anions follows the Hofmeister series, which for some common anions is $SCN^- < ClO_4^- < NO_3^- < Br^- < Cl^- < acetate^- < SO_4^{2-} < PO_4^{3-}$. Although phosphate would appear from this to be more effective than sulfate, in practice phosphate at neutral pH consists of a mixture of HPO_4^{2-} and $H_2PO_4^-$ ions, which are less effective than PO_4^{3-}. As far as cations are concerned, monovalent ions should be used, with $NH_4^+ > K^+ > Na^+$ in precipitation effectiveness [64]. Thus, of the common salts that are effective in causing precipitation, sodium, potassium and ammonium sulfates, phosphates, and citrates are attractive candidates. Further considerations of the suitability are the cost of kilogram quantities of sufficient pure salt, the solubility (several molar solutions are required), and the heat of solution. It is undesirable to get heating while dissolving in the salt; also, a large ΔH^0 of solution means large changes in solubility with temperature. Finally, the density of concentrated solution should be considered, since centrifuging a precipitate depends on the difference in density between the aggregate and the solvent (cf. Section 1.2).

Most potassium salts are ruled out on solubility grounds, though potassium phosphates can be used. Sodium citrate is very soluble, and can be used also, but rarely offers any advantage other than the possibility of use above pH 8—which is not possible with ammonium salts, because of the buffering action of ammonia. Citrate cannot be used so effectively below pH 7 for a similar reason. Sodium sulfate forms several hydrates and has a complex solubility phase diagram, but it is not highly soluble at low temperature. Except when it is required to operate at a high pH, one salt has all the advantages and no disadvantages for a typical protein: ammonium sulfate.

The solubility of ammonium sulfate varies very little in the range $0-30°C$; a saturated solution in pure water is approximately $4\,M$. The density of this saturated solution is $1.235\,g\,cm^{-3}$ (cf. Section 1.2). In contrast, $3\,M$ potassium phosphate at pH 7.4 has a density of $1.33\,g\,cm^{-3}$, which makes it impossible to centrifuge off protein, since this is also the density of protein aggregates in this solution. Because of this density effect, combining with the extra density caused by other salts and compounds present in crude extracts, precipitates in the range $3-4\,M$ ammonium sulfate may also be difficult to sediment.

Ammonium sulfate is available in sufficiently pure state to be used in large quantities economically. However, small contaminating amounts of heavy metals, especially iron, could be detrimental to sensitive enzymes. Metal-complexing agents such as EDTA should be present in the solution before adding the ammonium salt. It is worth noting that 1 part per million of a heavy metal converts to the biochemically active concentration of a few micromolar, when the ammonium sulfate concentration is $3\,M$. The amount of ammonium sulfate to add to reach a predetermined concentration can be calculated from a simple formula (Eqs. 4.4, 4.5). Because the addition of salt causes an increase in volume, allowance for this is made in the formula. Thus, although the solubility of ammonium sulfate is $533\,g\,liter^{-1}$ at $20°C$, $761\,g$ must be added to 1 liter of water to make a saturated solution (1.425 liters). The density of the saturated solution is thus $1761/1425 = 1.235\,g\,cm^{-3}$, as indicated above. It is usual to quote ammonium sulfate concentration as percent saturation, making the assumption that the starting material is equivalent to water in its ability to dissolve ammonium sulfate. For this, Eq. (4.5) is used:

Grams ammonium sulfate to be added to 1 liter of a solution at $20°C$:
(a) At M_1 molar, to take it to M_2 molar:

$$g = \frac{533(M_2 - M_1)}{4.05 - 0.3M_2} \tag{4.4}$$

(b) At S_1 % saturation, to take it to S_2 % saturation:

$$g = \frac{533(S_2 - S_1)}{100 - 0.3S_2} \tag{4.5}$$

The two equations are related by the assumption that 100% saturation = 4.05 M. Note that in practice, because of other salts present and the protein itself, the *actual* solubility of ammonium sulfate in the sample will be somewhat less than 533 g liter^{-1} at 20°C. The amount of ammonium sulfate required can be read from Appendix A, Table I for 5% intervals; other intervals can be interpolated from these figures. Although pH may be important in precipitation, in most cases small variations in pH from one experiment to another make little difference. It is best to operate at a neutral value (pH 6–7.5); ammonium sulfate has a slight acidifying action, so around 50 mM buffer (e.g., phosphate) should be present.

One of the best-studied systems for ammonium sulfate fractionation is rabbit muscle extract. Czok and Bucher [63] divided the extract into five fractions by ammonium sulfate precipitation; in each fraction certain enzymes were especially enriched. Using this work as a basis, all glycolytic enzymes plus a few others could be purified after an initial division into four ammonium sulfate fractions [65].

One great advantage of ammonium sulfate fractionation over virtually all other techniques is the stabilization of proteins that occurs. A 2–3 M ammonium sulfate suspension of protein precipitate or crystals is often stable for years, and it is a normal packaging method for commercial enzymes. The high salt concentration also prevents proteolysis and bacterial action. It is always useful to have a stage in a purification where the sample can be left overnight. An ammonium sulfate precipitate, either before centrifuging or as a pellet, is a good form in which to keep a sample.

After deciding to do an ammonium sulfate fractionation, the next decision is what percentage saturation to try. Material precipitating before

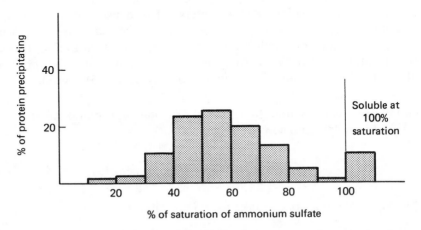

Figure 4.8. Typical ammonium sulfate precipitation ranges for proteins from a crude tissue extract.

Table 4.2. Trial Fractionations with Ammonium Sulfate

	Percent saturation range	Percent enzyme precipitated	Percent protein precipitated	Purification factor[a]
First trial	0–40	4	25	
	40–60	62	22	2.8
	60–80	32	32	1.0
	80 supernatant	2	21	

Conclusion: Enzyme precipitated more in 40–60% than in 60–80%; try 45–70%

Second trial	0–45	6	32	
	45–70	90	38	2.4
	70 supernatant	4	30	

Conclusion: Good recovery, but purification factor not as good as in first trial; if
purity important, try 48–65%:

Third trial	0–48	10	35	
	48–65	75	25	3.0
	65 supernatant	15	40	

[a] Specific activity of fraction divided by specific activity of original sample.
Source: Scopes [66]

25% saturation is generally particulate and preaggregated or very-high-molecular-weight protein. A typical distribution of protein precipitation from a crude extract is shown in Figure 4.8. An idealized set of results of trial fractionations with ammonium sulfate is reproduced in Table 4.2. Fractionations are always a compromise between recovery of activity and degree of purification. If the source material is valuable or difficult to obtain, or the objective is to get as much enzyme as possible without too much concern for absolute purity, then degree of purification may be sacrificed for recovery. On the other hand, if the fractionation is the first step from a readily available source material, recovery can be sacrificed for purity. In many cases the next step to be used will give a similar result regardless of the recovery and purity attained in the initial salt fractionation, in which case recovery of activity is more important than degree of purification. In that case broad ammonium sulfate cuts can be used, resulting in relatively little purification; the step is essentially a cleaning-up operation to get a messy extract into a suitable state for, e.g., adsorption chromatography.

The solution to which salt is to be added should be provided with a stirring system, and solid ammonium sulfate is weighed out. Any lumps should be broken up before slowly adding the solid. The first bit can be dissolved in quite quickly, but as the intended percent saturation is approached, the last increments of salt should be added more slowly.

Dissolved air may come out of solution and cause frothing, but no harm is done unless overvigorous stirring itself causes frothing, in which case surface tension effects on proteins caught in bubbles can cause denaturation. After the last bit of salt has dissolved, stirring should continue for 10–30 min to allow complete equilibration between dissolved and aggregated proteins; then the solution is centrifuged. Generally, around $10^5 g \, \text{min}^{-1}$ is sufficient, i.e., 10,000 g for 10 min or 3,000 g for 30 min; at higher salt concentrations somewhat more centrifugation may be needed. The supernatant is decanted, its volume noted, and the amount of salt required for the next cut calculated. The precipitate is dissolved in a suitable buffer. Note that the volume of buffer for dissolving the precipitate need be no more than 1–2 times the volume of the *precipitate*, since that will be enough to reduce the salt concentration to well below the precipitation point of the proteins present. If all the precipitate does not dissolve, the insoluble material is probably denatured/ particulate, and should be removed by further centrifugation. It is always best to keep solutions as concentrated in protein as possible. This improves stability, and it is always easier to dilute something later than it is to reconcentrate it.

The redissolved precipitate contains considerable amounts of ammonium sulfate, and often this must be removed before proceeding to the next step. Dialysis or gel filtration on a "desalting" column are the normal methods employed (cf. Section 1.4). Measurement of enzyme activity can be carried out on the redissolved precipitate or on the supernatant, but beware that in the latter case, the amount of ammonium sulfate added into the assay mixture with the sample could upset the assay. The precipitate should always be tested anyway, for although losses in ammonium sulfate fractionation are usually small, the activity recovered in a precipitate may not be as much as the difference between successive supernatants.

Rather than dissolve solid salt into the sample, it is possible to add a suitable volume of saturated solution, rather in the same way as adding organic solvents (cf. Section 4.4). This is particularly useful if the desired protein is all precipitated by the time 50% saturation is reached, as this would only involve a twofold increase in volume. On a large scale even this might be unacceptable, but on a small scale it can be an appropriate method. Higher percentage saturation needs still more increase in volume; dilution of the proteins leads to the requirement of a still higher percent saturation for equivalent precipitation, so the method is not advised for going much above 50%.

An important point to remember is that salt never precipitates *all* the protein, but just reduces its solubility. If the starting mixture contains an enzyme at, say, 1 mg ml^{-1}, and addition of salt causes a reduction in its solubility to 0.1 mg ml^{-1}, this means that 90% of it precipitates. This would be an acceptable recovery. On the other hand, if the enzyme was only at $0.1\,mg\,ml^{-1}$ to start with, the same concentration of salt would not cause any precipitation! In other words, the salt concentration range for "precipitation" is not an absolute property of the protein concerned, but depends on both the properties of other proteins present (coprecipitation) and on the protein concentration in the starting solution. Different salt concentrations may be needed at different stages in the purification procedure.

Finally, an important application of salting out procedures is not so much for fractionating samples, but for concentrating them. An eluate from an ion-exchange column or from gel filtration may need to be concentrated. Salting out by adding sufficient ammonium sulfate to precipitate all the protein is an efficient way of doing this—provided that the sample is not too dilute to start with. It is convenient to dissolve in $60\,g$ ammonium sulfate for every $100\,ml$, to give an approximately 85% saturated solution. Few proteins have solubilities in excess of $0.1\,mg\,ml^{-1}$ at this level of ammonium sulfate. But if the starting protein concentration is less than $1\,mg\,ml^{-1}$, one should try to increase this by, e.g., ultrafiltration (cf. Section 1.4) before salting out.

4.4   Precipitation with Organic Solvents

General Theory

The method of protein precipitation by water-miscible organic solvents has been employed since the early days of protein purification. It has been especially important on the industrial scale, in particular in plasma protein fractionation [67]. However, its use in laboratory-scale purifications has been less extensive; arguably it has been underutilized, considering there are some advantages compared with salting out precipitation; but there are also disadvantages. As the principles causing precipitation are different, it is not necessarily an alternative to ammonium sulfate, and can be used as an additional step.

Addition of a miscible solvent such as ethanol or acetone to an aqueous extract containing proteins has a variety of effects which, combined, lead to protein precipitation. The principal effect is the reduction in water activity. The solvating power of water for a charged, hydrophilic protein molecule is decreased as the concentration of organic solvent increases. This can be described in terms of a reduction of the dielectric constant of the solvent, or simply in terms of a bulk displacement of water, plus the partial immobilization of water molecules through hydration of the (organic) solvent. Ordered water structure around hydrophobic areas on the proteins' surface can be displaced by organic solvent molecules. This can result in a relatively higher "solubility" of these areas. Some extremely hydrophobic proteins, normally located in membranes, may be soluble in 100% organic solvent. But the net effect of miscible organic solvents on cytoplasmic and other water-soluble proteins is a *decrease* in solubility to the point of aggregation and precipitation.

The principle causes of aggregation are likely to be electrostatic and dipolar forces, similar to those occurring in the salting-in range in the absence of organic solvent (Section 4.2). Hydrophobic attractions are less involved because of the solubilizing influence of the organic solvent on these areas. It has been found that precipitation occurs at a lower organic solvent concentration at around the isoelectric points of the proteins, which supports the suggestion that the aggregation is similar to that occurring in isoelectric precipitation. A two-dimensional representation of proteins in a water–organic solvent mixture is shown in Figure 4.9. Here aggregation is occurring by interactions between opposite-charged areas on the proteins' surfaces.

Another feature affecting organic solvent precipitation is the size of the molecule. It has generally been found that the larger the molecule, the lower percentage of organic solvent required to precipitate it. Thus, if one could compare a range of proteins with different molecular sizes but similar hydrophobicity and isoelectric points, the precipitation order would be the reverse of the order of size. In practice, the different compositions and net charges of proteins in a given mixture make this an approximate rule at best. Large molecules aggregate sooner because they have a greater chance of possessing a charged surface area that matches up with another protein (Figure 4.9). Some experimental results showing

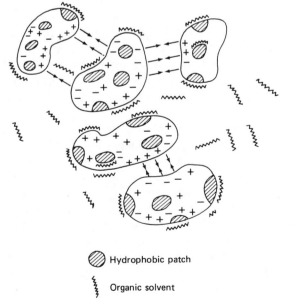

Hydrophobic patch

Organic solvent

Figure 4.9. Aggregation of proteins by interactions in an aqueous–organic solvent mixture.

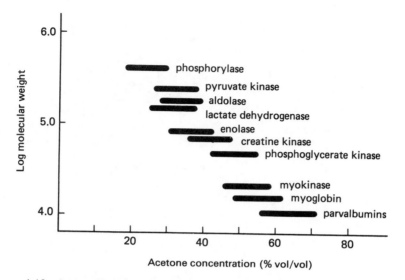

Figure 4.10. Approximate precipitation ranges in acetone at 0°C, pH 6.5, $I = 0.1$, of some proteins found in muscle tissue extracts. (From Scopes [66].)

both the size relationship and the isoelectric point effect have been presented previously [66] and are reproduced in Figure 4.10. In an early investigation of organic solvent fractionation, it was concluded that acetone, at pH 6.5, was the most suitable for rabbit muscle extracts [68]. The isoelectric points of the major components of such muscle extracts, in phosphate buffer, are in the range 6.0–7.5.

Choice of Solvent

The solvent used must be completely water-miscible, unreacting with proteins, and have a good precipitating effect. The two most widely used solvents are ethanol and acetone. Others that can be used include methanol, *n*-propanol, *i*-propanol, dioxan, 2-methoxyethanol (methyl cellosolve), and various other more exotic alcohols, ethers, and ketones. Safety should be a major consideration, both in terms of flammability and in terms of noxious vapors. For these reasons dioxan, 2-methoxyethanol, and other ethers should be eliminated from the list.

One advantage of organic solvent fractionation is that it can be carried out at subzero temperatures, since all the miscible solvents form mixtures with water that freeze well below 0°C. This is fortunate, since it is most important that the temperature is kept low. At above about 10°C denaturation effects become substantial. The reason for denaturation

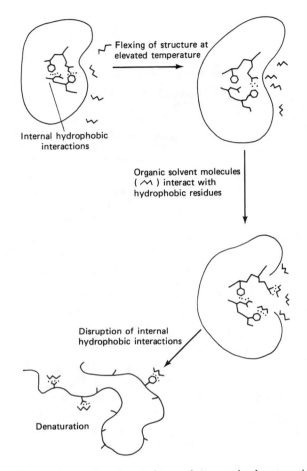

Figure 4.11. Effects of organic solvents in causing protein denaturation.

concerns the intramolecular hydrophobic interactions that help maintain protein structure. At low temperatures, the lack of conformational flexibility means that organic solvent molecules are unlikely to penetrate the internal structure and cause destabilization. But at higher temperatures, small organic molecules enter "cracks" in the surface that occur spontaneously due to natural flexing of the structure, and attach themselves through hydrophobic forces to internal residues such as Leu, Ile, Tyr, Phe, Val, etc. (Figure 4.11). At higher temperatures these internal hydrophobic forces in the protein molecule are stronger, and relatively more important in maintaining the molecule's integrity; loss of these interactions quickly results in denaturation.

Certain organic solvents are more effective at denaturing proteins than others. It has long been recognized that longer-chain alcohols are more denaturing than short-chain ones. An illustration of this is presented in Section 4.5. Although ethanol is used in standard procedures for plasma protein fractionation [67], plasma proteins as a group are unusually stable—few are enzymes—and they survive conditions which cause denaturation of more sensitive proteins. Acetone may have a lesser tendency to cause denaturation than ethanol, because slightly lower concentrations are needed to cause comparable amounts of precipitation at low temperatures. It is also more volatile, which enables it to be removed easily from redissolved precipitates under reduced pressure.

Operating Procedures

Because the temperature must be kept low, an ice–salt bath or other container that can be maintained below 0°C should be prepared and the sample undergoing fractionation chilled to 0°C. The protein concentration can be $5-50\,mg\,ml^{-1}$, but the salt concentration should not be too high. If the salt concentration is high (e.g., subsequent to an ammonium sulfate fractionation), then electrostatic aggregation is impaired, higher levels of organic solvent are needed, and denaturation is more likely. On the other hand, at very low salt concentrations a very fine precipitate may be formed that can be difficult to sediment. Ideally, the starting value of the ionic strength should be 0.05–0.2. Addition of the solvent to water causes heat evolution due to the negative ΔH^0 of hydration of solvent molecules. Consequently, slow addition with efficient cooling is the procedure to follow (use a glass container, not plastic, for good thermal conductivity; stainless steel beakers are even better). If the protein proves to be particularly unstable to organic solvent treatment, it may be best to allow the temperature to fall below 0°C almost at once, as the added solvent lowers the freezing point. But most proteins are much more resistant than this, and it is probably unnecessary to maintain the temperature much below 10°C while the solvent remains below 20% vol/vol. After this amount of solvent has been added, no further heating occurs, so a constant temperature can be maintained with ease. A typical temperature curve for continuous addition of (cold) organic solvent, with efficient cooling, is shown in Figure 4.12.

When the first cut, usually between 20 and 30% (acetone or ethanol), is taken, the temperature must be controlled closely at the desired value, e.g., 0°C, and left there for 10–15 min before centrifuging in a precooled rotor to maintain that temperature. The amount of precipitation is highly dependent on the temperature and is greater at lower temperature values. (This is a rare exception to one of nature's frustrating laws which states that a change of one thing makes another one worse. Decreasing temperature will not only lessen the chance of denaturation in itself, but it will also allow the use of a lower level of organic solvent.) Since

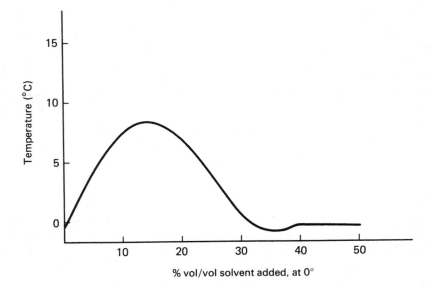

Figure 4.12. Temperature changes as cold organic solvent is added to an aqueous solution, with efficient cooling. After about 15% solvent is reached, the heating effect of hydrating more solvent is less than the external cooling. After 20% solvent there is little or no more heating effect.

subzero temperatures are difficult to maintain accurately, especially during centrifugation, a standard practice is to use "0°C ± 1°C" for organic solvent precipitations.

Most proteins precipitate with acetone in the range 20–50% vol/vol. There is a problem in defining percentages by volume; properly quoted, a percentage should add the statement "assuming additive volumes." For if one adds 50 ml acetone to 50 ml water, the result is only 95 ml of liquid. Describing this as "50% vol/vol acetone assuming additive volumes" at least defines the operation, although it is not mathematically correct. Similarly, a 50:50 mixture of ethanol and water loses 4% of the additive volume. The volume loss is due to the formation of hydrated solvent complexes that occupy a smaller volume than their constituent components. The volume change introduces a marginal difference between reaching an assumed percentage in one step (e.g., 0–50% by adding an equal volume of solvent) and reaching it stepwise; but this is unlikely to be of significance. Unless clearly indicated otherwise in an experimental protocol, take it that quoted percentages are "vol/vol assuming additive volumes."

After equilibrating the mixture at the given percentage solvent at a fixed temperature for 10–15 min, the precipitate can be centrifuged. Because most miscible organic solvents have lower densities than water, the sedimentation of aggregated protein can be very rapid. In fact, at high percentages, when volumes may also be large, sedimentation at 1 g (i.e., on standing) may be sufficiently fast to allow decantation of much of the supernatant without using a centrifuge. But centrifugation is usually necessary, and large-capacity centrifuges capable of generating only a few thousand g are quite satisfactory, provided that they are refrigerated. For reproducible results the need for constant temperatures both before and during centrifugation cannot be stressed too much.

The precipitate can now be redissolved, using *cold* buffer of suitable composition. Organic solvent precipitates are sometimes hard-packed and gluey, and may be quite difficult to redissolve. If a precipitate is not dispersing to make a clear solution, do not add more and more buffer—there is probably denatured protein present. It is usually easier to disperse protein precipitates in a small volume. Doubling the volume of the precipitate will halve the organic solvent concentration, which should allow solution of all undenatured protein. Once a solution has been excessively diluted it is difficult to concentrate again. Moreover, if excess organic solvent is to be removed by evaporation, the higher the starting concentration the more quickly solvent will be lost.

The presence of small amounts of residual organic solvent, e.g., up to 5% vol/vol, are unlikely to affect other fractionation methods, with the exception of hydrophobic chromatography (cf. Section 6.7) and perhaps salting out fractionation. With relatively stable proteins, ammonium sulfate fractionation can be carried out in the presence of this residual solvent, though the salt concentration required for precipitation is likely to be somewhat higher because the organic solvent molecules interfere with the hydrophobic aggregation at high salt concentration. Excessive amounts of solvent can be removed by placing the solution in a Buchner flask and evaporating at reduced pressure. The temperature may be allowed to rise to 25–30°C by immersing the flask in warm water during this process. However, some solutions froth excessively, and this method is not always satisfactory. The best way to remove all trace of organic solvent, plus any other low-molecular-weight compounds that may have precipitated with the protein, is to use a gel filtration column appropriate to the volume of sample (Table 1.2).

To the supernatant from the first organic solvent precipitation, more solvent is added to reach the next percentage value. A formula for calculating the amount of organic solvent to be added is

$$\text{Volume to add to 1 liter to take \% from } x \text{ to } y = \frac{1000(y - x)}{100 - y} \text{ml}$$

A table of amount of pure solvent to add to each level is presented in Appendix A (Table II).

At 50% solvent, only proteins of molecular weight less than 20,000, or proteins far from their isoelectric point are likely to remain in solution.

Precipitation of acidic proteins may be aided by inclusion of divalent metal ions such as magnesium, which form magnesium protein complexes with reduced net charge, encouraging aggregation. Alternatively, a pH change toward the isoelectric point may result in precipitation without increasing the organic solvent concentration.

> Note that pH values, as measured directly with a pH electrode standardized against aqueous buffer, will generally increase as organic solvent content increases. A simple way of explaining this is to consider an acid dissociation to be the sum of two processes:
>
> $$HA = H^+ + A^-$$
>
> $$H^+ + H_2O = H_3O^+$$
>
> The second process is affected because the water activity is reduced by organic solvent; thus, K_a for the overall process is smaller; pK_a is larger.

It has been stressed that denaturation can occur if the temperature is too high in organic solvent fractionation. But denaturation is selective; some proteins and enzymes are remarkably stable. In particular, extracellular proteins and enzymes from thermophilic organisms have to be more robust because of their natural environment. Such proteins often survive organic solvents at elevated temperatures. Use of organic solvents for selective denaturation is discussed further in Section 4.6.

4.5 Precipitation with Organic Polymers and Other Materials

Salts and organic solvents are not the only materials that can cause aggregation of proteins without denaturation. Polson et al. [69] investigated the ability of a variety of high-molecular-weight, neutral, water-soluble polymers to precipitate plasma proteins. Although several were effective in causing precipitation, the high viscosity of most of the solutions made their use as protein precipitants impractical. There was one exception, polyethylene glycol, which is available in a variety of degrees of polymerization. Solutions of this polymer up to 20% wt/vol are not too viscous, and many components of the plasma precipitate before 20% wt/vol is reached. Polyethylene glycol of molecular weight 4,000 or greater is most effective; the two types commonly used for protein precipitation have molecular weights of about 6,000 and 20,000. The behavior of the proteins is rather similar to their behavior in precipitation by organic solvents, and indeed the polyethylene glycol (PEG) molecule can be regarded as polymerized organic solvent, although the percentage required to cause a given amount of precipitation is lower. The first plasma protein precipitated by PEG is fibrinogen, a rather large, highly asymmetric molecule that is only just "soluble" in the plasma in the first place. Then (at neutral pH close to their isoelectric points) γ-globulins precipitate, followed by various other components. As with organic solvents, proteins become more soluble in PEG solutions as the pH moves away from their isoelectric points. The presence of divalent metal ions can also have a considerable specific effect on the solubility of various proteins in PEG [70].

PEG precipitation has been used with considerable success for intrinsically low-solubility proteins, such as "globulins." For other "albumins," such a high concentration of PEG may be needed to cause a precipitate that (a) the viscosity and density of the solution make problems in centrifuging, and (b) in complex protein extracts, phase separation can occur, with a protein-rich heavy phase separating from a lighter PEG-rich phase above (see Section 8.3).

PEG is not as easy to remove from a protein fraction as either salt or organic solvent; as a polymer it will not dialyze rapidly, and separation from proteins on conventional desalting columns may not be good—especially if MW 20,000 PEG was used. Nevertheless, a residual low level of PEG is not detrimental to many procedures; salting out, ion exchange, affinity chromatography, or gel filtration may be carried out without having to remove the PEG first.

Charged polymers (polyelectrolytes) have also been used for protein purification [67,71], especially in industrial applications, where their cheapness and lack of waste disposal problems make them attractive. This is because precipitation occurs at very low percentage, around 0.05–0.1 wt/vol. But their use is restricted to specific cases because of the mode of precipitation. Electrostatic complexes form between polyacrylic acid salts and positively charged proteins at somewhat low pH values, usually in the range 3–6; many enzymes would not withstand these low pH values. Moreover, many proteins are not sufficiently positively charged until the pH is at the lower end of this range. Sternberg and Hershberg [71] described the precipitation of various proteins by both linear and cross-linked polyacrylates. Basic proteins such as lysozyme, cytochrome C, protamines, and trypsin were precipitated effectively.

Other precipitants include caprylic acid salts (n-octanoic acid) and rivanol (2-ethoxy-6,9-diamino acridine lactate), principally in plasma protein fractionation [67]. It is probable that a wide variety of other organic compounds possessing dual functions of hydrophobicity and polarity could cause precipitation, but such techniques have not been extensively investigated.

4.6 Affinity Precipitation

Affinity methods in protein purification may be defined as procedures that make use of the protein's specific and selective interaction with a

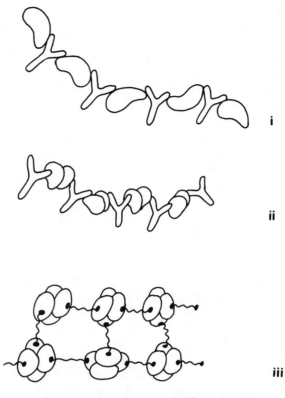

Figure 4.13. Formation of immunoprecipitates and affinity precipitates. (i) Chain-type aggregate with a polyclonal antibody and monomeric protein. (ii) Chain-type aggregate with polyclonal or monoclonal antibody and a dimeric protein. (iii) Three-dimensional aggregate formed with antibodies and tetrameric protein, or with bis-ligands and tetrameric protein.

ligand, usually a substrate or an inhibitor if the protein is an enzyme, which may be attached to an insoluble matrix or may be free in solution. Among the affinity methods included are affinity chromatography (Chapter 7), affinity electrophoresis (Section 8.2), and affinity phase partitioning (Section 8.3). Also in this category is a precipitation method in which addition of a compound that specifically binds to the enzyme or protein, causes that protein to aggregate and precipitate from solution.

Simple addition of ligand to a protein solution may cause precipitation in rare cases—especially if the ligand is a large molecule such as a nucleic acid. But in general, an enzyme's solubility properties are not altered sufficiently on binding to a ligand to make this a successful purification step. But aggregation and precipitation of protein by addition of antibody (a specific ligand) is a well-known phenomenon, and it occurs because the antibody is divalent and either (1) the antibody is polyclonal, so that more than one site on the protein surface is recognized or (2) the protein is oligomeric. In this way a giant molecular interaction aggregate is built up, and this aggregate forms the precipitate (Figure 4.13). Immunoprecipitation is a form of affinity precipitation, but the latter name is more often used for the technique where the precipitant is not an antibody, but a divalent specific ligand, created by linking chemically two ligand molecules together [Figure 4.13 (iii)]. The main successes of this technique have been using "bis-NAD" for precipitating dehydrogenases [72,73], but in principle other bis-ligands should work well, provided that the interaction between protein and ligand is strong enough. Recent developments with "bis-dyes," which are cheaper to make, show promise [74].

4.7   Precipitation by Selective Denaturation

General Principles

Almost every other section of this book stresses treating proteins, and especially enzymes, by gentle procedures to avoid denaturation. Yet there are a large number of enzymes, and many nonenzymic proteins, which are very robust and stand quite extreme conditions. One can use this exceptional stability by exposing an impure preparation to these conditions, whereby unwanted proteins denature and precipitate out of solution. Denaturation implies destruction of the tertiary structure of a protein molecule and formation of random polypeptide chains. In solution these get tangled together and aggregate physically, and to some extent chemically, through formation of -S–S- bonds. Nevertheless, denatured proteins in low-salt conditions, well away from their isoelectric point, may remain completely in solution and precipitate only when the pH is adjusted. This is because repulsion between the charged random polypeptides keeps them apart; closer to their isoelectric point they will aggregate. Alternatively, higher salt concentration can disrupt these intermolecular repulsive forces and also allow aggregation.

The objective of a selective denaturation step is to create conditions where the protein in question will not denature much, whereas many other components are denatured and precipitated by the treatment. The main stresses imposed are (1) temperature, (2) pH, and (3) organic solvents. These three are not generally independent since temperature denaturation depends strongly on pH and vice versa, and with organic solvents, temperature, pH, and ionic strength must be carefully defined. Ionic strength may make some difference to (1) and (2) also, especially if comparing behavior at low I ($\sim 0.01 \rightarrow 0.1$) with that at high salt concentration, e.g., $1\,M$ ammonium sulfate. The three types of denaturation are described separately below.

Temperature Denaturation

Denaturation of proteins by heat is a first-order process with the rate constant obeying most normal thermodynamic principles. Thus the rate constant k_{den} can be related to temperature by:

$$\frac{d(\ln k_{den})}{dT} = \frac{E_{act}}{RT^2}$$

where E_{act} = energy of activation = $\Delta H^{\ddagger} + RT$, and $\Delta H^{\ddagger}$ = enthalpy of activation. What makes heat denaturation of proteins so different from normal chemical reactions is the fact that E_{act} is so much higher than normal. Typical values for observed chemical reactions (including enzyme-catalyzed ones) lie in the range $40-70\,kJ\,mol^{-1}$. For heat denaturation, E_{act} is often in the range $200-400\,kJ\,mol^{-1}$. This is because of the large positive entropy change—visualized as an increase in disorder—when a protein starts to unfold.

The theoretical curve for the amount of a protein denatured at a given temperature in 10 min is shown in Figure 4.14. In this example, E_{act} is taken as $400\,kJ\,mol^{-1}$, and the 50% point at 50°C. Note that at only 5° above the 50% denaturation temperature, barely 1% remains, but at 5° below, as much as 8% is lost. In other words, the curve is asymmetric. Different proteins will have differing E_{act}, and so the steepness of their denaturation curves will vary. But more importantly, the position along the temperature scale at which denaturation takes place also varies widely. Consequently, it is possible to choose a temperature that completely denatures one protein, while leaving over 95% of another protein unaffected; a collection of proteins might have individual denaturation

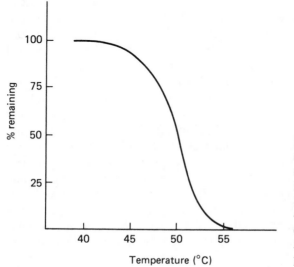

Figure 4.14. Theoretical diagram showing the percentage of protein remaining undenatured after 10 min incubation. E_{act} = $400\,kJ\,mol^{-1}$, and 50% denaturation occurred at 50°C. It is assumed that the process is irreversible.

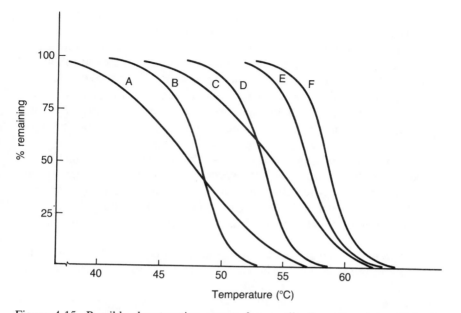

Figure 4.15. Possible denaturation curves for a collection of proteins; F is the most stable to heat; A and B the least stable (see text).

curves as illustrated in Figure 4.15. In this case, if one were attempting to purify protein D, heating for 10 min at 52°C would result in almost total denaturation of A and B, and a little of C. But 90% of D would remain, and all of E and F. Depending on the relative amounts of these proteins in the mixture being treated, this might be a worthwhile step, especially if it was otherwise difficult to separate A or B from D.

As with so many other empirical fractionation methods, it is highly unlikely that one would have any idea of the behavior of many, if any, of the proteins in the mixture—certainly nothing like the detail of Figure 4.15. But the behavior of the particular protein of interest can be determined by using small-scale trials at, say, 5°C temperature intervals. The time of incubation is only important for reproducibility; 1 min or 60 min will give the same *shapes* of curves, shifted along the temperature axis. But whereas 1 min might be very convenient on a 1-ml scale, it would be almost impossible to heat up, then cool rapidly a large volume within a short time. While approaching and leaving the incubation temperature, further denaturation will occur. Thus, a heat-denaturation step must have a reproducible temperature/time curve. Moreover, it is *essential* that the temperature of the protein solution itself be measured, not the liquid outside.

An important consideration with heat denaturation is the possibility of proteolysis. Proteases, like other enzymes, are more active at higher temperatures. They tend to be fairly stable, so even if the desired protein remains active, there is a chance of nicking or minor modification spoiling the final preparation. For this reason it may be preferable to do a heat step in the presence of ammonium sulfate; many proteases are inhibited by the high salt concentration. But the salt may stabilize the proteins further, so that still higher temperatures are required to cause denaturation. There are many documented examples where the presence of a substrate further stabilizes its enzyme, and so can protect it against denaturation. But destabilization by substrate is also possible.

The buffer composition of the sample to be heat-treated must be adjusted carefully in order to get reproducible results; the temperature profile must be reproducible, and above all, the pH of the solution must be carefully defined. As indicated in the subsection below, a very small difference in pH can make a large difference to the amount of denaturation.

Heat denaturation trials can be made in a range of pH values; what one is looking for is the maximum loss (precipitation) of unwanted proteins while getting a good recovery, say 90%, of the desired protein. At the extremes of pH, temperatures need to be lower, and this is more properly described as pH denaturation.

pH Denaturation

The further one moves away from the physiological operating pH of the enzyme, the less stable it becomes. Note that it is not always a matter of being away from neutrality; pepsin is very stable and normally active at pH 1–2, but denatures quickly at pH 7 or higher. Nor is it necessarily true that the optimum stability coincides with the optimum pH of an enzyme—the optimum pH is frequently not relevant physiologically, since activity under experimental conditions of substrate saturation does not necessarily correspond to physiological conditions. Nevertheless, it is expected that an enzyme will be maximally stable at or close to its physiological environmental pH.

Extremes of pH cause denaturation because sensitive areas of the protein molecule acquire more like charges, causing internal repulsion, or perhaps lose charges which were previously involved in attractive forces holding the protein together (Figure 4.16). The most stable pH is not necessarily related to the isoelectric point; indeed, many bacterial and plant proteins are isoelectric in the pH range 4–5 but unstable in that range; they exist at a pH of 6–7 in the cell. But the susceptible *areas* illustrated in Figure 4.16 might be effectively "isoelectric" at the physiological pH.

The rate of denaturation in acid/alkaline conditions is not the protonation step illustrated in Figure 4.16; protonation/deprotonation of charged

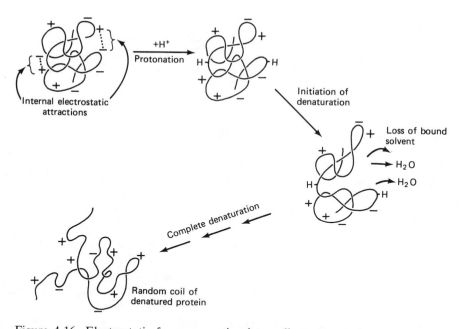

Figure 4.16. Electrostatic forces operating internally on a protein molecule are abolished by protonation, leading to opening up of molecule and complete denaturation.

residues is virtually instantaneous. It is the opening-up step which determines the rate, and this process is essentially the same as occurring in "heat" denaturation. Thus temperature is all-important, and in these more extreme pH conditions, temperatures around 0–10°C rather than 50–60°C may be denaturing. In view of the great difference that a few degrees makes, pH denaturation should be carried out at a carefully specified temperature. Now pH adjustment can be made quickly down to the desired value, and up again (or up and down again), so there is no problem of slow approaches to conditions when scaling up the procedure. Also, pH adjustment should be carried out with an appropriate acid or base—this is discussed further in Section 12.3. Strong acids and bases should be avoided unless necessary. Tris and acetic acid can be used for adjusting pH values in the range 4.5–8.5; stronger acids or bases are needed only to go outside this range. Lactic acid is suitable down to about pH 3.5, after which phosphoric or even sulfuric acid might be needed. If the enzyme is stable at pH 2, then brief exposure to a drop of strong acid is less likely to cause harm than the same exposure of a protein that denatures at about pH 5. For high pH, diethanolamine (to pH 9) or sodium carbonate (to pH 10.5) can be used; sodium or potassium hydroxide will be needed to get above pH 11. In these extremes of pH it is likely that a substantial proportion of denatured proteins will remain in solution, even if the salt concentration is moderate. But on readjusting to neutrality, denatured proteins will normally precipitate out. It is advisable, after neutralization, to incubate briefly at 25–30°C to encourage completion of the aggregation of the denatured proteins before centrifuging off the precipitate.

Denaturation by Organic Solvents

In Section 4.4 the use of organic solvents as protein precipitants at low temperatures was described. The temperature is maintained around 0°C to avoid denaturation. In this section the use of organic solvents as denaturants is described; the temperature is usually kept at 20–30°C or even higher to encourage the process, and the concentration of organic solvents used is usually lower than that needed to cause precipitation of native proteins.

The principle rests on the differential sensitivity of proteins to this treatment. Some proteins are remarkably stable, and this property can be exploited, allowing other, less stable proteins to be denatured. As examples, rabbit muscle creatine kinase is stable in alkaline conditions to 60% vol/vol ethanol at 25°C, and this was used as a step in the original purification of this enzyme [75]. Yeast alcohol dehydrogenase can be purified easily after an initial treatment with 33% vol/vol ethanol at 25°C [76].

A study of the effect of alcohols was carried out on yeast glyceraldehyde phosphate dehydrogenase. Using different alcohols, it was found that the longer the aliphatic chain, the more denaturing the alcohol was (Figure 4.17). The percentage of n-alcohol required to cause 50% denaturation in 30 min at 30°C decreased by a factor of *exactly* 2 for every methylene added to the alcohol. Thus the percentages for methanol, ethanol, propan-1-ol, butan-1-ol and pentan-1-ol were 34, 17, 8.5, 4.2, and 2.2 respectively; lack of miscibility of longer-chain alcohols prevented this from being extended further. Branched chain alcohols were less denaturing (and more water-miscible), indicating that the length of aliphatic chain, i.e., the hydrophobicity, is the important feature. Acetone gave very similar results to ethanol. It would appear from these results that it might be safer to use methanol for organic solvent precipitation if denaturation is a problem.

The pH and temperature must be carefully defined for organic solvent denaturation, since the three stresses are all operating together on the proteins. With so many possible variables, one is unlikely to ever reach the ideal procedure, i.e., maximum recovery of the required enzyme combined with maximum loss of other proteins. On the other hand, it is possible to make a large number of small-scale tests in a relatively short time so that a near-optimum set of conditions can be chosen.

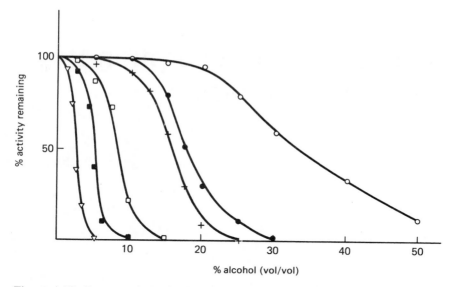

Figure 4.17. Denaturation of yeast glyceraldehyde phosphate dehydrogenase by alcohols. Enzyme at $2\,mg\,ml^{-1}$ was incubated at 30°C for 30 min, cooled, and any precipitate removed by centrifugation. A sample of the supernatant was assayed for enzyme activity remaining. (○), methanol; (●), ethanol; (□), propan-1-ol; (■), butan-1-ol; (∇), pentan-1-ol; (+), propan-2-ol. (Author's unpublished results.)

A frequently used example of organic solvent precipitation, but this time with an immiscible solvent, is chloroform denaturation of hemoglobin (e.g., [9,77]). Purification of erythrocyte enzymes is complicated by the fact that well over 90% of the total protein in the red blood cell is hemoglobin. Shaking a hemolysate with ethanol-chloroform causes complete denaturation of the hemoglobin; a small percentage of chloroform mixes with the aqueous phase and has a strong, specific denaturing effect on the hemoglobin molecule. Thus the first stage in isolating an erythrocyte enzyme is often to carry out chloroform treatment to remove the main contaminating protein hemoglobin.

Chapter 5
Separation by Adsorption I: General Principles

Proteins adsorb to a variety of types of solid phases, usually in a selective manner. Consequently, adsorption techniques, especially when adopted in column chromatography, have become widely used; they frequently result in purification steps that give the greatest increase in protein purity, and, in the case of an enzyme isolation, the greatest increase in specific activity. Although column chromatography is the ideal way of getting optimum resolution, batchwise adsorption methods should not be forgotten, since they can be very rapid, and so are valuable techniques when speed is a priority.

The principal adsorbents for proteins are ion exchangers, hydrophobic materials, inorganics such as calcium phosphate gels, chemically synthesized ligand adsorbents in which a ligand of mixed functional characteristics is attached to a matrix, or biological compounds such as substrates, enzyme inhibitors, or antibodies, which constitute affinity adsorbents. The matrix used has classically been of cellulose, agarose, or other carbohydrate polymer. But there has been a rapid development of alternative, mainly synthetic matrices by companies in the field of protein isolation, and these will be briefly discussed in Section 5.5. The basic principles of using adsorption techniques are discussed in this chapter, and uses of the particular adsorbents are described in Chapters 6 and 7.

It should be mentioned at this point that the principles of chromatography are the same whether one is referring to the traditional "low performance," or "open column" systems, or the more fashionable "high performance liquid chromatography" (HPLC). The "high performance" refers more to the instrumentation than to either the theory or the results. For HPLC columns can give very poor performance (for example, if the recovery of activity is zero!), just as more traditional systems can give very high performance if the conditions are right. Further differentiation and specific discussion of HPLC will be given later.

5.1   General Chromatographic Theory

Partition Coefficients

The behavior of small molecules on chromatographic columns is now well understood. The interaction of a small molecule with the adsorbent can be expressed mathematically, and can be described by simple parameters. With proteins, things are more complicated, because not only may the protein interact with the adsorbent at more than one site on its (the protein's) surface, but also the types and strengths of interaction at these sites may be different. Perhaps even more important is the fact that the adsorbent itself presents a heterogeneous array of binding sites, even assuming that the interacting areas are randomly distributed (Figure 5.1). Thus, no single parameter can describe adequately the protein–matrix interaction (except in highly specific affinity sites; see Chapter 7), and chromatographic theory for proteins must be a compromise of approximations and assumptions.

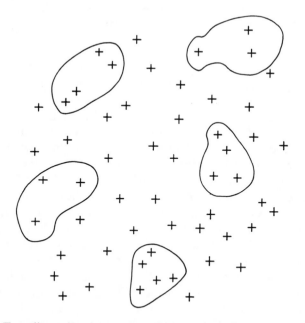

Figure 5.1. Two-dimensional representation of the polydispersity of adsorbent sites in an ion exchanger. Protein molecules are shown interacting with three, four, and five positive charges (+) on the exchanger. The charged groups may be randomly distributed, or heterodisperse due to inhomogeneities in the adsorbent.

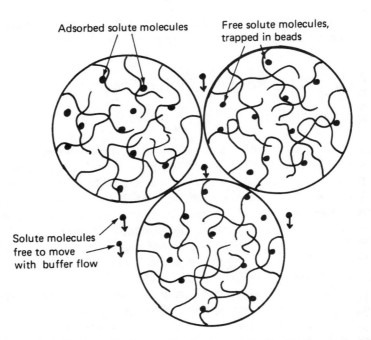

Figure 5.2. Distribution of solute molecules in an adsorbent column consisting of packed beads. Molecules are either adsorbed (mostly within the beads), non-adsorbed within the beads, or freely moving between the beads. Only the latter fraction moves down the column.

True chromatography describes the separation of solutes according to their different partitioning between two phases (in most cases solid and liquid), resulting in differing mobilities down a column (or sheet, layer) of solid particles in the presence of flowing phase. We can say that a solute of partition coefficient 0.8 (defined here as that *fraction* of the solute that is adsorbed on the solid) will move at 20% of the speed of the liquid flow, because at any instant 80% is adsorbed and 20% is free in solution. The individual molecules equilibrate rapidly between adsorption and solution, so each molecule travels at the same speed averaged over a period of time. By measuring the rate of movement of the band of solute compared with the liquid flow, we can determine this partition coefficient. R_f, the mobility of the solute relative to the liquid, is $1 - \alpha$, where α is the partition coefficient. To relate this to genuine interactions between the solute and the adsorbent we have to consider the solute as occupying not two, but three different states. These are (1) adsorbed, that is, inter-acting noncovalently and reversibly with the adsorbent; (2) not adsorbed, but trapped within the solid particles; and (3) present in the flowing liquid between the adsorbent particles (Figure 5.2). In most forms of chromatography the major part of the adsorption occurs within the particles, not on the outside surface. Within the particles there is no solvent flow; all the net movement down the column of both solvent and solute occurs in the spaces between the particles.

Provided that transfer of solute from inside to out of the particles is rapid, we can consider that equilibrium is established, and can define a more rigorous partition coefficient α', which relates to the equilibrium between adsorbed and desorbed solute molecules within the particles:

$$\alpha' = \frac{[\text{adsorbed molecules}]}{[\text{adsorbed molecules}] + [\text{desorbed molecules}]} \qquad (5.1)$$

The volume occupied by the space exterior to the particles is generally referred to as the void volume, V_0 (see Section 8.1). The total effective volume, inside and out, is V_t. The proportion of solute molecules *outside* is then $(1 - \alpha') \times V_0/V_t$. If we call the flow rate of liquid outside the particles F_{out}, the rate of movement of the average solute molecule is:

$$\text{Flow rate of solute} = F_{out} (1 - \alpha') \times V_0/V_t \qquad (5.2)$$

However, since the liquid distributes itself throughout the particles, at any given moment only the fraction V_0/V_t of the liquid is moving. Thus, to compare the rate of flow of the solute with the actual flow rate of liquid through the whole column, F, the expression in Eq. (5.2) must be multiplied by V_0/V_t, and we arrive at the simple expression:

$$\text{Flow rate of solute} = F \times (1 - \alpha') \qquad (5.3)$$

As this is the same as the expression using the experimentally observed partition coefficient, it follows that $\alpha = \alpha'$, so we can use the same value of α both experimentally and theoretically.

Note that certain assumptions have already been made. It was assumed that transfer of solute (and solvent) molecules from inside to outside of the particles was rapid compared with the speed of liquid flowing past. As protein molecules have low diffusion coefficients, this assumption is rarely completely valid. Nonequilibrium deviations from ideal behavior are important simply because most protein chromatography is carried out at flow rates much faster than is needed for equilibrium to be reached.

Zone Spreading, Resolution, and the Plate Height Concept

In a chromatographic column, many factors contribute to the spreading of the zones containing specific solutes. Apart from diffusion, there is the theoretical concept of chromatographic spreading, first described by Martin and Synge in 1941 [78]. In this pioneering paper they introduced the concept of the "height equivalent to a theoretical plate"—derived from similar concepts in fractional distillation in which "plates" are the structures on which vapors condense. A "plate" in column chromatogra-

phy can be considered to be the largest uniform zone able to accommodate the solute, although this does not necessarily correspond with experimentally determined values. In preparative chromatography this is likely
to be determined by the amount of sample loaded, which may occupy a
significant portion of the column, but in analytical work or where otherwise dealing with very small loadings compared to the capacity of the
column, other features of the column become the determinants of plate
height. The concept of plate height can be used to describe chromatography as occurring in a number of discrete steps, transferring solute
from one "plate" to the next in proportions determined by the solute's
partition coefficient. The smaller the plate height, the more discrete steps
there are in a given length of column, so the more chance there is of
getting separation between similar components. In Figure 5.3, theoretical
values are plotted for the distribution of solutes (disregarding diffusion,
although this may contribute to the value of the plate height, as shown
later) after a given number of plate transfers (that is a set volume of buffer flow through the column) has taken place. Solutes with partition coefficients of 0.8, 0.5 and 0.2 have been illustrated.

It should be stressed at this point that the definition of the partition
coefficient α that I am using here is the *fraction* of protein adsorbed,
giving values between 0 and 1, rather than the more common definition
of the *ratio between* adsorbed and free solute, which gives values between
0 and infinity.

Examination of these theoretical curves tells us several things. First,
the zone spreading increases with number of plate transfers rather than
distance travelled down the column. So at $\alpha = 0.2$, after 100 plate transfers, the sample has traveled much further but has not spread any more
than the $\alpha = 0.8$ sample. And by the time the $\alpha = 0.8$ sample reaches the
equivalent position in the column to the 0.2 sample, it has spread very
much more. If the end of the column were at 80 plate heights, then the
high α value solute would emerge in a larger volume, and very much
later, than the low α value solute. The center of the peak emerges after
$1/(1 - \alpha)$ times the column volume of buffer has passed through the
column.

The second thing to be noticed is that a complete separation of the
three solutes (had they been applied simultaneously) has occurred after
100 plate transfers, but not after 20. This is understandable, since the
separation depends on the number of steps. Nevertheless the $\alpha = 0.2$ and
0.8 solutes are well separated even after 20, and still not much spread, so
if the task is only to separate these two, a column with length of only
20 plates would be best: it would be quicker to finish, and the samples
emerging would be more concentrated than if using a longer column.

Resolution, theoretical plate heights, and number of plates can be
deduced experimentally for a given situation. Resolution, R is defined
experimentally as the distance between two neighboring peak maxima

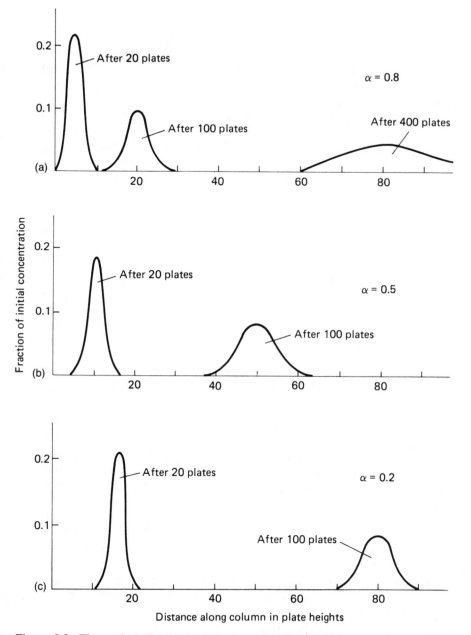

Figure 5.3. Theoretical distribution of solute concentration along a column after the buffer front has travelled a defined number of plate heights. In (a), the partition coefficient is 0.8; in (b) it is 0.5 and in (c) it is 0.2. Distributions were calculated according to the equations of Martin and Synge [78].

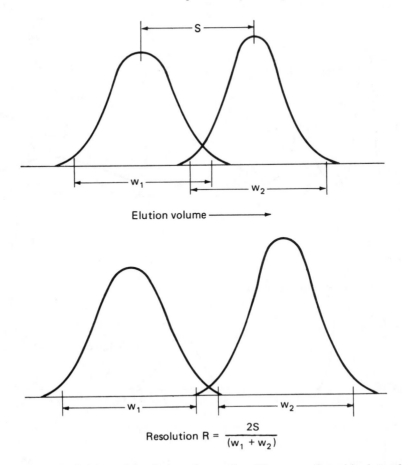

Figure 5.4. Definition of resolution. Separation S between the midpoints of the two peaks is clearly definable. The peak width (W) is defined according to requirements of allowable cross-contamination, since the trailing edges of the peaks prevent a definition based on *total* width.

(the separation, S) divided by the mean of the peaks' widths, W. We must include a factor to describe the degree of resolution we regard as satisfactory, so that the figure $R = 1$ is obtained for, e.g., 98% separation (or more or less as required). The peak width for each solute is then described in terms of the width (volume) containing 98% of the solute, 100% being theoretically impossible (Figure 5.4).

The plate height, H, is strictly defined as the variance of the Gaussian concentration profile divided by the migration distance which, as we only observe the results after emergence from the column, is the column length:

$$H = \frac{\sigma^2}{L} \tag{5.4}$$

But it is easier to calculate this value using an approximation relating the *total* volume of solvent that has passed through the column as the peak emerges (V_e), and the value of the peak width defined above:

$$H = \frac{L \, W^2}{16 \, V_e} \tag{5.5}$$

As mentioned above, if the value of H is not mainly determined by a high protein loading, factors contributing to its value include, diffusion, A, which increases with time, and therefore is inversely proportional to flow rate; nonequilibrium considerations, B, which increase with flow rate; and a general "column quality" factor, C, independent flow rate, which is determined by the quality of packing, of particle shapes, and particle size distribution. These three factors are combined in the Van Deemter [79] equation:

$$H = \frac{A}{v} + Bv + C \tag{5.6}$$

Graphically, this plots H against flow rate, v, as shown in Figure 5.5.

This shows that there is an optimum (minimum) value for H in terms of linear buffer flow rate, and that the optimum flow rate is much slower for high-molecular-weight solutes such as proteins. More extensive treatment of these parameters is given in [80], and further discussion in Section 5.4.

The above descriptions of idealized chromatography are based on isocratic operation, that is, when the buffer conditions do not change. Most protein chromatography is operated either with a gradient elution system, which results in continuous changes in partition coefficients of each adsorbed component, or with stepwise changes in buffer compositions. In the latter case, concepts of plate heights and theoretical optimum conditions are not very relevant; other considerations are more important. It is also not often possible (or necessary) to determine the parameters for each protein, but the theoretical description does give a useful idea of what might be happening in real situations.

The chromatographic theories have been developed principally for low-molecular-weight solutes, and are not always appropriate for proteins and other high-molecular-weight compounds. Among the differences encountered with high-molecular-weight polymers are matrix exclusion effects, large ranges of interaction strengths with different parts of the adsorbent, and often high adsorption capacity (in g cm^{-3}, despite the deceptively low value when this is expressed in molar terms). Biochemists are accustomed to thinking of molecular interactions in terms of dissociation constants—which often can be related to K_m values in enzyme kinetics. By using the same concept, of a dissociation constant for the interaction between adsorbent and protein, a simple theory of protein

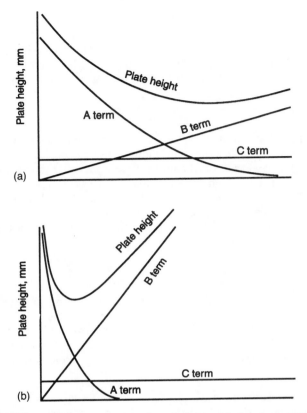

Figure 5.5. Effect of flow rate on plate height: (a) for low-molecular-weight solutes, and (b) for high-molecular-weight solutes such as proteins. Calculated from the van Deemter equation (5.6). Note that the A term, due to diffusion, is unimportant for large molecules except at very low flow rates.

chromatography can be built up; this is particularly useful for describing affinity techniques, in which this dissociation constant can be related to the dissociation constant between free enzyme and substrate.

Adsorbent matrices for proteins have requirements different from those used for low-molecular-weight compounds. The adsorbent must have a very open structure so that the protein can penetrate the particles and reach the binding sites; if not, then only the few sites on the particle surfaces are available, and the capacity of the adsorbent is very low. Second, the nature of the adsorbent must not be harmful to proteins, nor should the backbone matrix material interact with the protein strongly —the adsorptive power is usually created by adding substituents to a supposedly inert backbone material. These two factors make ion exchanger adsorbents designed for low-molecular-weight compounds useless for protein work. For example, highly cross-linked substituted polystyrene resins do not allow penetration by proteins, and they interact by hydrophobic as well as ionic forces on the exterior, which can make any adsorbed protein difficult to elute. It has also been suggested that strongly acidic, e.g., sulfonic acid, or strongly basic substituents are harmful. However, this is a confusion of terminology; the charged form used in ion-exchange work is the complementary weak base (e.g., sulfonate groups) or weak acid (e.g., quaternary ammonium groups), and can be no more harmful than charged groups complementary to weak acids and bases.

The first successful materials used in column chromatography of proteins were the cellulose-based ion exchangers developed by Peterson and Sober [81]. Many other adsorbents had been used for batch techniques, but few were suitable for column work because of poor flow characteristics. Cellulose powder, suitably treated, opens up to a porous structure that protein molecules can penetrate. Moreover, substitution of the cellulose with charged groups increases and stabilizes the porosity as a result of mutual repulsion between the charges. But agarose, the carbohydrate that gels at low concentration and is so widely used by microbiologists as agar, has been adopted as the most suitable matrix for protein purification (with the exception of high pressure systems). Agarose gel beads, sometimes chemically cross-linked for stability under high flow rates, are still the matrix of choice for moderate- to large-scale protein chromatography. They are inert, nonadsorbent in themselves, but capable of a high degree of derivatization to create ion exchangers or affinity-type materials, while retaining permeability to (and so capacity for) most proteins below 10^6 molecular weight. In addition, they can be formed with a range of porosities suitable for gel filtration (Chapter 8).

More recent developments have been towards adsorbent matrices that are capable of being used under higher pressures and faster flow rates than agarose, while retaining those desirable characteristics of high binding capacity, inertness, and adaptability that agarose has. Unfortunately, one other characteristic of agarose, its relatively low cost, has not been considered a primary objective to be reproduced in the new materials.

The Dissociation Constant for Protein–Adsorbent Interaction

In order to describe the behavior of protein on a chromatography column in simple terms that relate to familiar features such as dissociation constants and protein concentration, some further equations can be developed. These are based on an assumed dissociation constant between protein and matrix binding site, from which can be calculated the partition coefficient. This is the important parameter in determining elution position of individual components.

The interaction between a protein and an adsorbent is rarely one simple combination like that between an enzyme and its substrate. Many possible configurations of binding sites on the matrix present themselves to a protein, which itself may have an oligomeric quaternary structure and so offer more than one surface for binding. Figure 5.1 is a two-dimensional representation of how a protein might distribute itself in an adsorbent that has randomly distributed sites; it might interact with two, three, or more sites, giving a polydispersity of interaction strengths. Consequently, an array of possible "dissociation constants" may be expected, not necessarily evenly distributed over the range. Is it possible to average out the spread of values and use a single figure that approximates the interaction of most molecules? If not, the mathematical complexity of the situation and the necessity of defining a function to describe the distribution of constants would make this approach a near-impossible exercise. For semiquantitative descriptions, one can use a single "apparent" dissociation constant, and deviations from ideal behavior due to this approximation can be explained by the fact that this single constant is only an approximation.

Simplified Theory of Adsorption

If we call the concentration of protein in free solution at equilibrium with the adsorbent p, the concentration of free effective binding sites m (effective binding sites means sites that can be occupied by protein without steric hindrance from adjacent bound protein, and whose affinity for protein is sufficiently high to have an observable effect), and the concentration of matrix-bound protein q, then the dissociation constant for the protein–matrix interaction K_p is defined as:

$$K_p = \frac{m \cdot p}{q} \tag{5.7}$$

In order to define these terms properly, the volume occupied by a given amount of matrix must be clarified. For a sensible definition we should assume that the adsorbent matrix is settled to a volume such as it would occupy when packed in a column. In these circumstances, a typical, modern adsorbent based on agarose would have an effective liquid volume of 85–90% of the total, the adsorbent plus bound water occupying the remainder. Thus the *amount* of protein in free solution around the adsorbent is somewhat less than its concentration times the *total* volume.

Having defined this dissociation constant, we can now use it in a way that usefully relates to adsorption chromatographic procedures. Adsorption represents a partitioning of solute (protein) between the two phases, liquid and solid. A partition coefficient can be defined as being the amount of solute adsorbed at any instant as a fraction of the total. Alternatively, it could be defined as that fraction in solution, in which case it is identical with R_f, the mobility of the solute relative to the buffer front. The former will be used here and called $\alpha (\alpha = 1 - R_f)$. This partition coefficient α will be referred to frequently during the rest of this chapter. It should be noted that reference to a *low* α value implies zero to about 0.5, whereas a *high* α implies a value close to unity.

The relationship between α and K_p can now be determined:

Let the total effective concentration of adsorption sites (i.e., capacity for adsorbing the protein in question) be m_t. Let total protein concentration (free + bound) in the adsorbent be p_t. Then:

$$m_t = m + q \tag{5.8}$$

$$p_t = p + q \tag{5.9}$$

$$\alpha = \frac{q}{p_t} \tag{5.10}$$

Substituting $q = p_t \alpha$ and m, p from Eqs. (5.8) and (5.9) into Eq. (5.10) we get:

$$K_p = \frac{(m_t - p_t \cdot \alpha)(p_t - p_t \cdot \alpha)}{p_t \cdot \alpha} = \frac{(m_t - p_t \cdot \alpha)(1 - \alpha)}{\alpha}$$

So:

$$p_t \cdot \alpha^2 - \alpha(m_t + p_t + K_p) + m_t = 0 \tag{5.11}$$

Now α is the solution of this quadratic lying between 0 and 1. Note that in conditions where $m_t \gg p_t$, i.e., there is a large excess of binding sites compared with protein to bind, then:

$$m \approx m_t, q \approx p_t, \quad \text{and} \quad p = p_t - q = p_t(1 - \alpha)$$

So:

$$K_p = \frac{m \cdot p}{q} = \frac{m_t \cdot (1 - \alpha)}{\alpha} \tag{5.12}$$

Now, α is a value that can be experimentally determined very simply; p_t and m_t are also easy to measure in specified conditions. Thus, the validity of these arguments can be experimentally tested, and values of K_p determined. Typical concentrations of *effective* binding sites would vary between about 0.01 mM for affinity adsorbents to over 1 mM for ion exchangers (depending on protein size), and total protein concentrations would be in a similar range. Some solutions to Eq. (5.11) are presented in Table 5.1, using a variety of values of K_p. For a useful adsorption on column chromatography, α needs to be at least 0.8, from which it can be seen that the value of K_p generally needs to be 0.1 mM or less. This has important ramifications when considering affinity adsorption techniques.

The conventional description of chromatographic processes makes use of the Langmuir isotherm, which expresses the amount bound (q) in terms of the concentration of free solute (p). Using the symbols defined

Table 5.1. Solutions for α from the Equation
$p_t\alpha^2 - \alpha(m_t + p_t + K_p) + m_t = 0$

K_p (mM)	P_t (mM)	m_t (mM)	α
0.1	0.1	1.0	0.90
0.1	0.5	1.0	0.85
0.1	1.0	1.0	0.73
0.1	0.1	0.1	0.38
0.1	0.5	0.1	0.16
0.1	0.01	1.0	0.91
0.01	0.01	1.0	0.99
0.01	0.5	1.0	0.98
1.0	0.01	1.0	0.50
0.1	0.1	10	0.99
1.0	0.1	10	0.99
0.1	0.1	0.2	0.59
0.1	0.1	0.5	0.81
0.1	0.1	2	0.95
0.1	0.1	5	0.98

above, this expression can be deduced from the dissociation constant K_p:

$$K_p = \frac{m \cdot p}{q} = \frac{(m_t - q) \cdot p}{q} \tag{5.13}$$

$$q(K_p + p) = m_t \cdot p, \quad q = \frac{m_t \cdot p}{K_p + p}, \quad \text{or} \quad \frac{k_1 p}{1 + k_2 p}$$

This is the Langmuir isotherm. Note that the concentration term p is the *free* concentration at equilibrium, not the total. Thus, the Langmuir isotherm does *not* represent a plot of amount adsorbed against amount *added*, as is sometimes thought. It is much more difficult to measure p than p_t. Using some convenient values, the Langmuir isotherm expression is compared with the amount adsorbed per amount added and with α in Figure 5.6.

Preparative column chromatographic techniques typically involve applying a relatively large sample—often in a volume greater than the volume of the column—in a buffer chosen so that the protein of interest adsorbs. In the starting conditions, there are three possible fates for a particular protein. Either it adsorbs totally and immovably ($\alpha = 1$), it does not adsorb at all ($\alpha = 0$), or it adsorbs partially ($0 < \alpha < 1$) and moves down the column, emerging when $1/(1 - \alpha)$ column volumes of liquid following the start of application have passed through. Thus, if the applied sample is 5 times the column volume, and the column is to be washed with a further 5 vol, for this protein to remain in the column, α would need to be greater than 0.9. (We are assuming here that the total adsorptive capacity of the column is much greater than the amount of protein in the sample.) As another example, if the sample contains two proteins with very different α values, and only a small volume is applied to the column, a very convenient separation might be achieved (see, for example, Figure 6.12). If the protein of interest has an α value of 0 or only a little greater than zero, then it passes through the column, along with all other proteins of $\alpha = 0$. This may be a good purification step if the majority of proteins are left behind adsorbed to the column. No true chromatographic principles are involved in this case, since there is no subtle binding and desorption of proteins as they move down the column that is responsible for the separation. The enzyme in the nonadsorbed fraction would be obtained rapidly, since the column does not need to be developed with an elution scheme; it should be obtained in 100% yield, and in a volume not much bigger than before application to the column. Thus, although the full benefits of chromatography can be obtained only by first adsorbing all of the desired protein ($\alpha > 0.8$), then eluting with some scheme involving changing buffers, some partial purification may be achieved even if it is not adsorbed.

Even if all adsorption sites were equivalent, the α value would depend on protein concentration [see Eq. (5.10), Figure 5.6b], with higher protein concentrations leading to a smaller *proportion* (though larger amount) adsorbed. The fact that there is in truth a whole range of K_p values further aggravates this effect. Stronger binding sites (low K_p) are occupied first; higher protein concentrations need more binding sites, which means that weaker binding sites must also be used, leading to a higher average K_p and a lower α. Some quantitative values for an enzyme, rabbit muscle

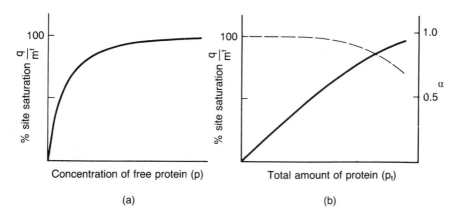

Figure 5.6. (a) The Langmuir isotherm plot, expressing the percentage of site saturation (amount adsorbed) against *free* solute concentration. (b) Percentage site saturation (*solid line*) and the partition coefficient α (*dashed line*) against the *total* solute added.

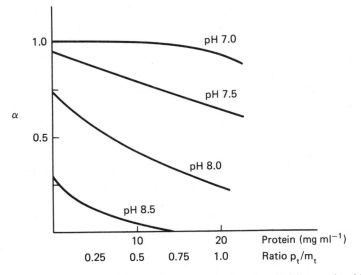

Figure 5.7. Variation of α with protein concentration for rabbit muscle aldolase adsorbing to CM-cellulose, at several pH values. (Experimental results from Scopes [82].)

aldolase adsorbing to CM-cellulose, have been presented elsewhere [82]. These results are illustrated in Figure 5.7, in which the variation of the measured value of α with the total amount of protein present is plotted for four different pH values. If we consider a band of aldolase on a column at pH 7.5, the total protein concentration in the middle of the band may approach m_t, with an α value of about 0.6–0.7. Any protein that moves out ahead of the main band, whether due to diffusion or to uneven flow in the column, becomes diluted, so its α value is higher. At $p_t/m_t = 0.1$, $\alpha = 0.9$ in this case. Consequently, it mostly absorbs, and so is retarded relative to the main band behind it. As a result, there is a self-sharpening effect on the leading edge of the protein band. However this increase in α also occurs on the trailing edge: decreasing protein concentration causing an ever-increasing strength of binding. So as α approaches 0.9, the rear part of the protein band moves with only 10% of the rate of the buffer flow, compared with 30–40% for the main band. As a result, with constant buffer conditions the enzyme would elute from the column with a sharp, concentrated leading edge, but a very long tail (Figure 5.8). This is an occurrence commonly observed in isocratic conditions, i.e., when the buffer composition remains constant. It can be overcome using a continuously changing gradient of buffer conditions (see Section 6.2).

It can be calculated from Figure 5.7 that, in this case, the measured average K_p values change by a factor of about 5 from very low site occupancy to 50% occupancy. The value for the total available sites, m_t, may not be the same in all conditions. The total effective capacity of an ion exchanger may well be greater when the binding is very tight, and less when the binding is weaker, despite the fact that the adsorbent itself presents the same pattern and distribution of possible adsorbing sites. A protein with 50 net charges may be able to bind on a "weak" site, whereas when it has only 30 net charges, it may not have a strong enough interaction.

Another significant complication here is that there is a proportion of adsorbing sites which take quite a long time to be occupied. Indeed, the time dependence is such that total saturation may take as long as an hour to be reached. In these cases equilibrium treatment is clearly not appropriate. These slow, poorly accessible sites may become occupied during the long process of loading a column. If it takes a long time for the protein to go on, it usually follows that it must also take a long time for it to come off. Thus a rapid elution scheme will leave behind some small proportion of protein, eluting later than the main part—a further aggravation of the tailing effect already described in terms of equilibrium consideration of partition coefficients and variable dissociation constants. Some protein may never come off, especially if a change in buffer conditions leads to a shrinkage of the adsorbent, finally trapping the protein in a site that was already accessible only with difficulty.

It has been proposed, and evidence presented to confirm this, that adsorption occurs mainly on the outer portions of bead particles. As proteins bind, especially large proteins, they tend to block access to sites further buried in the beads. So when saturated, the beads may not have a homogeneous penetration of protein, but rather be like an onion, with the outer layer truly saturated, and successive layers less so (Figure 5.9). Smaller proteins are more capable of penetrating the outer layer, and so

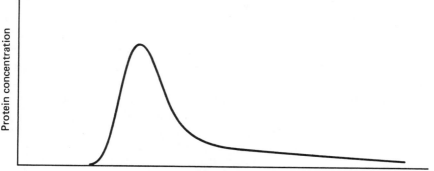

Figure 5.8. Shape of elution peak in constant buffer conditions expected for rabbit aldolase at pH 7.5 (from Figure 5.7). The long "tail" is caused by increasing value of α as the protein becomes more dilute.

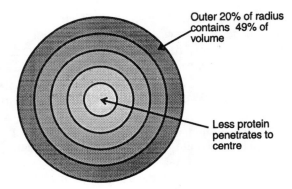

Outer 20% of radius
contains 49% of
volume

Less protein
penetrates to
centre

Figure 5.9. Cross section of spherical bead after protein has adsorbed. The darker sections contain higher concentrations of protein. The outer 20% of the bead, as measured by radius, constitutes 49% of the total volume, and as it is here that the protein is most concentrated, it is likely to contain 70% or more of the total protein.

may be selectively adsorbed inside. Even if only the outer one-fifth of the bead can be occupied, this represents close to 50% of the total bead volume, so it is advantageous *not* to have proteins going right inside to the center of the beads, considering the longer diffusion times. This penetration factor is obviously important when trying to quantitate and model chromatographic behavior, and has been proposed as the reason why actual adsorption kinetics are faster (and so better) than theory would suggest [83].

The final complication that upsets theoretical predictions is the occurrence of inhomogeneity of adsorbents; this is clearly a principal factor in causing the range of K_p values, but it also means that the concentration of binding sites, m_t (Eq. 5.8), is only an average value, as it is calculated from the concentration of bound protein. In some adsorbents, it may be that large volumes of matrix are so poorly substituted that there are few effective binding sites present, but where substitution does occur it is so extensive that virtually all the protein that binds does so in this relatively small fraction of the total volume of adsorbent. Consequently, the true, effective value of m_t to put into Eq. (5.11) may be much higher in local areas than the average value calculated from protein binding per cubic centimeter. This would result in higher α values than if the sites were distributed evenly through the matrix. Further discussion of this point occurs later under affinity adsorption (see Section 7.1).

In summary, the description of adsorption effects by using a dissociation constant for the protein–matrix adsorption can be useful, bearing in mind that in most cases a unique value for the dissociation constant does not exist, and the average value will vary depending on the percentage of adsorption sites occupied. Similarly, the partition coefficient, which is the experimentally important parameter, can decrease both directly due to approach to saturation, and indirectly because increasing site occupancy results in a lower average affinity, giving a lower α value.

5.2 Membrane Adsorbents; Radial Flow Columns

In order to overcome restrictions on flow rates that are imposed both by viscous forces as liquid flows around beads, and the rates of diffusion in and out of beads as used in columns, a new type of adsorbent based on membranes has been developed. Permeable membranes with small pores can have a relatively large surface area per unit volume. The unit volume is small, as the definition of a membrane implies thinness, but a stack of many membranes builds up an equivalent of a column, but without individual particles. Whereas flow rates in bead columns are restricted by the diffusion of solute into and out of the beads, with membranes, the ligands are on the surface of the matrix, and there is no diffusion of solute away from the buffer flow (Figure 5.10). Liquid flows through relatively easily under low pressure, the resistance determined mainly by the diameter of the pores traversing the membrane. The pressure required increases proportionately with the stack thickness. As there is no significant time delay for binding, the flow rates can be very fast ($100-500\,\mathrm{cm\,h^{-1}}$), limited only by the need for a protein molecule passing through a pore to have sufficient chance to encounter an adsorbing ligand on the pore's surface. A typical capacity of a membrane adsorbent is between 0.1 and $0.3\,\mathrm{mg\,cm^{-2}}$. A stack of 50 membranes is about 1 cm thick, so capacities of the order of $10\,\mathrm{mg\,cm^{-3}}$ are achieved. This is 5 to 10 times less than conventional column beads (ion exchange). Although the capacities are

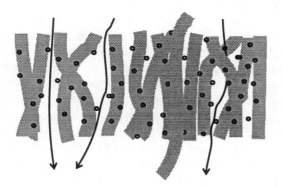

Figure 5.10. Cross section of a membrane, indicating how buffer flows through the pores. The active protein adsorbing areas, indicated by circles, are exposed on the surface of the membrane material, in immediate contact with the passing proteins.

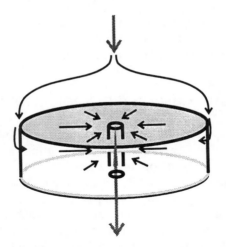

Figure 5.11. Principle of radial flow columns: the buffer solution passes in at the sides of the cylinder and moves in to the central core, from which it flows to the recorder and fraction collector.

lower, the advantages of high flow rates are attractive, particularly in a cyclic process which can be automated. Membranes for ion exchange and a variety of affinity applications are available commercially, e.g., MemSep™ from Millipore (Milford, MA). Recent examples of the use of membranes include the large-scale purification of formate dehydrogenase using cation exchange, anion exchange and red dye membranes [84], and purification of growth factors using immunoadsorbent (antibody) membranes [85].

Another way of increasing flow rates in column-type chromatography is the radial flow concept. In this system, the flow is perpendicular to the normal direction in columns, being from the outside of the column cylinder to the inside, where a collection tube receives the buffer flow and passes it to outside (Figure 5.11). The flow rate can be high because the cross-sectional area is large at the commencement of contact with the adsorbent (which is a conventional bead-type matrix), and although this decreases rapidly as the flow approaches the center, there is not much pressure drop. Thus the linear flow rate increases towards the center; proteins moving "down" the column, i.e., from the outside to the inside, are initially separated at a low linear flow rate, which is a good thing, but towards the end are rushed through quickly. Although focusing into the center portion concentrates the bands, this is counteracted by the fast linear flow at this point, so elution volumes are comparable with conventional columns. As with the membrane stack columns described above, radial flow columns (ZetPrep™ and Zetaffinity™, Cuno Life Sciences, Meriden, CT) are available in a variety of chemistries for ion exchange, activated, and precoupled affinity materials.

5.3   Batch Adsorption

General Principles

Some of the simplest steps in protein purification make use of batch adsorption, in which a flocculant material is added to a solution, often the initial extract, and proteins are adsorbed to it. Any adsorbent can be used; it should be capable of rapid recovery and washing by flocculation or filtration; on smaller scales centrifugation may be appropriate. The chief requirement for a successful batch method is that the partition coefficient α of proteins being adsorbed should be very close to 1, and that a significant number of other proteins should have lower α values. The reason for this is indicated in Figure 5.12 and Table 5.2, where it is seen that the *concentration* of free protein is the same throughout the volume, so the total *amount* not adsorbed can be substantial if α is even fractionally below 1.

Figure 5.12 allows the following calculations. For a two-phase system: free liquid and settled adsorbent volumes are V_s and V_a, respectively. The concentration of free protein, p, in V_a is the same as in V_s. The amount of protein adsorbed is qV_a, where q is the concentration of adsorbed protein. The amount of protein not adsorbed is $p(V_a + V_s)$. The partition coefficient α is defined as $\alpha = q/(p + q)$. This allows the concentration of free protein to be written as $p = q(1 - \alpha)/\alpha$. Therefore, the fraction protein adsorbed, f, is:

$$f = \frac{qV_a}{p(V_a + V_s) + qV_a} = \frac{V_a\alpha}{V_a + (1 - \alpha) V_s} \tag{5.14}$$

Table 5.2 gives an idea of how close α must be to unity to obtain a good adsorption yield of 80–90%. With most adsorbents, K_p values smaller than $10^{-6} M$ are needed to give α values of 0.98 or greater. If α is over 0.99, very small amounts of adsorbents can be used, provided the value of m_t is sufficiently large to accommodate the adsorbed protein. A

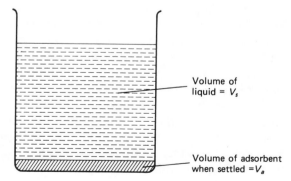

Figure 5.12. Definition of volumes in batchwise adsorption procedure.

Volume of liquid = V_s

Volume of adsorbent when settled = V_a

Table 5.2. Fraction of Protein (f) Adsorbed in
Different Batch Conditions, from Eq. (5.14)

V_a/V_s	α	f
0.05	0.9	0.30
0.1	0.9	0.45
0.2	0.9	0.60
0.5	0.9	0.75
0.1	0.95	0.63
0.2	0.95	0.76
0.5	0.95	0.86
0.05	0.98	0.70
0.1	0.98	0.82
0.1	0.98	0.89
0.1	0.99	0.99
0.02	0.998	0.91
0.05	0.998	0.96

value of V_a/V_s of 0.02 would not be unreasonable if the protein concerned is present only in low amounts. However, it is uncommon for
general adsorbents to be so specific; large amounts of other proteins are
likely to bind, reducing the effective capacity of the adsorbent. Affinity
adsorbents are much more likely to be selective, but, as mentioned in
Section 7.1, K_p values smaller than $10^{-6} M$ are unlikely to be achieved
without substantial nonspecific interactions. Immunoadsorbents should
be particularly suitable for batch applications, since they select out the
antigen and bind it tightly. Dye ligand adsorbents fulfill these conditions also, but may be relatively nonspecific; consequently, quite large
amounts may be needed before all the enzyme wanted is adsorbed.
As an example, glucose 6-phosphate dehydrogenases from yeasts and
bacteria can be 95% adsorbed onto selected dye adsorbents from a crude
extract at $V_a/V_s = 0.05$. From Table 5.2 this implies an α value of at least
0.99.

Sometimes one must judge whether a batch treatment is better than a
column. Batch methods offer the advantages of speed and the possibility
of treating large volumes of material, but unless α is well above 0.98,
some losses are inevitable. Columns can result in 100% adsorption even
at α values of 0.95, but are more trouble, slower, and may clog if dealing
with crude extracts. Consider the situation of having 1 liter of material
and adding $100 \, cm^3$ of adsorbent, resulting in 63% of the protein being
adsorbed ($\alpha = 0.95$, Table 5.2). If this α value is applicable to a column,
and the capacity (m_t) is far greater than the amount of protein present,
then running the whole sample on to a $100 \, cm^3$ column will result in the
leading edge of nonadsorbed protein being located halfway down [move-
ment rate = flow rate $\times (1 - \alpha)$, volume passed in = 1 liter, so movement =
$50 \, cm^3$]. Washing with a further 500 ml of buffer will still not cause elution

Table 5.3. Purification of Phosphoglucose Isomerase from a Yeast Extract Using Batch Adsorption of Unwanted Proteins

	Protein (mg)	Enzyme activity (units)	Specific activity
Crude extract	30,500	173,000	5.7
After Zn(OH)$_2$ treatment	6,630	115,000	17.3
After bentonite treatment	905	109,000	120

Source: Ref. 86

of the protein, and a stepwise or gradient elution procedure now should result in an excellent chromatographic separation. But if the procedure is to be scaled up to, say, 20 liters, then columns may not be economical. The question is whether 63% adsorption by a batch treatment, with more lost on washing, is a sufficiently good recovery. If not, then α must be increased somehow, or the amount of adsorbent increased. Using twice as much adsorbent does not improve the recovery very much; an increase in α by changing the buffer conditions is preferable and simpler—if possible.

Even if the enzyme does not adsorb at all, batch processes can be extremely useful for removing other proteins. As an example, treatment of crude yeast extracts with first zinc hydroxide and then bentonite adsorbs much of the protein material, but leaves all the phosphoglucose isomerase ([86]; Table 5.3). There are many other examples, sometimes going under the category of "clarification," but incidentally involving adsorption of unwanted material. Even if the degree of purification is not great (e.g., only 10–20% increase in specific activity), the protein removed may otherwise have been an important contaminant at a later stage.

Practical Approaches

The simplest type of batch adsorption involves adding the adsorbent (pre-equilibrated with the appropriate buffer), stirring for a few minutes, allowing to settle, and then filtering under suction. The pad on the filter funnel is then washed with buffer until little nonadsorbed protein remains. It is sucked to damp-dryness, then the elution buffer mixed in (temporarily removing the suction) and allowed to stand with the adsorbent for a few minutes. It may be better to transfer the adsorbent with the elution buffer to a beaker and stir for a few minutes, before returning the slurry to the filter funnel. Meanwhile, the collecting flask is washed out thoroughly and drained before replacing it and sucking the elution buffer through. The process is then repeated, and finally, without mixing the pad, a further volume of buffer passed in to ensure that all the desorbed enzyme is

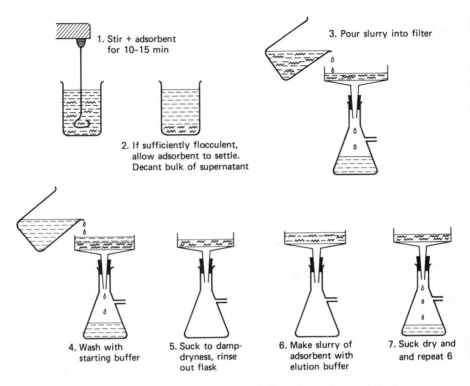

Figure 5.13. Operational procedure for batchwise adsorption and elution.

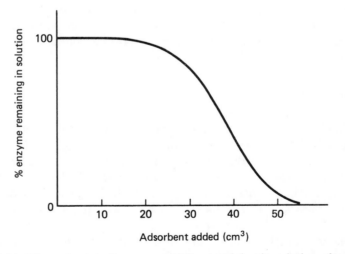

Figure 5.14. Measurement of enzyme activity remaining in solution after adding successive batches of adsorbent indicates optimum protocol. In this example, $20\,cm^3$ of adsorbent should be added first and then removed. Then a further $30\,cm^3$ is added to adsorb the enzyme.

washed out. Each batch of elution buffer should be kept to only 1–2 times the volume of the adsorbent pad. The process is described diagrammatically in Figure 5.13.

More complex protocols may lead to better purification. For instance, Figures 5.14 and 5.15 illustrate an example in which there are proteins present which adsorb more strongly than a required enzyme. Consequently none of the enzyme binds to the first bit of adsorbent added. In this case, the first batches of adsorbent should be removed before adding the batch that takes out the enzyme. A small-scale trial to find the optimum batches of adsorbent can be carried out first.

Elution can also be stepwise, removing some unwanted material in an initial elution before the buffer to elute the wanted enzyme is added. Again, a small-scale trial can find the optimum (Figure 5.15).

Typical batch adsorbents are:

Hydroxyapatite
Ion exchangers (especially phosphocellulose)
Affinity adsorbents
Dye ligand adsorbents
Hydrophobic adsorbents
Immunoadsorbents

Because of the speed and potential large-scale use of batch methods, they should be seriously considered in any preparation where an adsorbent is used with a large volume of material. With so many different types

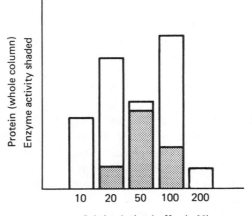

Figure 5.15. Stepwise elution to determine optimum procedure for extracting enzyme. In this example enzyme is eluted mostly above 20 mM salt. Probably the optimum protocol would be (1) wash with 20 mM salt to remove unwanted proteins, (2) wash two times with 50–60 mM salt to obtain most of enzyme without too much nonspecific protein.

of adsorbent being available, e.g., a multitude of dye ligand possibilities, it becomes more likely that one exists with the appropriate characteristics for any particular enzyme. If it can be applied to a crude extract, there is a great saving in time, and requirements for larger-scale equipment may disappear.

5.4   High-Performance Liquid Chromatography

General Principles

Originally, HPLC stood for "high pressure liquid chromatography." The separation of low-molecular-weight compounds by column chromatography was limited in resolution by the diffusion of the compounds during the chromatographic run; diffusion is time-dependent, so a faster process would allow higher resolution. Faster flow rates require smaller beads, otherwise there is insufficient time for the solutes to pass in and out of the beads and attain (or approach, see below) chromatographic equilibrium. But small beads create high back pressures, necessitating the development of high pressure pumps and columns to support the media and pressures. Whether it was the successful development of rigid, small, high quality bead resins, or of high-pressure pumps, or simply the requirement of higher resolution that was the main force behind the introduction of HPLC is a matter of history, but the hardware for HPLC was introduced in the 1970s principally for the separation of small molecules. The adsorbents used were mostly of the hydrophobic type, using organic solvents such as methanol or acetonitrile in gradients with water, and often acidic solutions ("reverse-phase" chromatography). Protein chemists saw this as an opportunity to get higher resolution separation of protein mixtures, at least in an analytical context, but it was many years before HPLC achieved the sorts of results for proteins that were routine for small molecules. This is because it was necessary to develop alternative resins that had pores sufficiently large to be permeated by the protein molecules; also, the flow rates had to be reduced somewhat because of the lower diffusion coefficients of large molecules such as proteins. But the main problem was that, in the conditions used in reverse phase, such as 0.1% trifluoroacetic acid and up to 50% organic solvent, most proteins are denatured, and so even if good resolution was achieved, bioactivity would often be irreversibly lost. HPLC had by now come to stand for "high performance liquid chromatography." This recognizes that high pressures were not as important as high quality of the adsorbent material in the column, especially in the development of high-performance gel filtration materials for proteins.

To this day, HPLC means different things to different people. To some, it means only reverse-phase columns operated with organic solvents and completely automated equipment. To others it is the equipment itself, the pumps and mixers, columns and detectors and collectors, regardless of the principles of separation. But realistically, it should be the *performance* that is high, even if achieved with simple equipment. For, often, the use of "HPLC equipment" has resulted in relatively poor performance, but with the implication that this is the best that could be achieved. The medium should not be the message—it is results we want.

When high-performance liquid chromatography, generally abbreviated to HPLC, was first applied to proteins for preparative purposes, results were somewhat equivocal and of dubious advance compared with more conventional techniques. The problems may be listed as follows:

1. To obtain a reasonable capacity, it is necessary for most adsorption to occur within bead particles; thus, they must be open enough to be freely penetrable by protein molecules. It has been difficult to manufacture beads that are both sufficiently porous and sufficiently strong not to collapse under high pressure gradients.
2. Diffusion coefficients of molecules are inversely proportional to their linear dimensions. Thus a spherical protein molecule of 100,000 Da has a diffusion coefficient 5 times less than a similarly shaped molecule of 1,000 Da. This means that a column with ideal parameters for small molecules would need to be operated about 5–10 times slower to deal with averaged-sized proteins.
3. To get high resolution, very small plate heights are needed, which means that only a very small segment of the column can be occupied by a particular fraction; consequently, loadings must be very low if the full benefits of the quality packing are to be realized.
4. Scaling up, to overcome restriction (3), is possible, but extremely expensive owing to the costs of packing materials.

It was not until equipment and materials specifically designed for proteins were developed that HPLC became a routine procedure for isolating proteins, especially bioactive ones. The Pharmacia FPLC "Fast Protein Liquid Chromatography" (Uppsala, Sweden) was introduced; its basic principles were those of ion exchange, but based on highly refined top-quality beads (Monobeads^{TM}, $10 \mu m$ diameter) that are able to produce extremely high resolution. Capacity is limited (mainly by cost; few researchers can afford Monobead columns larger than 1 ml, with optimum capacities no more than a few mg), but as a final step in a protein purification, the Monobead columns, and similar adsorbents developed since by many other manufacturers, have approached a theoretical optimum in resolution.

Because of limitations in scale, HPLC of proteins is suited mainly for the following:

1. The last step in a purification which results in less than 5 mg of final product (but repeated application cycles can increase this figure);

2. When dealing with minute amounts of mixtures (and where only microgram quantities of product are required);
3. When the budget is virtually unlimited, allowing large preparative columns to be used;
4. For analytical work, comparing peaks in one sample with another.

Many protein isolations do come into one of these categories. Although the enzymologist or protein crystallographer may want much more, a protein's natural occurrence may be quantitatively so low that the final step may only yield a few micrograms. In work with recombinant DNA, to detect a structural gene requires an oligonucleotide probe which can be synthesized, provided that a short stretch of amino acid sequence of the protein is known. Such is the sensitivity of modern sequencing techniques that given a pure sample, it is possible to get sufficient sequence from as little as 5–10 pmol, which, for a polypeptide of 50,000 Da, is less than 1 μg. And newer sequencing techniques are predicting a further factor of 10 reduction in amount needed. HPLC techniques not only give ideal separations, but may be the only way of handling such small amounts of protein. Capillary columns, working with microliter volumes of sample are now available (see also Chapter 10).

Relationships Between Bead Size, Flow Rate, Pressure, and Optimum Performance

Before modern HPLC equipment was introduced, the main problem in getting small solutes of closely similar properties separated was the rapid diffusion and coalescing of adjacent fractions during the periods of hours of the chromatography. In order to get sufficient plate numbers, columns needed to be long, but this increased the time of operation. Another way of reducing plate height is to decrease the size of the particles in the column, and make them spherical beads to optimize the packing. But the smaller particles meant that flow rate decreased, so the shorter column took just as long to elute. The answer was to increase the pressure, and this meant the development of pressure-resistant columns, joints and connectors, pumps to generate the pressure, and all the control systems that are now routine on HPLC apparatus. Ever smaller beads meant even higher pressures, but the results justified the means. The relationships between these various parameters are as follows:

Rate of attainment of equilibrium is inversely proportional to the diffusion coefficient of the solute, and also to the bead diameter. Thus, halving the bead diameter halves the time to reach, say, 90% of an approach to equilibrium for a specific molecule.
Flow rate is proportional to the square of the bead diameter under a given pressure, assuming that the beads are packed ideally.

Resolution is proportional to the square root of the column length. To double the resolution requires a column 4 times longer, operated in the same conditions of flow rate, loading etc., and it thus takes 4 times longer to complete the chromatography.

Optimum performance is attained at the smallest value of plate height, which (as shown in Figure 5.5) depends on flow rate; for proteins this optimum flow rate is generally very slow.

The term "reduced velocity," a dimensionless number, has been used [87,88] to relate these properties to determine an ideal flow rate, for a given column:

$$V_{red} = \frac{v \cdot d_p}{D} \tag{5.15}$$

Where v = linear flow rate
d_p = bead diameter
D = diffusion coefficient of solute.

It is suggested that a value of V_{red} in the range 3–10 is optimum. Using values for typical proteins of sizes 50,000–200,000 Da on columns with bead diameter of 100 μm (e.g., conventional, low-pressure chromatography), the flow rates suggested are less than 2 cm h^{-1}. For beads of 10 μm, the common size used in HPLC of proteins, 20 cm h^{-1}, is indicated. On a standard 7.5 mm diameter HPLC column, 20 cm h^{-1} represents 0.15 ml min^{-1}, some 10 times slower than the columns are usually operated. This same factor of 10 applies to the large-bead conventional chromatography, where flow rates are usually 20–30 cm h^{-1} rather than the 2 cm h^{-1} indicated above. We clearly have a compromise here, experimentally determined, between theoretical optimum operation and convenience of speed; slight loss of resolution by operating at about 10 times V_{red} appears to be acceptable considering the substantial reduction in time.

Note that the transition from low pressure operation to full-blooded HPLC occurs very sharply, at bead sizes around 20–40 μm. The optimum flow rate *increases* inversely with bead size, and the pressure to obtain even the *same* flow rate increases with the inverse square of the bead size. Thus, halving the bead size leads to an eightfold increase in the pressure needed to achieve a theoretical optimum flow rate (Table 5.4; Figure 5.16).

Note also that plate number (other things being equal) is proportional to bead size, so for a given number of *plates*, column length must increase proportionately with bead size. The total time for one column volume to pass through the column, for columns of the same number of plates, increases fourfold for a doubling of bead diameter, if one is to obtain the same resolution. In gel filtration (Chapter 8), one column volume

Table 5.4. Relationships Between Bead Size in Chromatographic Columns and Various Operating Parameters. The theoretical flow rate has been calculated for a perfect column, according to Eq. (5.15). The compromise flow rate is taken as 10 times this value, except where the pressure required exceeds 150 atm, for which lower rates have been used. The flow time is taken as the time for 1 column volume to pass through a column of length equal to 10,000 times the bead diameter, which represents columns of theoretically equivalent resolving power. Note that for the large beads, the very slow flow rates suggested could be faster, because there is not full penetration of the beads (see Figure 5.9)

Size of beads (μm)	Theoretical optimum flow rate [$cm\,h^{-1}$; Eq. (5.7)]	Compromise flow rate ($cm\,h^{-1}$)	Flow rate at 1 atm pressure ($cm\,h^{-1}$)	Approx. pressure for comprm. flow rate (atm)	Flow time at comprm. flow rate (see above)
5	40	100	0.6	150	3 min
10	20	150	2.5	65	4 min
20	10	100	10	10	12 min
40	5	50	50	2	50 min
80	2.5	25	200	.25	3 h

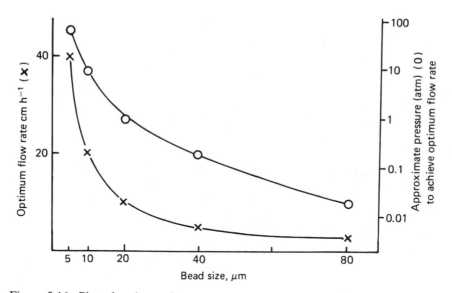

Figure 5.16. Plot of optimum flow rate (×) against bead size, and of pressure (○) required to achieve that flow rate. Note that the pressure scale is logarithmic.

completes the run, but in adsorption chromatography many column volumes may pass through before the desired solute is eluted.

Resolution depends on these various factors, and experimentally we find that optimum resolution can be approached with a column that has about 5×10^8 beads in it. More than this number does not greatly improve the performance, but resolution will fall away with columns containing significantly less than this number. For $10 \mu m$ beads, a $0.5 ml$ column contains this number. Half a billion $80 \mu m$ beads occupy about $200 ml$.

HPLC columns suitable for protein work are available from many different suppliers. Standard columns may be used on routine equipment; however, operators of HPLC equipment are often concerned about nonvolatile buffer salts that must be used in processes other than reverse-phase, which can cause damage by corrosion, or blockage by crystallization if the system is not washed out well after use. Equipment designed especially for use with buffer salts for protein work is available from most major equipment manufacturers. Prepacked columns contain adsorbents based on silica particles or synthetic organic polymers with hydrophilic groups. The standard size is in the range $1-20 ml$ total volume. Theoretical plate heights as small as $0.05 mm$ are possible. Although this size plate could only accommodate a few tens of micrograms of protein, the actual capacities are higher because the plate "height" is not a physically real concept except when operating in isocratic (nongradient) mode. Operation at $0.5-2 ml min^{-1}$, approximately 10 times the ideal flow rate (see above), results in elution times in the range 5–60 min, the higher figure for strongly retarded proteins.

The principles of separation, with the exception of reverse-phase, are the same as in "low-performance," or conventional low-pressure chromatography. Thus, ion exchangers are most commonly used, followed by hydrophobic adsorbents, high-performance gel filtration columns, and other miscellaneous adsorbent types. Discussion of these principles is reserved for the next two chapters.

It must be stressed that the carefully manufactured packing materials for HPLC operation are very much more expensive than the larger bead-sized conventional adsorbents, and economics dictates that the HPLC column must be capable of multiple reuse, tens if not hundreds of times. To this end, it is essential that the applied sample (1) contains no particulate material, and (2) contains no component that will be unstable on the column, which could result in precipitation and clogging. Buffers should be filtered and degassed before use, and high quality reagents used at all times. The sample should be allowed to equilibrate to the pH and temperature, to ensure that any aggregation in the conditions of the column occurs beforehand, and then it should be forced through a 0.2 μm filter before application. Preliminary fractionation procedures (from a crude extract) are essential; a conventional chromatographic step on a similar type of adsorbent to precede the HPLC column may be advisable. Even completely clear samples may contain non-protein components which adsorb irreversibly to the column. Although columns can be cleaned with various detergents, and with care, alkalis or acids, it is better not to allow this to be necessary. Each column type should have a recommended clean-up protocol supplied with it.

Undoubtedly, the principal use of HPLC is in the final cleaning up of small amounts of protein—often just prior to its intended use—rather than in large-scale early stages of isolation procedures. Not only does it have the edge in resolution over more conventional chromatographic methods, but it presents the possibilities of handling very small amounts of protein with little loss and rapid processing.

5.5   Types of Adsorbent Used in Protein Chromatography

Adsorbents used in protein chromatography have two essential features: the nature of the matrix (i.e., what the beads are made of), and the type of functional group that carries out the adsorption.

Nature of the Bead Matrix

The matrix is an insoluble polymer which is cast into bead form. (Some of the cheaper matrices, especially those based on cellulose, are morpholog-

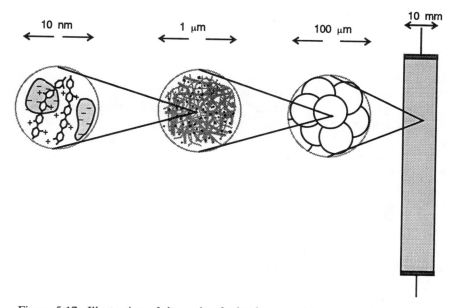

Figure 5.17. Illustration of the scale of adsorbent particles and protein molecules in a typical chromatographic column. Each magnifying stage represents 100 times: one million times between the column on the right and the molecular representation on the left of protein molecules and positively charged polysaccharide chains.

ically fibrous or chunky, rather than spherical.) With the exception of some very small beads (~5 μm) designed for surface adsorption only, they must be sufficiently porous so that proteins can penetrate within. The network of pores creates a total surface area many times greater than the outside surface of the beads. These pores may be just large enough to accommodate proteins of around 10^6 Da, for example, those matrices based on agarose and other carbohydrate polymers, or still larger pores, which generally means that the solid structure has to be more rigid than carbohydrate polymers can achieve. Synthetic organic polymers (see Appendix D), inorganic matrices such as silica, and composite structures have been developed. Simultaneously with the development of these newer matrices has come the requirement for high flow rate/high pressure resistance. This has meant that the two mutually opposite requirements, porosity and rigidity, have to be balanced; not surprisingly there are different matrices for different tasks.

The scale of thee matrix chains and proteins, relative to the column itself, can be gauged from Figure 5.17. Each step in magnification is 100-fold; the width of the column is about 1,000,000 times the width of a protein molecule.

For operating at low pressures (up to 2 atm), bead sizes in the range 50–150 μm are needed. Most companies sell such material in loose form, both derivatized as ion exchangers, etc., and underivatized for use in gel filtration or as a base matrix for derivatization by your own method (see, e.g., Chapter 7). These large-sized beads are not monodisperse; some include a two- to 3-fold range of diameters. HPLC adsorbents are sold prepacked in columns of 1- to 20-ml volume, and generally have bead sizes in the range 5–15 μm. Each column has a very narrow bead diameter range for optimum packing and chromatographic behavior; they are operated at pressures of 10s of atmospheres. Between these extremes are the 30–40 μm beads (narrow size range), some sold only in prepacked columns, others as bulk material for self-packing. These adsorbents can be operated at moderate pressures (2–10 atm), and represent a true scaling up from HPLC, as they are most suited for columns of volumes 20–500 ml. They give plate numbers comparable with the HPLC columns which have smaller beads and smaller volume. The largest-sized beads are generally less sophisticated materials used at low pressure, and are more suited to very large-scale processes. But these are widely used in laboratory enzyme purification protocols, in which simple apparatus can be used. A not insignificant point to note is that the cost of adsorbents is approximately constant *per bead*! This means that 1 ml of 10 μm beads costs about 1,000 times as much as 1 ml of 100 μm beads.

As has been explained, the small beads take less time to equilibrate during protein binding because of shorter diffusion distances. However, it has been found in practice that large beads have a faster response than expected theoretically. This has been put down to the possibility that only the outer part of the beads is functional; adsorption occurs preferentially in this area as it is the first to be accessed, and the bound proteins tend to block the pores and so leave the centers relatively free. Only the smallest proteins get into the center, and as they diffuse faster their behavior does not necessarily slow things up. As mentioned previously, it can be calculated that the outer 20% of the radius of the bead contains close to 50% of the volume, and so even if blocking of pores is not a problem, much of the adsorption and desorption takes place quite close to the surface (see Figure 5.9).

The ligand, e.g., DEAE-, or an aliphatic hydrophobic, see below, is normally attached by simple chemical reaction to the bead matrix. The ligands are distributed evenly, as far as is possible, but natural fluctuations in density of the matrix backbone will lead both to higher densities of ligands in some areas, and to a lower accessibility for larger proteins. The mode of attachment can be more important with affinity ligands (Chapter 7), in which spacer arms are introduced to hold the ligand away from the matrix backbone; this is not normally done with simple materials such as ion exchangers or hydrophobics. A high-performance adsorbent based on spacing multiple ion exchange groups on a flexible arm has been introduced, which has advantages of rapid kinetics for adsorption and desorption [89]; these "tentacle" adsorbents (Millipore) have given very satisfactory results in our hands.

For batch adsorption, the size of beads is less important for diffusional effects than it is for sedimentation rate. A batch process may take 20–60 min to operate, which is ample time for proteins to equilibrate even with large beads. If the beads are large and dense they will sediment quickly under gravity. An ideal bead for batch processing would be one made of inorganic material with a high density, and in the 100–200 μm diameter range.

Summary of Adsorbent Types

The detailed operating conditions for various types of adsorbent are described in Chapters 6 and 7. Briefly, this section will indicate the variety available and in what circumstances they are most useful.

Ion Exchangers

These are the most widely used of all adsorbents. Ion exchangers exploit the different net charges on proteins at a given pH, and interact with the proteins principally by electrostatic attraction. There are essentially two classes: anion exchangers (positively charged matrix) and cation exchangers (negatively charged matrix). Within these two classes are many types which differ in chemistry of substituent, degree of substitution, and mode of attachment of the substituent. The commonest anion exchangers are diethylaminoethyl- (DEAE-) and quaternary amino ethyl- (TEAE-, QAE-) substituents attached directly to hydroxyl groups on the matrix (Figure 5.18). In addition, polyethylene imine exchangers have been used, but mainly for chromatofocusing (Section 6.1).

Each of these adsorbents has a single positive charge on a nitrogen atom; in the case of the quaternary substituents, this is nondissociable, but with the DEAE- adsorbents, the proton on the nitrogen dissociates at high pH to leave a totally uncharged material above pH 9.5. What is more, DEAE- adsorbents are partly uncharged at neutral pH, and do not become fully protonated even at pH 6 [90]. Although this means that at the pH they are usually operated in (7 to 8) the adsorbents are only partly charged, they act as strong buffers, and this may have a positive role in separations, which would not be occurring with the quaternary substituents. Otherwise, the nature of the matrix, and the presentation, accessibility, and density of the substituent is more important than its chemical structure.

Cation exchangers fall into three chemical categories: "weak" carboxymethyl substituents (CM-), "strong" sulfonate groups (S- and SP-), and

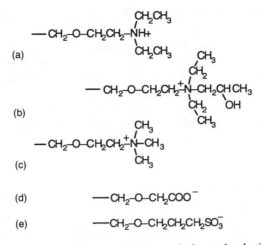

Figure 5.18. Structures of the commonly used charged substituents on anion exchangers. (a) diethyl aminoethyl-, DEAE-; (b) hydroxypropyl diethyl aminoethyl-, (quaternary) Q-; (c) trimethyl aminoethyl-, TMAE- or Q-; (d) carboxymethyl-, CM-; (e) sulfopropyl-, SP- or S-.

phosphates. The latter, although undoubtedly acting as cation exchangers, also have certain affinity actions, in that they rather specifically interact with enzymes that bind multiply-phosphorylated substrates such as nucleic acids [91], and with enzymes that bind sugar phosphates [92,93]. Because no bead-type phospho- adsorbent has yet been made available commercially, this sort has largely gone out of use as a simple cation exchanger. Carboxymethyl- adsorbents are widely used, and are fully charged above pH 4.5. But for operation at lower pHs than 4, the "stronger" sulfo-types are needed, as carboxymethyl groups progressively protonate and become uncharged as the pH falls below 4.5. Sulfonate groups are attached to the matrix through propyl and hydroxypropyl chains (Figure 5.18).

Hydrophobic and Reverse Phase Adsorbents

Hydrophobic adsorbents (hydrophobic interaction chromatography, HIC) exploit the variability of external hydrophobic amino acid residues on different proteins, and interact with the proteins by virtue of the fact that in aqueous solvents, hydrophobic patches on proteins seek out other hydrophobic surfaces preferentially. Hydrophobic interactions are strengthened by high salt concentrations and higher temperatures, and are weakened by the presence of detergents or miscible organic solvents (Section 6.4). The commonest hydrophobic adsorbents used in protein chromatography consist of short aliphatic chains (C_4 to C_{10}), or a benzyl (phenyl) group attached through simple chemistry to hydroxyls on the matrix (Figure 5.19). In addition, there are mixed-type hydrophobics which include, for instance, a positive charge. The extent of binding of a hydrophobic protein depends on the type and density of substitution of the matrix, as well as the nature of the buffer conditions. Hydrophobic adsorbents are much more variable in behavior than ion exchangers. Because of the nature of the interactions, resolution is generally poorer than in ion exchange chromatography.

Reverse phase adsorbents are essentially similar to the hydrophobics described above, containing aliphatic chains between 8 and 18 carbons long. They are used exclusively in HPLC, and represent the direct extrapolation of HPLC methods from small molecules to proteins. The term "reverse phase" also derives from small-molecule chromatography. "Normal phase" for such systems involves use of a hydrophilic (polar) matrix such as silica, and an organic solvent (hydrophobic). In reverse phase, the matrix is hydrophobic, and the solvent, at least in the starting conditions, is polar—mainly water. Reverse phase chromatography for proteins has been highly successful when dealing with small, structurally sturdy proteins, up to about 30,000 Da in size. Because of the operating conditions (acids, organic solvents), larger proteins and most enzymes are not suited to this technique. It has its greatest application in the separation of peptides, such as those generated by enzymatic digestion of proteins. Reverse phase chromatography of proteins operates by denaturing the polypeptide chains and exposing the internal hydrophobic residues to the adsorbent, followed by elution by weakening the interaction forces with organic solvent. Recovery of bioactivity depends on the reversibility of the denaturation. Nevertheless, the high resolving power of the technique, combined with the fact that bioactivity of the product is not always needed, makes this a valuable final step in protein purifications.

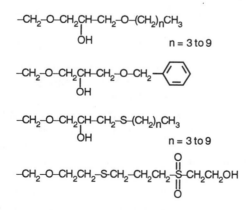

Figure 5.19. Structures of some commonly used hydrophobic adsorbents (see text).

Inorganic Adsorbents

Although several different inorganic oxides and phosphates have been used for adsorption of proteins, they have in the main not been suitable for column chromatography, but have been used in the batch mode. Alumina, titania, and zinc hydroxides have all been used to good effect from time to time, as have less defined media such as powdered silicaceous rocks such as bentonite. But the main inorganic adsorbent is hydroxyapatite, a form of calcium hydroxide-phosphate. Hydroxyapatite can also be formulated for use in columns, and as such is available from several manufacturers. Its mode of action is discussed in Section 6.3.

Affinity Adsorbents

There are many affinity adsorbents available, some highly specific for particular proteins, others being "group adsorbents" which bind to classes of proteins rather than an individual one. The strict definition of affinity chromatography involves the description of the interaction with the adsorbent as mimicking a natural protein–ligand interaction. Full details are described in Chapter 7. Before choosing to use an affinity adsorbent, it is worth checking on the adsorption capacity and the price of the material, as costs per milligram of protein bound are generally high. Although high-performance affinity adsorbents are available, the separate advantages of HPLC and affinity chromatography tend to be mutually incompatible, unless speed of operation is of prime concern.

Pseudo-affinity adsorbents cover a large range of types of material, and show selectivity which does not obviously or necessarily involve a natural binding site on the protein. These adsorbents include those called "biomimetic," in which the binding site *does* interact, e.g., dyes, relatively nonspecific "multifunctional" adsorbents which operate by a variety of types of interactions, and nonspecific but selective materials such as metal chelates. These will all be discussed further in Chapters 6 and 7.

Immunoadsorbents and Protein A/G

Antibody biotechnology has become a major part of diagnostics in the clinical and food industries, as well as in research. It is possible to purchase, but more often it will involve making, an antibody adsorbent for a particular protein that is being isolated. Generally, activated gels ready for attachment of your antibody can be purchased (Appendix D). More generally useful are columns that specifically bind antibodies, which are in great demand for antibody purification. These are mostly based on the proteins isolated from *Streptococcus* sp. that interact strongly and specifically with the invariant portion of immunoglobulins. Protein A, which binds IgG immunoglobulin molecules, is the most widely used; Protein G, which also binds IgG, is better for certain subclasses of IgG molecules, and for immunoglobulins from larger domestic animals. These materials are discussed further in Chapter 9.

5.6   Operating Conditions for Column Chromatography

Sample Application

As has been indicated immediately above, samples used in column chromatography should be clarified and free of particulate matter that would clog the column and render it useless. Ideally, whether using a low-pressure, large-bead system, or HPLC, the sample should be filtered to remove particulates, and should not be in a state (of pH, temperature, salt conditions) in which aggregation is likely to occur during the time of the run. The protein concentration (and so the volume) may be adjustable before application, either by ultrafiltration concentration (Chapter 1), or by dilution. The optimum value for this will depend on circumstances, but generally should not be much greater than $10\,\mathrm{mg\,ml^{-1}}$. At higher concentrations, the proteins themselves become significant ionic species, and may affect the adsorption behavior. In addition, with soft gel adsorbents, the osmotic pressure of the protein can cause bead shrinkage at high protein concentration. When the value of α of the desired protein (Section 5.1) is significantly less than unity, a sample volume many times the column volume will result in the protein being adsorbed through much of the column, even beginning to elute before all has gone on. So, for low α values, a high protein concentration/low volume is most suitable. Large volumes have many disadvantages, but may be unavoidable as a result of the previous operation.

The capacity of adsorbents for proteins varies considerably, and will be an important factor in deciding the size of column for a particular sample. If the separation depends critically on optimization of chromatographic behavior, maximizing resolution, then loading is limited to the maximum that still allows acceptable resolution. If, on the other hand, resolution is good, or chromatographic behavior is less important, as in affinity techniques (see Chapter 7), then high loadings can be used (also see below). With ion exchangers, a good average value for loading is in the range 1–10 mg per ml of adsorbent. Smaller loadings run the risk of high losses, and using a larger column than necessary has other disadvantages (e.g., speed, cost). Similar loadings are generally suitable for most other protein adsorbents.

Overload and Displacement Chromatography

The sample being applied will contain a mixture of proteins; there may be several major components and many minor ones. These components each interact with different strengths, and during sample application several events occur. If one considers the first few moments when only a very small part of the sample has run on, then the top surface of the column will contain each component, with proportions bound and free according to their partition coefficients (Figure 5.20). As the next part of the sample runs in, the more weakly bound components (lower α values) partly desorb and move to occupy the next section of the column. The sites revealed now become occupied with the fresh sample. The more strongly bound components from the first stage remain on the top; gradually, the top part of the column becomes saturated with these strongly bound components, while the lower α components move into lower parts of the column. When the top part is saturated, all newly applied proteins pass through until they reach an unoccupied site which has just been vacated by a low α component. In this way, during sample application the components tend to separate out into regions of the column according to their partition coefficients. This can frequently be observed directly when applying a sample containing several colored proteins. The efficiency of this "striation" at the top of the column depends on the flow rates and relative α values; a better separation of components occurs at slow flow rates. The phenomenon can be described as displacement, in which the more strongly binding protein components displace the weaker ones. Displacement chromatography is a technique in itself, in which proteins are successively removed from a column using displacers which are either ampholytes [94,95] or displacer proteins that are not part of the sample [96,97]. Although useful theoretically, particularly for obtaining sharp separations between very similar components, it has not yet been used *purposely* to a great extent. But displacement chromatography in which the indigenous proteins are displacers, is a phenomenon which occurs widely in many important column procedures, and can explain why excellent separations are achieved even at very high loadings, when theoretically, plate height and, so, resolution, should be much reduced.

If the loading as illustrated in Figure 5.20 is continued indefinitely, then weakly bound components start to appear in the eluate. Gradually, the protein concentration in the eluate increases, as it includes not only the unbound proteins, but displaced fractions from the column. Ultimately, only the most strongly bound proteins remain on the column; in ideal conditions all binding sites on the column are occupied with only these, all weaker ones having been displaced. If the most strongly bound proteins form only a small part of the total protein, then it may be possible to apply a very large amount of protein to the column, and still retain these proteins. If your desired protein is among them, then this is an effective and economical way of purifying it. This process can be called "indigenous protein displacement chromatography," or, alternatively, "overload chromatography." In Figure 5.21 is illustrated some results of an overloaded column of a dye adsorbent to which a crude bacterial extract was applied, and the appearance of certain enzymes in the eluate monitored. The overloading technique is similar to using batch adsorption; in both cases, insufficient adsorbent is used to bind all proteins, but

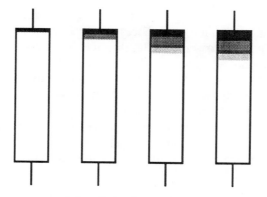

Figure 5.20. Appearance of banding of proteins as a column is loaded: the weaker-binding proteins are displaced further down the column by strongly binding proteins.

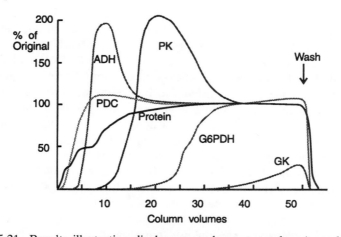

Figure 5.21. Results illustrating displacement chromatography. A crude extract of the bacterium *Zymomonas mobilis* (11 mg protein ml^{-1}) was continuously applied to a column containing the dye adsorbent Procion Yellow H-E3G until 50 column volumes had gone on. Protein and enzyme activities in the eluted fraction were measured, and expressed as a percentage of the value in the original extract. PDC (pyruvate decarboxylase) was not significantly adsorbed. ADH (alcohol dehydrogenase II) was adsorbed at first, but then displaced at 5 column volumes in a peak 2 times higher than the applied level, the peak representing all that had been previously bound. PK (pyruvate kinase) was bound more strongly, and may be regarded as one of the proteins that displaced the ADH; it too was diplaced after 15 column volumes, and the peak is equivalent to the PK previously bound. G6PDH (glucose 6-phosphate dehydrogenase) binds strongly, but was overloaded by 25 column volumes; the adsorbed enzyme was just beginning to be displaced toward the end. GK (glucokinase) was still binding to the column and displacing other proteins at the end of the experiment. This is an ideal column for purifying glucokinase, since it could all be eluted at the end by a simple buffer change; most other major proteins had already been displaced and washed off.

the desired protein is among the most strongly binding. For application of the overload technique, it is necessary to find an adsorbent that the desired protein binds to very tightly; this may require screening of affinity and multifunctional adsorbents to be described in Chapter 7.

Displacement effects are also very important during elution, when there are high loadings and large amounts of each component being eluted. The same basic phenomenon is occurring, as each component competes for binding sites as it moves down the column. Resolution of closely similar components can be higher than theoretical levels because these competitive displacements are not taken into account in normal theory [98,99].

Flow Rates

The flow rates used in a chromatographic process are always a compromise between conditions for ideal adsorption and resolution, and time spent for completion of the process. We have already mentioned the concept of an idealized "reduced velocity" (Section 5.4), and the empirically deduced compromise that flow rates about 10 times higher than this theoretical concept have proved to be optimum. Nevertheless, there has been a tendency in the past few years to carry out chromatography at ever faster speed, especially in large-scale commercial applications where time is money. Faster flow rates mean higher pressures, and it has been necessary to develop new adsorbents that can withstand these pressures without collapsing. These developments have been made, and we now have available adsorbents that can be used at flow rates 100 or even 1,000 times the optimum values suggested from the reduced velocity (V_{red}) calculation. Although these flow rates cannot achieve the resolution (or the adsorption capacity, see below) that slower rates can, the fast flow compromise may still be operationally and/or economically worthwhile. In separations where the desired protein component is well resolved from impurities, a higher speed separation will save time without necessarily sacrificing purity of the product (see Figure 5.22).

The capacity of a column for any particular protein, which is determined by watching for the breakthrough point of the protein when overloading, is reduced at higher flow rates. This is termed the "dynamic capacity" of the column, and can be explained simply in terms of the time needed for proteins to penetrate all parts of the beads and find the binding sites; faster flow rates carry some of the protein through before it has a chance to find an unoccupied binding site. At zero flow rate (e.g., in batch-type adsorption), the total capacity can be determined. In the column, the dynamic capacity describes how much of the protein binds before significant breakthrough occurs. In Table 5.5 are some figures made available by Pharmacia on the dynamic capacity of their Protein G

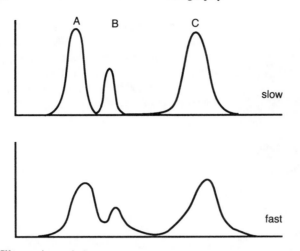

Figure 5.22. Illustration of the use of different flow rates on the separation of three protein components. At slow flow rate, all three are well separated. At fast flow rate, components A and B are not well separated, but both are still completely separated from component C. If it is C that is required, the fast flow rate is suitable.

Sepharose TM Fast Flow. It can be seen that flow rates up to $100 \, \mathrm{cm \, h^{-1}}$, corresponding to about 50 times V_{red}, result in a loss of about half the potential capacity; faster rates greatly reduce this to levels that would not be viable with such an expensive adsorbent.

To illustrate why the dynamic capacity decreases with flow rate, Figure 5.23 shows an idealized picture of adsorption to beads in a column at high flow rate. Adsorption occurs first on the outer part of the beads, and this

Table 5.5. Influence of flow rate on dynamic capacity: experiments with Protein G Sepharose^{TM} Fast Flow. V_{red} (see text) for IgG is in the range $1-3 \, \mathrm{cm \, h^{-1}}$. At 10 times V_{red}, only a little loss of capacity occurs, but at higher flow rates, the loss is substantial

Flow rate, $\mathrm{cm \, h^{-1}}$	mg IgG bound per ml of adsorbent
0[a]	23
30	18
100	12.3
200	7.1
300	4.7
500	2.3
1,000	0.83

[a] Total capacity determined by batch equilibrium.
Data: Courtesy of Phamacia.

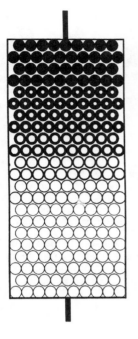

Figure 5.23. The effect of a fast flow rate during column loading. The top beads become fully saturated with protein, but farther down the column, only the outer portions of the beads have a chance to bind the protein (see also Fig. 5.9), as it passes quickly down to lower parts of the column.

tends to restrict access to the inner parts. As the flow is too fast to allow equilibration through the beads, protein passes down to the next section, where it can bind to the outer part of fresh beads. As loading continues, protein has time to penetrate the top beads fully, but by then, beads near the bottom are receiving protein molecules, and soon, protein begins to emerge from the column. The dynamic loading capacity has been reached long before the lower beads are saturated. At very low flow rates, on the other hand, each section of the column is saturated with protein before molecules pass through to the next section, so much more can be applied to the column before there is any emergence from the bottom; the dynamic capacity approaches the total capacity.

There are circumstances when high flow rates, despite loss of capacity, are very useful. An example is the reduction of the ionic strength of a bacterial extract by dilution with water to a low enough value so that the desired enzyme would bind to a cation exchanger [100]. The volume so obtained was very large, but using a fast flow adsorbent, the sample (710 liters) could be applied at a flow rate of $315\,\mathrm{cm\,h^{-1}}$ on a column of $1,500\,\mathrm{cm^2}$ cross section, which took about 1.5 h. A loss of dynamic capacity was made up in convenience of operation.

Aside from these possibilities of faster than optimum flow rates, the most suitable flow rates have already been presented (Table 5.4). For large beads in the $80-150\,\mu\mathrm{m}$ size range, $20-30\,\mathrm{cm\,h^{-1}}$ is appropriate at

20°C, while about half of this rate is best in the cold room. For 30–40 μm beads, flow rates in the range 50–100 cm h^{-1} at 20°C are appropriate. For HPLC columns with 10 μm beads, flow rates of 2–4 cm min^{-1} are optimal, which is about 1 ml min^{-1} on a typical 1 ml volume column.

Chapter 6

Separation by Adsorption II: Ion Exchangers and Nonspecific Adsorbents

6.1 Ion Exchangers—Principles, Properties, and Uses

General Principles

Although Chapter 5 was a general one describing the principles of chromatography, ion exchangers were mentioned frequently, and much of the theory has been tested using ion exchange adsorbents. Proteins bind to ion exchangers by electrostatic forces between the proteins' surface charges (mainly) and the dense clusters of charged groups on the exchangers. The substitution level of a typical diethylaminoethyl (DEAE-) cellulose or carboxymethyl- (CM-) cellulose may be as much as $0.5\,mmol$ cm^{-3} (packed, swollen adsorbent), that is $0.5\,M$ of charged groups. If such a concentration were evenly distributed in three dimensions, the average distance between each charged group would be $1.5\,nm$. A compact globular protein of MW 30,000 has a diameter of $4\,nm$. The charges are of course balanced by counterions such as metal ions, chloride ions, and sometimes buffer ions. A protein must displace the counterions and become attached; generally, the net charge on the protein will be the same sign as that of the counterions displaced—hence "ion exchange." The protein molecules in solution also are neutralized by counterions; the overall effect in a given region of the adsorbent must be electrically neutral. This is illustrated in Figure 6.1. In this diagram it has been assumed that a protein with a net negative charge is preequilibrated in Tris-chloride buffer; the counterions associated with the protein are thus $HTris^+$. The adsorbent (DEAE-cellulose) also equilibrated with the buffer has Cl^- counterions.

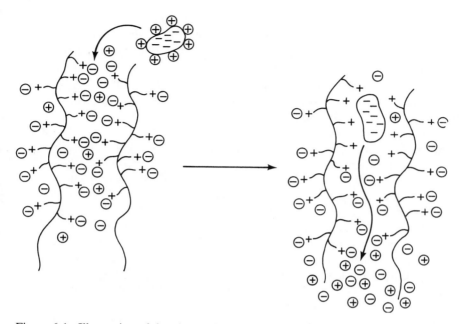

Figure 6.1. Illustration of the "ion exchange" occurring when a negatively charged protein adsorbs to an anion exchanger. Seven positively charged ions (e.g., HTris$^+$) associated with the protein molecule are displaced, together with seven negative ions (Cl$^-$) from the exchanger.

The protein then displaces the chloride ions, occupies a site in the adsorbent, and Tris-Cl is discharged. It may be noted that this Tris-Cl is the (acid) salt of Tris, and if the displacement is rapid, it could lead to a lowering of the effective pH in the column ahead of the band of adsorbed protein.

In addition, the pulse of Tris-Cl represents an increase in the ionic strength of the buffer flowing through the column. Typically, for every $1\,mg\,ml^{-1}$ protein that is adsorbed, it is replaced by approximately $1\,mM$ of extra buffer salts. The pH of nonadsorbed fractions can be shown to be lower than that of the preceding buffer wash (or higher, when referring to a cation exchanger) if the protein concentration of the applied sample is high. To avoid these pH and ionic strength changes, which can result in less adsorption than expected, the applied protein concentration should not be too high—especially if a large proportion of it is expected to adsorb to the column. This also implies that there should be adequate buffering. In the absence of any buffer, quite extreme pH changes could occur and cause denaturation. Generally $10\,mM$ of a buffer within 0.3 unit of its pK_a is the minimum desirable level (see Table 6.1 and Section 12.3) and protein *being adsorbed* should not be less than $5\,mg\,ml^{-1}$. (Proteins that are not being adsorbed have less influence.)

Two forms of interaction have been suggested for the adsorption of proteins to ion exchangers. One is termed "irreversible," the other "reversible." As has been pointed out by Peterson [101], the term *irreversible* is unfortunate in implying that protein, once adsorbed, will not come off again. This would not be very useful. A smooth change of buffer conditions to cause the transition from so-called irreversible to reversible adsorption must, in fact, be a gradual process. The probability that a protein molecule will dissociate in 1 second may be 99% for "reversible" adsorption and 0.001% for "irreversible" (it can never genuinely be zero), but where the dividing line comes is really a matter of opinion. These probabilities could be calculated from adsorption/pH curves such as illustrated in Figures 5.7 and 6.2, but ultimately, one ends up with just another way of expressing the partitioning between protein and ion exchanger. Experiments similar to those illustrated demonstrate the transition from absorbed (partition coefficient close to unity) to nonadsorbed (partition coefficient zero) as the pH of the buffer changes [82]. Calculations from these curves using the equations in Section 5.1 have given some quantitative estimates of the magnitude of dissociation constants (assuming quasiequilibrium conditions) between protein and adsorbent. In these examples the adsorbent studied was CM-cellulose. Not only was it possible to calculate the values for α and K_p, but also, from the rate of change of these with pH, a value for the energy of interaction per charge on the protein was obtained.

In practice, this is the rate of increase in the energy of interaction for each additional positive charge, since asymmetries of charge distribution and uncertainties about exact isoelectric points make it difficult to ascertain exactly how many charges on the protein molecule are really contributing to the interaction with the adsorbent. The derivation of these parameters is shown below and in Figure 6.2.

From the pH/titration curve and a knowledge of the protein's isoelectric point, the number of charges on the molecule at each pH can be calculated. A plot of log K_p against the number of charges on the protein molecule, z, gives a slope of s (see Figure 6.2), where $s = (d \log K_p)/dz$.

The energy of interaction per mole is given by $\Delta G^0 = -2.3RT \log K_p$.

Therefore, the rate of change of energy of interaction per mole per charge z is:

$$\frac{dG^0}{dz} = -2.3RT\frac{d\log K_p}{dz} = -2.3RTs \qquad (6.1)$$

Values for dG^0/dz for various proteins are listed in Table 6.1. It can be seen that the values show little definite trend; they are mostly in the range $1-2\,\text{kJ}\,\text{mol}^{-1}$. However, the smallest protein, myokinase (MW 21,500), had the largest value, indicating that the extra positive charges that the protein acquired as it titrated through the range pH 7.5–6.0

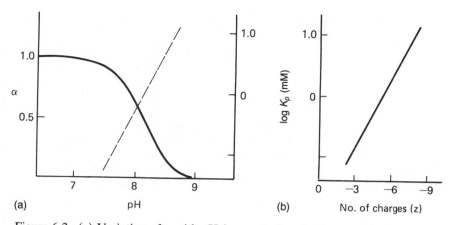

Figure 6.2. (a) Variation of α with pH for yeast phosphoglycerate kinase on CM-cellulose, $I = 0.01$ (*solid line*). Values for K_p have been calculated using Eq. (5.12) (*dashed line*). (b) log K_p plotted against number of charges on the enzyme molecule, calculated from a titration curve and taking the isoelectric point as 7.1. The value of s (see text) is 0.4.

were interacting more strongly with the ion exchanger than was the case with larger protein molecules. This is due to either the fact that the smaller molecule is able to pack more tightly into the exchanger (confirmed by capacity experiments, see below), or that in a small molecule, each charge is inevitably closer on average to the protein molecule's surface, and therefore forms a stronger electrostatic attraction with the ionized carboxymethyl groups of the adsorbent. To estimate the number of these ionized groups that are interacting with each charge in the protein, one can use appropriate values for likely distances between the average residue in a protein and the ionized carboxymethyl groups, use the dielectric constant, D, for water, and use a value of $2\,kJ\,mol^{-1}$ per charge in Eq. 6.2:

Table 6.1. Change in the Energies of Interaction of Enzymes with CM-Cellulose for Each Positive Charge on the Enzyme

Enzyme	$dG^0/dz(kJ\,mol^{-1})$
Yeast phosphoglycerate kinase	2.0
Rabbit muscle AMP kinase	4.2
Rabbit muscle creatine kinase	1.5
Rabbit muscle aldolase	0.9
Rabbit muscle pyruvate kinase	1.0
Yeast pyruvate kinase	1.7
Rabbit muscle lactate dehydrogenase	0.7
Beef heart mitochondrial aspartate aminotransferase	0.9

$$E = \frac{-z_1 z_2}{Dr} \qquad (6.2)$$

where z_1, z_2 = charges on attracting components: z_1 for protein; z_2 for adsorbent; r = distance between them; taken as 5 nm:

$$E(\text{per charge } z_1) = -\frac{z_2}{Dr} = 2\,\text{kJ}\,\text{mol}^{-1} \qquad (6.3)$$

This gives z_2 = approximately -5, i.e., on average about 5 carboxymethyl groups are exerting their influence on each positive charge.

These calculations are based on extremely broad assumptions but at least the value for z_2 is reasonable when considering the probable three-dimensional density of charge clusters of carboxymethyl groups in relation to the sizes of proteins.

Adsorptive Capacities of Ion Exchangers

The amount of protein that can be bound per unit volume of packed ion exchanger can be very high indeed. However, for most exchangers it depends greatly on the size of the protein molecule, with the smaller molecules adsorbing to a greater extent. Expressed in molar mass terms the differential is still greater. Some published values are given in Table 6.2. Molecules with molecular weights greater than 10^6 are likely to be excluded from most cellulose-based ion exchangers, the exclusion effect being essentially the same as is occurring in gel filtration chromatography (cf. Section 8.1). Larger molecules can bind only to the surface of the ion exchanger particles, and so the capacity for these is very low.

Table 6.2. Capacity of CM-Cellulose (Whatman CM 52) for Various Proteins at $\alpha = 1.0$ (pH 6.0, $I = 0.01)^a$

Protein	MW	Capacity (mg cm^{-3})	Capacity m_t (mM)
Hen egg white lysozyme	14,300	130	9.0
AMP kinase	21,400	100	4.6
Yeast and rabbit muscle phosphoglycerate kinase	45,000	70	1.55
Phosphoglycerate mutase	60,000	40	0.67
Creatine kinase	82,000	35	0.43
Enolase	88,000	48	0.55
Lactate dehydrogenase	140,000	21	0.15
Glyceraldehyde phosphate dehydrogenase	145,000	26	0.18
Aldolase	160,000	22	0.14
Pyruvate kinase	228,000	12.5	0.055

^{a} See Figure 6.3 for a plot of these values.
Note: All enzymes are from rabbit muscle unless otherwise indicated.
Source: Scopes [82].

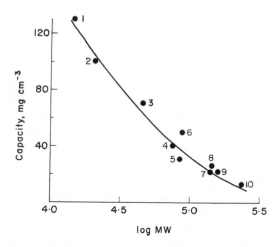

Figure 6.3. Capacity of CM-cellulose for a variety of proteins (in conditions where $\alpha = 1.0$). Proteins used: (1) hen egg white lysozyme; (2) AMP kinase; (3) phosphoglycerate kinase (both rabbit muscle and yeast preparations gave the same capacity); (4) phosphoglycerate mutase; (5) creatine kinase (at 0°C); (6) enolase; (7) lactate dehydrogenase; (8) glyceraldehyde phosphate dehydrogenase; (9) aldolase; (10) pyruvate kinase. (From Scopes [82].)

The total capacity relates to the amount of the internal volume of the particles that is accessible to the protein, and as with gel filtration, this is simply a hole-size problem. As shown in Figure 6.3, the capacity, in mg cm^{-3}, increases with decreasing molecular weight in such a way that capacity against log MW is nearly linear.

It is interesting to reflect that an adsorbent already "saturated" with a large protein molecule is still capable of taking up small molecules, provided that the "pores" are not sterically blocked up by the large molecules. In an experiment to confirm this, using the conditions described in Figure 6.3, with CM-cellulose at pH 6.0, 1 cm^3 of CM-cellulose was treated with an excess of rabbit muscle pyruvate kinase (MW 228,000); 13 mg was adsorbed. The CM-cellulose was washed with buffer, then 50 mg lysozyme (MW 14,500) was added. All the lysozyme adsorbed to the CM-cellulose, without displacing the pyruvate kinase. This illustrates that overloading of columns is more likely to result in the emergence of the larger molecular size proteins first; there may be sufficient capacity to adsorb all the smaller molecules—depending on the relative proportions of large and small molecules in the applied sample. Similar results have been presented for HPLC reverse phase columns [102]. Most synthetic ion exchangers have higher exclusion limits, for example Trisacryl™ is claimed to be able to take up molecules up to 10^7 in molecular weight [103].

Types of Ion Exchangers

A large variety of types of ion exchangers have been developed over the years, but many of them have found limited use or have been superseded by superior materials. The beginner need have only two types, a cation exchanger (e.g., carboxymethyl-) and an anion exchanger (e.g., DEAE-). Carboxymethyl and diethylaminoethyl groups have been attached to cellulose, agarose, dextran, silica, and synthetic polymers to provide suitable materials for protein ion exchange chromatography. The more recent trend is to attach charged groups that remain charged at extremes of pH, such as quaternary amino groups (Q-type) and sulfopropyl (SP-) (see Figure 5.18). Early materials were coarse cellulose particles that had some desirable features such as rapid flocculation and good flow rate in columns, but because of their microstructure they were not very porous, and so had relatively few binding sites for proteins—especially large proteins. Their capacity was limited. An improved "microcrystalline" cellulose material was developed by Whatman, Maidstone, U.K., and their two exchangers, CM 32 and DE 32 (dry powders), or CM 52 and DE 52 (damp, preswollen versions), found widespread use, and are still competitive in price with the more sophisticated materials developed since, though not able to provide such good resolution.

With the development of methods for producing spherical beads of sufficient porosity and rigidity, high-capacity adsorbents with reliable, reproducible properties, fast flow rates, and high resolving power have become generally available. Many companies produce ion exchange materials of varying grades (bead size); the finer beads are generally only available in prepacked columns. A list of some of the products currently available (1992) is in Appendix D.

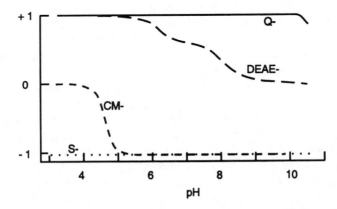

Figure 6.4. Ionization curves for different types of ion exchangers: DEAE-, Q-, CM- and S-. The "strong" exchangers Q- and S- are fully charged at all usable pH values.

The two main decisions for choosing an adsorbent are: (1) the charge, plus or minus, and the nature of the group resonsible for that charge, and (2) the nature of the matrix particles, in terms of bead size, flow rate required under pressure, capacity, and often, cost per unit volume. The second choice is a matter of choosing from ranges offered by the manufacturing companies; the first choice, the nature of the substituent, may require some further consideration. For cation exchangers, the only significant difference between carboxymethyl- and sulfo- groups is when operation below pH 5 is envisaged. Because carboxymethyl- groups begin to become protonated below pH 5, they give a fairly sharp titration with a pK_a of 4.5, and below pH 4, few groups remain charged. So for low pH operation, the "stronger" acidic group of sulfonate is needed. Sulfo-groups remain fully charged right down to pH 1. There is a region of operation of carboxymethyl ion exchangers around pH 4.5 where the titration of the groups can actually benefit the separation of proteins. Since the exchanger itself acts as a strong buffer, resolutions of closely similar p*I* proteins can be achieved in this pH range, in a similar manner to the operation of chromatofocusing.

With anion exchangers, the situation is more complex. The quaternary amino groups are fully ionized up to pHs higher than needed in protein chromatography, and so behave in a simple fashion, i.e., fully charged at all times. But the most used ion exchanger, with DEAE- groups, becomes uncharged at high pH, and is not suitable for use above pH 8.5. Unlike carboxymethyl-, however, DEAE- does not titrate with a simple pK_a, but titrates over a wide pH range from 10 down to 6 (Figure 6.4). Anion exchange columns are usually operated in this pH range, the buffering of the exchanger itself is an important feature in separations attained. It should be noted that whereas proteins acquire more negative charges as the pH goes up, DEAE- exchangers lose positive charges. Consequently, it is possible for a protein which does not titrate much in the pH range 6–8, to bind more strongly at the lower pH because the exchanger has a higher charge density. Use of anion exchangers at pHs less than 7 has been unjustifiably neglected; in fact it is advisable for low p*I* proteins, provided that they are stable at slightly acid pHs.

pH and Donnan Effects

The pH in the microenvironment of an ion exchanger is not exactly the same as that of the applied or eluting buffer because Donnan effects can repel or attract protons within the adsorbent matrix. In general, the pH in the matrix is up to 1 unit higher than that in the surrounding buffer in anion exchangers, and 1 unit lower in cation exchangers (Figure 6.5). The lower the ionic strength of the buffer, the larger this difference is. This has important ramifications when considering the stability of enzymes as a

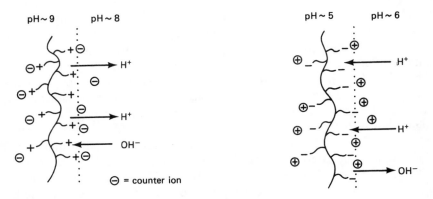

Figure 6.5. The Donnan effect on pH in the microenvironment of ion exchangers.

function of pH; an enzyme that is perfectly stable at pH 5.5 may denature rapidly in solution at pH 4.5. If such an enzyme is adsorbed on a cation exchanger at pH 5.5, it experiences a microenvironmental pH lower than 5.5, and is likely to denature on the adsorbent. Because it leads to a shift toward extreme pH values, the Donnan effect limits the operational pH range of ion exchangers, especially in the mildly acid range that is needed on cation exchangers. In general, enzymes tend to be more stable in the mildly alkaline range, e.g., pH 8–10, than in the mildly acid, pH 6–4, so loss by high pH denaturation is less of a problem when using anion exchangers. To minimize these effects, buffering power should be maximized, which means selection of a buffer which gives the highest buffering power for a given ionic strength (see Table 6.4).

Elution of Adsorbed Protein

In theory, two general methods for eluting proteins are available (plus affinity elution; cf. Section 7.4). These are (1) to change the buffer pH to a value where binding is weakened—lower pH for an anion exchanger, or higher pH for a cation exchanger; and (2) to increase the ionic strength, thereby weakening the electrostatic interaction between protein and adsorbent. In practice, method (1) is not generally very successful. This is because, unless there is a very high buffering capacity, sudden large pH changes happen as proteins become eluted, so there is little separation of individual components. At low ionic strength, buffering capacity is by definition low, and the attempt to change the pH by applying a pH gradient is frustrated by the buffering power of the proteins adsorbed to the column and especially in the case of DEAE-adsorbents, the buffering of the adsorbent groups themselves. A typical result is shown in Figure 6.6. Only if the proteins required are adsorbed very strongly in the first place, can a pH gradient be used successfully. In this case, strong buffering at ionic strength 0.1 or greater can be employed.

By employing buffering systems that maximize buffering power, and using strongly buffering ion exchangers, it has been possible to achieve high resolution of proteins by pH elution. The technique has acquired its own name, "chromatofocusing" [104,105,106]. In order to minimize the ionic strength, polymeric buffers called ampholytes (also see Section 8.2) are used, together with an anion exchanger that has a very broad titration over the pH range 5–10. The ampholytes have both positive and negative charges, and contribute very little to ionic strength overall. Consequently, they can be used at relatively high concentration, and so control the pH on the column very closely. Proteins are eluted according to their isoelectric point, starting at high pH and working down. The operating procedure is to start by equilibrating the column at a pH high enough to bind the protein of interest, in ampholyte buffer, then to apply to the column the ampholyte adjusted to a pH lower than the protein's pI. A pH gradient develops on the column, and a steady drop in pH eluting from the column is attained. The proteins should be eluted close to their isoelectric points. In ideal conditions, resolution of proteins differing in pI by as little as 0.05 units is possible on a chromatofocusing column. The anion exchanger used has polyethylene imine substituents; the charged imino groups interact with each other to create a continuum of pK_a values. Chromatofocusing is successful with proteins that are stable and soluble at their isoelectric point, as they are eluted in a buffer close to their pI. Optimum results also require that the protein being isolated have a good buffering around its isolelectric point; the best examples are proteins isoelectric at pH 6–7, with a high histidine content. It is necessary to be

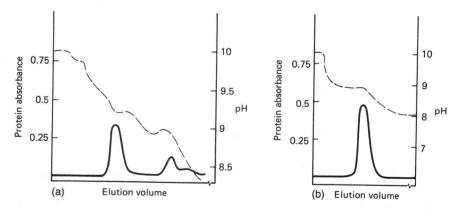

Figure 6.6. Examples of irregular pH gradients obtained from columns of DEAE-Sephadex. After starting at pH 10.1, buffers at pH 8.0 (a) or pH 7.5 (b) were applied to the columns. Elution pH values fell irregularly and dropped sharply as the protein (solid line) was eluted. (From Sluyterman and Wijdenes [106].)

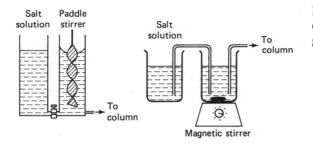

Figure 6.7. Simple gradient mixers (linear gradients).

able to remove the ampholytes following chromatofocusing before any further treatments.

Salt gradients are the commonest means of eluting proteins from ion exchangers. Whereas pH changes can be "held up" by buffering action of proteins and column substituents, salts pass through the column relatively unhindered. Usually either potassium or sodium chloride is used to generate the gradient. A linear gradient can be formed by a simple mixing device (Figure 6.7). The action of the salts can be considered in one of two ways. The salt can directly displace the protein; the ions (e.g., chloride ions on DEAE-adsorbents) occupy the positively charged sites and block reattachment by protein. Alternatively, the system can be regarded as an equilibrium in which even strongly bound proteins spend some time not adsorbed; the presence of the salt ions between the unattached protein and the adsorbent greatly weakens the attraction between the two. In either case, the desorbed proteins are replaced by counterions, and so "ion exchange" is a true description.

As salt concentration increases, the value of α decreases. On a column, this means that proteins begin to move down, and as they will be moving

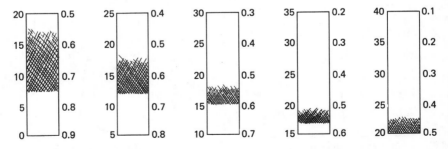

Figure 6.8. Elution of a protein with a salt gradient. On the left of each column diagram the salt concentration in the column is given (mM); on the right is the corresponding α value (for simplicity, variation in α with protein concentration has been ignored). The protein band moves faster as the salt increases, and emerges in this case at $\alpha = 0.5$.

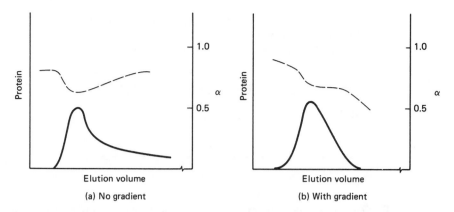

Figure 6.9. Elution of a protein (*solid line*) with a salt gradient. The variation of α with protein concentration means that without a gradient (a) considerable "tailing" of the peak occurs. With a gradient (b) the "tail" is accelerated and eluted quickly.

at a slower rate than the salt passing by, each part of the protein band is continuously experiencing an increasing salt concentration. This causes a sharpening of the bands. The leading edge is sharpened as a result of protein concentration effects, as described earlier. In addition, it is also sharpened by the fact that the gradient of salt in the column means that α is higher at the front of the band than at the rear. Consequently, the rear tries to catch up with the front, causing a sharpening, or decrease of the depth of the band. This sharpening effect of a gradient is illustrated in Figures 6.8 and 6.9. The usefulness of stepwise increases in salt concentration to elute proteins is described in the next section.

6.2 Ion-Exchange Chromatography— Practical Aspects

The isoelectric point of a protein depends on the proportions of ionizable amino acid residues in its structure. Positive charges are provided by arginines, lysines, and histidines (the latter positive below pH 7). Any free N-terminal amines will also bear a positive charge below pH 8. Unless one is operating at an unusually high pH, e.g., above 8.5, all arginines and lysines can be considered positively charged. Negative charges are principally on aspartate and glutamate residues, and above pH 6 virtually all these residues are ionized, as are C-terminal carboxyls. At high pH values (>8) cysteines may become ionized also. A few residues, usually associated with the active site of an enzyme, may prove exceptions to these generalizations, with pK_a values several units away from

Table 6.3. Choice of Ion Exchanger for Purification of
a Protein with a Known Isoelectric Point (Assuming
That the Protein Is Stable Only in the pH Range
5.5–8.5)

Isoelectric point	Ion exchange	Buffer pH
8.5	Cation	$\leqslant 7.0$
7.0	Cation	$\leqslant 6.0$
	Anion	$\geqslant 8.0$
5.5	Anion	$\geqslant 6.5$

normal. For example, the enzyme creatine kinase has an ionized cysteine residue at its active site with pK_a 7 [107]. In the normal pH range of enzyme stability on ion-exchange columns, namely 5–9, the variation in net charge is due mainly to histidine residues, since these have pK_a values in the range 6–7. Consequently, a protein with few histidines will have a very similar charge over quite a wide pH range, so the pH used for ion-exchange chromatography may not be critical (but see comment above for DEAE- exchangers). Conversely, proteins with high histidine content vary considerably in net charge through this pH range, and the precise pH for adsorption or elution needs to be more carefully defined. Above pH 8.5 and below pH 5.5, titration of other groups on proteins is extensive, and a very small variation in pH can make a large difference to the adsorption behavior (K_p and α values). Often the limited range of protein stability restricts the pH that can be used, and if the isoelectric point is outside the stability range, only one class of ion exchanger can be used, as indicated in Table 6.3. There are three pH-related features of enzymes: the isoelectric point, the optimum pH for activity, and the pH-stability range. These are *not* necessarily related; if an enzyme has an isoelectric point of 5, this does not mean it is active at pH 5, nor is it necessarily stable enough to handle at that pH. Table 6.3 assumes that all proteins behave ideally and do not adsorb on the "wrong side" of their isoelectric point, and that useful values of α (>0.9) are not reached until 1 or 1.5 pH units of the "right side." There are several documented cases where this simple concept proves to be incorrect—proteins adsorbing to cation exchangers even 1 pH unit above their isoelectric point [108], or vice versa for anion exchangers. There are "sticky" proteins that will bind to both types of exchangers in the same buffer, and "slippery" proteins that are difficult to get to bind to any adsorbent within a stable pH range. These effects are due partly to nonelectrostatic interactions, e.g., hydrophobic, and partly to uneven distribution of charges over the protein surfaces. A cluster of like-charged groups, for instance at an active site that binds a multiple-charged substrate, can cause adsorption despite an overall average charge on the protein of the other sign [66].

Trials to Determine Ion-Exchange Behavior

In practice, the conditions for adsorption will be found empirically. Provided that you have a simple method of detecting the protein you are trying to purify, for instance if it is an enzyme that can be stained for activity, then a preliminary experiment of analytical isoelectric focusing, combined with the specific enzyme stain can reveal its isoelectric point. But a simple experiment using ion exchangers will normally suffice equally well. A small sample of the preparation containing the protein can be passed through a gel filtration column to change the buffer. One milliliter can be processed through a 10-ml column in about 10 min (see Table 1.2). For a trial anion exchange adsorption, the buffer should be 20 mM Tris pH 8.0 (see below for buffers). A small column of 1–2 cm^3 of DEAE-cellulose or similar material should be sufficient, preequilibrated in the Tris buffer. If the protein does not adsorb, then probably the process has resulted in a good deal of purification, since many other proteins are likely to have adsorbed to the DEAE-cellulose; moreover, the protein will probably adsorb to a cation exchanger. A second trial, using 20 mM N-morpholino-ethanesulfonic acid (Mes) buffer, pH 6, and a small column of CM-cellulose will settle this point. If the protein adsorbs to either exchanger (or at least disappears from the nonadsorbing fraction), the column should at once be treated with 1.0 M NaCl. This will almost always be a sufficiently high ionic strength to elute all proteins; then a check is made to ensure that the desired protein is washed off. Sometimes "adsorption" is really "inactivation," so before going ahead on a larger scale, the percentage recovery should also be checked.

The next stage depends on the result of the previous trial. If the protein adsorbed to neither column, a high degree of purification is already probable. The sample can be passed through columns of both adsorbents in a compromise buffer, e.g., pH 7.0. The two columns can be coupled together; alternatively, an intermediate collection and pH adjustment can be used. It is best not to have to change the buffer itself. For instance, it is particularly convenient to use Tris buffer adjusted to pH 8.0 with Mes for passing the sample through a DEAE-cellulose column. This may be expected to remove much unwanted protein. Then more Mes (acid form) is used to adjust the nonadsorbed fraction containing the enzyme to pH 6.5 [109]. The Mes now becomes the buffering species with HTris$^+$ merely a counterion, and the enzyme sample is immediately run on to CM-cellulose, where in most cases it will be adsorbed. If it is not, the eluted protein fraction will be many times purified compared with the starting sample.

If the protein adsorbed to one of the columns and was eluted successfully by salts, it may be worthwhile seeing whether it would adsorb at a more exacting pH, i.e., higher for CM-, lower for DEAE-, since this would lessen the quantity of other proteins binding simultaneously.

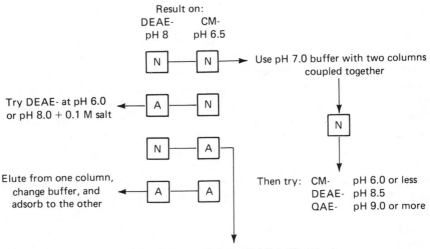

Figure 6.10. Flow chart illustrating possible trials of adsorption and elution from ion exchangers. N, not adsorbed; A, adsorbed.

Suitable buffers are described below. Alternatively, an increase in ionic strength at the same pH would have a similar effect, e.g., adding $50\,mM$ or $0.1\,M$ KCl to the buffer. A flow chart of progressive trials is suggested in Figure 6.10.

Buffers for Use in Ion-Exchange Chromatography

In all protein work, the correct choice of buffer can be crucial to success. Maintenance of pH in ion-exchange chromatography is of paramount importance, since the ionic interactions are so dependent on pH, and the buffers themselves are ionic and may be taking part in the ion exchange process. It is generally advised that the buffering ions should not themselves interact with the adsorbent, that is, that the charged form(s) of buffer should be of the same sign as the substituents on the adsorbent. If this is not so, undesirable and unpredictable pH changes may take place in the microenvironment adjacent to adsorbed proteins, causing uncontrolled changes in strengths of interaction, and possibly, loss by denaturation. Nevertheless there are many examples in the literature where this rule has been successfully broken; it is less important to observe when higher concentrations of buffer can be used. The commonest nonobservance of the rule is in using phosphate buffers in anion-exchange chromatography; sometimes the presence of phosphate is essential for the stability of an enzyme. In the absence of any reason to do otherwise, one is advised to follow the rule and allow only simple anions to be present in anion exchange experiments (e.g., Cl^-, acetate$^-$) and simple cations with cation ion exchangers (e.g., K^+, Na^+, Mg^{2+}, $HTris^+$ for pH < 7). Remember that polyanionic complexing agents such as EDTA bind to anion exchangers, accumulate on the adsorbent, and may compete with proteins for binding sites.

Often it is necessary to maximize the buffering power at the least possible ionic strength to ensure adsorption of a weakly binding protein. To do this two other rules must be followed. First, the pK_a of the buffer should be not more than 0.5 unit, and preferably not more than 0.3 unit away from the pH being used. Second, one of the buffering species should be uncharged, and so not contribute to the ionic strength. For anion exchangers at pH 8.0, a Tris-chloride buffer obeys all three rules: the counterion is Cl^-, the pK_a of Tris is 8.1 (at 25°C cf. Section 12.3), and the buffering species are $HTris^+$ (noninteractive) and Tris (neutral). For cation exchangers at pH 6.5, a K-Mes buffer is suitable: the counterion is K^+, the pK_a of Mes is 6.2, and the buffering species are HMes (neutral) and Mes$^-$ (noninteractive). On the other hand, phosphate breaks two of these rules when used with DEAE-cellulose since it is the counterion, and neither buffering form is neutral (HPO_4^{2-} and $H_2PO_4^-$). On CM-cellulose, phosphate breaks the latter rule only. Table 6.4 gives some values of the buffering power at various pHs around the pK_a values of several buffers. Note that a divalent buffer such as phosphate has the same buffering power (millimoles per millimoles of buffer) for pHs above its pK_a as does the monovalent acetate or Mes, but its buffering power per unit of ionic strength is much less, by a factor of 4 or so. This means that using a divalent buffer instead of a comparable monovalent buffer requires up to 4 times as high an ionic strength to get comparable buffering power. Alternatively, using phosphate instead of, say Mops, at a specified ionic strength results in 4 times weaker buffering (Mops, N-morpholinopropane sulphonic acid, is a monovalent buffer with a pK_a similar to phosphate). Another point to note about Table 6.4 is that with monovalent buffers at pHs which require more of the neutral than the charged form, the buffering power per ionic strength remains constant (and maximal). Thus acetate, even at pH 3.7 with an ionic strength of 0.01, still has a buffering power of 10 mM, as defined in Table 6.4; the amount of undissociated acetic acid would be 0.1 M. A list of other possible buffers is given in Appendix C, and further lists are to be found elsewhere (e.g., [110,111,112]).

Suggested buffer compositions for ion exchange use are given in Table 6.5. The buffering strength, as defined above, should be *at least* 5 mM. This means that, for example, if using Tris-HCl with an anion exchanger at its pK_a or higher (pH 8.0+), there should be at least 5 mM HCl with Tris base to make up to the desired pH. But if using Tris below its pK_a, the amount of HCl must be increased to maintain the buffering power.

Table 6.4. Buffering Power of Commonly Used Buffers in Relation to the Ionic Strength of the Buffer Solution. The buffering power is defined as the amount of acid (for pH $<$ pK_a) or alkali (for pH $>$ pK_a) required to completely titrate the buffering species, in units of mM. Note that the buffering power per mM of buffer depends only on the difference between the pH and the pK_a. Values for other buffers can be calculated using their pK_a values and ionic species (for n value, see Chapter 12)

Buffer	pH	Buffering power for $I = 0.01$	Total buffer concentration for $I = 0.01$	Buffering power per mM of buffer
Acetate	4.2	10	42	0.24
	4.4	10	30	0.33
$n = -1$	4.7[a]	10	20	0.50
	5.0	5	15	0.33
	5.2	3	13	0.24
Imidazole	6.5	3	13	0.24
	6.7	5	15	0.33
$n = +1$	7.0[a]	10	20	0.50
	7.3	10	30	0.33
	7.5	10	42	0.24
Phosphate	6.5	1.7	6.7	0.24
	6.7	2	6	0.33
$n = -3$	7.0[a]	2.5	5	0.50
	7.3	1.4	4.2	0.33
	7.5	0.9	3.0	0.24
Tris	7.6	3	13	0.24
	7.8	5	15	0.33
$n = +1$	8.1[a]	10	20	0.50
	8.4	10	30	0.33
	8.6	10	42	0.24

[a] The pK_a for this buffer; actual value will depend on temperature and total ionic strength, see Chapter 12.

To get a buffering power of 5 mM at pH 7.0 using Tris, it is necessary to have 50 mM HCl (and 55 mM total Tris buffer). Clearly, this might be too high an ionic strength for binding of your protein. This problem is solved by using a buffer with pK_a closer to or less than 7.0, e.g., imidazole or histidine.

It is convenient to prepare the buffer for an anion exchanger by starting with 10 mM of HCl and adjusting it to the desired pH with the uncharged base (e.g., Tris base, imidazole base). This ensures that the ionic strength is accurately known (e.g., using 10 mM HCl will give $I = 0.01$ regardless of which (monovalent) base is used or what pH the solution is adjusted to. For cation exchangers, the converse strategy is used, e.g.,

Table 6.5. Suitable Buffer Compositions for Ion-Exchange Chromatography[a]

For anion exchange[b]			For cation exchange[c]		
Temperature (°C)	pH	Buffer	Temperature (°C)	pH	Buffer
20	8.5–9.2	Diethanolamine, or 2-amino-2-methyl-1,3-propanediol	0–30	3.6–4.3	Lactic acid
0	9.0–9.7		0–30	4.3–5.2	Acetic acid[e]
20	7.8–8.5	Tris [tris(hydroxymethyl)-aminomethane]	0–30	4.7–5.4	Pivalic acid[e] (trimethylacetic acid)
0	8.2–8.9				
20	7.4–8.0	Triethanolamine	0–30	5.2–5.8	Picolinic acid[g]
0	7.8–8.4		20	5.8–6.6	Mes[h]
20	6.6–7.3	Imidazole[d]	0	6.0–6.8	
0	7.0–7.7		20	6.3–6.9	Ada[h]
20	6.2–6.8	Bis-tris [bis-(2-hydroxyethyl)imino-tris-(hydroxymethylmethane)]	0	6.5–7.1	
0	6.5–7.1		20	6.8–7.5	Mops[h]
			0	7.1–7.8	
20	5.6–6.3	Histidine[d]	20	7.2–7.8	Tes[h]
0	5.9–6.6		0	7.6–8.2	
20	4.9–5.6	Pyridine[ef]	20	7.8–8.5	Tricine[h]
0	5.2–5.9		0	8.2–8.9	

[a] All buffers listed are 0.01 ionic strength. The usable pH range extends no more than 0.4 unit either side of the pK_a.
[b] 10 mM HCl, adjusted to pH indicated with base form of buffer.
[c] 10 mM KOH, adjusted to pH indicated with acid form of buffer.
[d] Complexes divalent metals, especially histidine.
[e] Volatile.
[f] Poisonous.
[g] UV absorption.
[h] See Appendix D.

titrating $10 \, mM$ KOH with a monovalent acid such as acetic, morpholinoethane sulfonic (Mes), or Tricine to the appropriate pH. The effects of temperature and ionic strength on pK_a values should be borne in mind (cf. Section 12.3), especially if the buffer is made up at room temperature but used in the cold. Unfortunately, when following published procedures, it is rarely clear whether the buffer was adjusted to the stated pH in the cold or at room temperature.

If the protein of interest adsorbs strongly at low ionic strength, it may be appropriate to increase the ionic strength of the application buffer, either by adding a neutral salt (KCl, NaCl) or by increasing the buffer strength. The latter choice is preferable up to a buffer concentration of about $0.1 \, M$, but higher values could have an interfering effect on the chromatography. Some buffers may be rather expensive to use at high concentration, and may absorb in the ultraviolet wavelengths. On the other hand, a higher buffering power makes the maintenance of exact pH more certain and may enable an elution protocol by change of pH.

Conditions of Adsorption

Once an appropriate buffer is chosen, one must consider how to load the column, what the column dimensions ought to be for a given amount of protein, and what other operating procedures are required. It is normal for the applied protein mixture to be in exactly the same buffer used to preequilibrate the column. If this is not necessary, because the protein of interest is strongly bound, so much the better, since time is saved not having to transfer the sample into a precisely determined buffer. Nevertheless, the pH should be identical with the column buffer and, preferably, the ionic strength similar. There are two usual ways of attaining the correct buffer composition: dialysis or gel filtration. Dialysis is the traditional method, and although it takes a long time, there are few manipulations. More frequently used these days is "desalting" by gel filtration. These processes have been described in Section 1.4.

The protein sample is now in an appropriate buffer for applying to the ion-exchange column. The only remaining adjustment that might be needed is the protein concentration itself. Even after manipulations to remove salts, etc., the concentration could be as high as $30 \, mg \, ml^{-1}$, and this is too high, especially if a substantial proportion is going to adsorb to the exchanger. Ion exchange involves buffer ions, and a rapid displacement of counterions by protein molecules could result in sharp local changes in pH and salt concentrations (see above). Any carefully defined conditions should be reproducible; whereas it is always possible that applying the sample at high protein concentration may work well in a particular case, it may not be readily reproducible. It is the common experience that conditions that are reproducible from day to day are very difficult to genuinely reproduce from laboratory to laboratory, so it is preferable to have methods that are not critically dependent on precise values of any parameter. The more dilute the protein sample is during application, the more ideal and reproducible the result will be. But very dilute solutions are impracticable, and might cause other problems if the α value of the desired protein is less than 0.95 (see below). A maximum concentration of adsorbing protein of $5\,\mathrm{mg\,ml}^{-1}$ is allowable, with a further $10\,\mathrm{mg\,ml}^{-1}$ of nonadsorbing protein. At first it may seem strange that any restriction be put on the latter. But remember that nonadsorbing protein carries counterions, which increase the ionic strength, weakening the proper interactions required during sample application.

Usually one assumes that in the conditions for adsorption, the protein will stick to the top of the column and stay there until buffer conditions are altered for the elution procedure. This is so if $\alpha = 1.0$ or very close to that value. But it may not be possible to reach such a state; the pH required may be outside the stability range, or one may have chosen to operate in less strongly adsorptive conditions to lessen the total protein adsorbing to the column. Consider the situation of $\alpha = 0.90$. At any point on the column reached by the protein, 10% remains in solution and passes down to the next point. Moreover, a fraction of the adsorbed protein is constantly desorbing and so being carried by the buffer flow to a lower point. On the average, each protein molecule moves down the column at 10% of the speed of the buffer, so after 10 column volumes of starting buffer have been applied, the protein will begin to emerge from the bottom.

If there are more strongly adsorbing proteins in the mixture being applied, the $\alpha = 0.90$ protein will be displaced more rapidly, since on desorbing from initial binding sites at the top of the column, it will find most other sites around it occupied irreversibly and will have to move down with the buffer to an area where these other proteins have not yet reached (see Section 5.5). So its average speed of movement down the column will be somewhat greater than 10% of the buffer flow. If the protein you want has an α value less than 0.90, it is clear that the applied volume should not be too much greater than the column volume, nor should much washing with starting buffer occur before the elution scheme is commenced.

Size and Dimensions of the Column

The protein concentration and the relative volumes of column and sample size are clearly important in deciding the size of column to be used. More important are the actual capacity of the column for the proteins being adsorbed, and the length of chromatographic separation needed, i.e., the plate number, see Section 5.1. For some applications, particularly stepwise elutions and affinity elution (Section 5.5), a short column can be used with up to one-half of its volume being used for the initial adsorption. As the full chromatographic development is not required in stepwise procedures, it is better to minimize the "unused" part of the column to lessen band spreading. But stepwise procedures are rarely true on–off situations ($\alpha = 1.0$ to $\alpha = 0$), so partial retention in the "unused" lower part of the column will in fact affect the elution pattern. For more subtle elutions involving gradients, a reasonable column length is needed, from 5 to 20 times the amount needed for adsorption.

The adsorptive capacity of ion exchangers for proteins can be very high, but it depends on molecular size. Small proteins bind in amounts in excess of $100\,\text{mg}\,\text{cm}^{-3}$, but very large proteins ($>10^6\,\text{MW}$) may not penetrate the exchanger and bind only to the surfaces of the particles, resulting in very low capacities (see Figure 5.9). Thus, it is difficult to predict in advance how much of the adsorbent is involved in the initial uptake of proteins. An average figure for complex mixtures of $30\,\text{mg}\,\text{cm}^{-3}$ might be used, but the distribution through the column may well be asymmetric. In addition to the banding effect alluded to above, relative exclusion of larger molecules would cause a large protein to occupy a greater proportion of column than the same amount of a small one. If the molecular size composition of the applied fraction is known (e.g., a fraction from gel filtration, Section 8.1), then a better estimate can be made of the likely capacity of the ion exchanger. Using the general figures quoted above, if one has 1 g of protein and expects about half of it to adsorb, it would be reasonable to apply this in 100 ml of buffer to a column of $100-200\,\text{cm}^3$ volume. Some $20-40\,\text{cm}^3$ at the top of the column would be used in adsorbing proteins. These calculations are intended for "strongly adsorbing" proteins, i.e., those with α values very close to 1.0; if there are proteins with α values above zero but below 0.95 in the mixture, they will be only partially retained, and distribute further down the column during the application phase.

In the early days of protein ion-exchange chromatography, the dimensions of columns were made very long and thin. Although ideal for true chromatography, the low flow rate necessitated by such a configuration means a long process time, often with harmful effects on the protein. Shorter, squat columns are more common these days. Theoretically, with perfectly even application and flow through the column, the shape would appear irrelevant, unless one is operating at very low protein loading. In that case, the plate number (see Section 5.1) is determined by column length and particle size, so a long column would result in better resolution. But at the high loadings commonly used in protein purification, the plate number is likely to be determined by the ratio of unoccupied adsorbent sites to occupied, which is relatively independent of column shape (Figure 6.11).

Figure 6.11. Columns of identical volume but different proportions. The long, thin column (a) with protein adsorbed to the top 20% has a relatively slow flow rate, but uneven flow is not usually a problem. The squat column (b) can have a very rapid flow rate, meaning a short processing time, but the flow must be even through the whole cross section, otherwise resolution is lost.

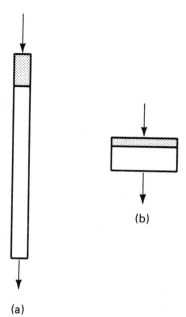

(b)

(a)

In practice, perfect flow through a column is difficult to achieve; longer columns even out imperfections better but take longer to run. For routine use, columns 4–5 times longer than their diameter are recommended. But for stepwise elution, ion-exchange columns should be still squatter, with heights being only 1–2 times the diameter. On scale-up, it is common to use columns with diameters greater than their height (see Chapter 10).

Procedures for Elution

Before describing the different methods of desorbing proteins from ion exchangers, it is useful to repeat two situations—the trivial one of $\alpha = 0$, where the protein passes straight through the column, and the intermediate situation where α is substantially greater than 0 but significantly less than 1. When $\alpha = 0$, all similar proteins (those that have the same charge as the adsorbent) pass through and, apart from boundary effects due to different levels of molecular exclusion, are collected in a volume not substantially larger than that applied. If the protein required is in this fraction and most proteins have been held back on the column, this is an excellent method for purifying it. Recovery is usually 100%. Although the full potentialities of chromatography have not been realized, such a

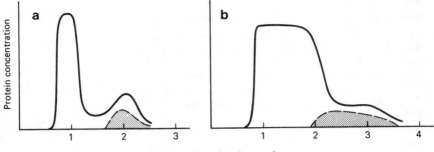

Figure 6.12. Elution of nonadsorbed ($\alpha = 0$) and retarded ($\alpha = 0.5$, shaded) proteins from an ion exchanger: (a) sample volume < 0.5 column volume; (b) sample volume $= 1$ column volume. Note that any tailing of nonadsorbed proteins with α values slightly greater than zero will cause a considerable overlap in (b).

procedure of nonadsorption is often used to decrease the total amount of protein being handled.

In the intermediate cases, α from 0.5 to 0.9 (values below 0.5 can be classed with $\alpha = 0$, since most of the protein will be only slightly retarded and will pass through the column quickly), significant retardation of the applied protein may lead to complete separation from the nonadsorbed fraction, but elution will occur eventually in the starting buffer. Two examples of possible results for $\alpha = 0.5$ are shown in Figure 6.12; the separation will depend greatly on the relative volumes of the column and the applied sample. Note that the total volume of buffer in which the enzyme is eluted will be greater than the volume of applied sample, even if there is no tailing. This is due both to chromatographic spreading, and the fact that in reality, because of a range of effective K_p values, α is a variable parameter, 0.5 being only an average. Nevertheless, if there is no problem in handling a large volume of eluted enzyme, and the sample can be applied in a small volume, this simple method of chromatography without buffer change can give a very good separation of proteins with α values between 0.9 and 0.5.

In most cases, the experimental conditions are such that the protein required is tightly bound in the starting conditions, and a buffer change is needed to elute it. There are three ways of decreasing α values on ion exchangers: (1) by changing the pH (up for cation exchangers, down for anion exchangers), (2) by increasing ionic strength, and (3) by affinity methods. Method (3) is described in detail in Section 7.4; (2) is the most commonly employed, and may be used in either stepwise or gradient form. Method (1), pH change, as a gradient, is not generally advisable other than in chromatofocusing columns. For stepwise applications, pH elution can be quite successful; the pH change in the eluate will be delayed compared with the new buffer front because of protein and exchanger titrations, but eventually a protein peak emerges, coincident with a rapid pH change. A rare, successful example of the use of a pH gradient is the separation of the two isoenzymes of yeast hexokinase on DEAE-cellulose in the pH range 5.6–5.2 [113]. Many aspartate and glutamate residues become protonated as the pH falls by only a fraction of a pH unit; consequently, the gradient, in pH terms, can be very gentle. In this pH range, the DEAE- adsorbent has almost no titration, and so does not significantly interfere with the development of the pH gradient.

Ionic strength gradients do not have any problems of irregular behavior, and increasing salt strength should emerge from the bottom of the column in much the same form as the gradient was applied at the top. The salt concentration needed to elute proteins will depend mainly on isoelectric points; a few proteins remain reversibly adsorbed to ion exchangers above $I = 0.5$; some require $1\,M$ salt to remove them, but rarely more. On anion exchangers, nucleic acid contaminants bind and are eluted in the salt range 0.5 to $1\,M$, almost independent of pH, since the charge on nucleic acids is pH-independent through the normal range. If your protein elutes in this salt range, and you want to separate it from nucleic acids, it would be worth trying the chromatography at a much lower pH, so that your protein elutes at lower salt concentration.

With cation exchangers, it is uncommon for a protein to bind at salt concentrations higher than $0.5\,M$, mainly because the pH used is such that even high-isoelectric-point proteins do not have a very high net charge. Phosphocellulose is different and behaves somewhat in the fashion of an affinity adsorbent; quite high salt concentrations are often needed to break the pseudospecific binding between the phospho- groups and phosphate-binding sites on the proteins.

Gradients (of ionic strength or pH) can be formed using a simple apparatus commercially available in which two containers are connected at the bottom, and an efficient mixing paddle is in the container from which the buffer is extracted. Simpler still is the two-beaker method with a magnetic flea to stir the sampled buffer (see Figure 6.7). By arranging the gradient mixing device above the column, it is possible to run the column under gravity feed, but a peristaltic pump to deliver a constant flow is preferable. More complex, but now routine equipment, are the electronically controlled mixers that can produce any gradient profile required. Pumps come in many shapes, sizes, and, above all, costs. An expensive pump should be capable of a continuously variable flow rate, minimum pulsation, and capability of operating against a fair pressure without decreasing the flow rate. However, many of these capabilities are not always necessary—in particular the latter, since it may be better for the pump to cease flowing against an increased pressure due to a blockage, than to burst the tubing and continue to pump liquids all over the bench. In general, pressure should *not* increase during the run. If it does, there is probably something wrong. Besides the excellent pumps available from many major companies specializing in biochemical equipment, there are simpler and much cheaper versions which are adequate for many purposes. HPLC is of course another matter; in this case, all pumps and gradient formers are part of the equipment.

Compact peaks containing the required protein can be obtained by a stepwise application of a buffer with higher ionic strength, and this can be a rapid and efficient way of purifying a protein. However, because of other factors affecting the α value, namely protein concentration and weak and strong binding sites (see Section 5.1), it is unlikely that a simple on–off system will operate well unless the step in salt concentration is quite large. With a large increase in salt concentration, many other proteins may be eluted in addition to the desired one. A sudden decrease in the α value, due to the salt, brings off much of the enzyme, but tailing effects are large, and a further increase in salt may be needed to bring off the rest of the enzyme (Figure 6.13). The second peak is *not* a different enzyme form; it would be most unwise to deduce the occurrence of isoenzymic forms from a stepwise elution procedure. The tailing of a peak is lessened by having a gradient. Indeed, this is one of the chief reasons for using gradient elution so extensively.

A few further points are worth noting about ion-exchange chromatography. Whereas simple theories imply that conditions for elution are exactly the inverse of those for adsorption, in practice some hysteretic

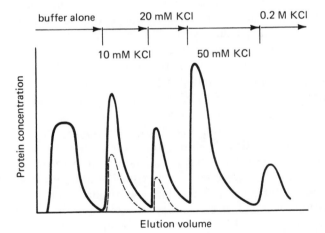

Figure 6.13. Stepwise elution from ion exchange column by increasing salt concentration. Enzyme (*dotted line*) has been distributed between two peaks, but the enzyme does not necessarily exist in two forms. Note the sharp leading edges and long trailing edges to the eluted protein peaks (see text). (From Scopes [66].)

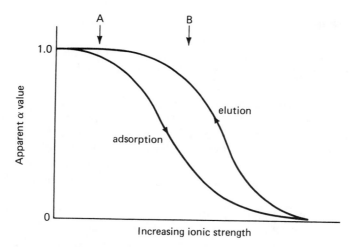

Figure 6.14. Hysteresis in ion-exchange chromatography. Due to nonequilibrium operation, conditions for adsorption appear to be slightly different and more demanding than conditions for elution (see text).

behavior is often noted. Hysteresis means, for example, that the α value effective when a particular ionic strength buffer is used for adsorption will be different from the α value when the same buffer is used for elution. Thus, in Figure 6.14, ionic strength A may be suitable for adsorption, and ionic strength B will not cause elution. But ionic strength B would not be suitable for adsorption. These effects must be explained in terms of nonideality, nonequilibrium processes, and other factors not taken into account by simple theory.

When operating cation-exchange chromatography, it is assumed that basic (i.e., positively charged) proteins will bind to the adsorbent. But it may be that the fraction being applied to the column contains nucleic acids or other negatively charged polymers. These may compete with the cation exchanger by acting as soluble cation exchangers themselves; protein–nucleic acid interaction is not only common, but to be expected if their charges are opposite, and the ionic strength is low. This problem can be remedied by removal of the nucleic acids using cationic polymers such as protamine or polyethylene imine either at an earlier step (Section 2.3), or immediately before application to the cation-exchange column. The amount of precipitant needed should be determined by trials (follow the ratio of UV absorption at 260 and 280 nm), and the pH carefully controlled by adjusting the precipitant to the pH value of the sample being treated. Many times have people concluded that their protein "does not stick" to cation exchangers despite a high isoelectric point—the reason being interference by nucleic acids.

6.3 Inorganic Adsorbents

There is a range of inorganic substances that have been used for protein adsorption, mainly oxides, insoluble hydroxides, and phosphates. Chief among these is calcium hydroxyphosphate, which in crystalline form is known as hydroxyapatite. The use of hydroxyapatite will be described below; the gelatinous form "calcium phosphate gel" is appropriate for batch adsorption, see Section 5.2. Meanwhile, a short listing of adsorbents will give an idea of the usefulness of inorganic materials. One particular advantage they have, especially in large-scale and industrial applications, is their cheapness; it is often not worth the trouble to clean them up after one use.

Unlike ion exchangers or affinity adsorbents, the inorganic materials do not have a readily explainable mode of action. Crystalline surfaces are made up of charged ions with associated water of hydration, so undoubtedly an electrostatic interaction is an important component of the adsorption of proteins to them. However, every positive area on the crystal surface has negative areas very close to it, and vice versa. Consequently, the interaction could be a polar dipole–dipole bonding (Figure 6.15a). Proteins have the greatest probability of having adjacent positive and negative groups in the neutral pH range; thus, whatever the isoelectric point may be, adsorption is most likely between pH 6 and 9.

Despite this simple interpretation, the practice is more complex. Buffers are always present, usually phosphate, and the buffer ions themselves adsorb to the inorganic materials. Thus, it is possible for hydroxyapatite to present an entirely negatively charged surface to a protein (Figure 6.15b). Increasing buffer concentration causes competition between protein and buffer ions for the charged sites on the adsorbent, resembling typical ion exchange behavior. Detailed discussions of these effects have been presented [114,115,116] with some quantitative data on the adsorption of certain proteins to both hydroxyapatite and TiO_2.

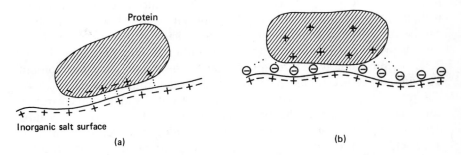

Figure 6.15. Dipolar interactions of proteins on inorganic salt surface: (a) simple interaction, (b) with intervention of buffer ion.

Table 6.6. Some Inorganic Adsorbents That Have
Frequently Been Used for Protein Purification

Alumina gel C_γ (gel and crystalline)
Bentonite (a silicaceous powder)
Calcium phosphate: Aged gel
                    Brushite
                    Hydroxyapatite
Titanium oxide
Zinc hydroxide gel

As with many other protein purification procedures, the only way
of finding out whether the method will be suitable is to try it and see.
Adsorption will be encouraged by low salt concentration, but sufficient
buffering should be present to avoid undesirable local pH changes in the
highly polar environment of the adsorbent surface. Elution of proteins
occurs at higher salt concentrations, which can be added directly or, in
the case of column chromatography, applied as a gradient.

Some of the inorganic adsorbents that have been used are listed in
Table 6.6. One often comes across the use of these materials not as chromatographic materials or even in the normal adsorption–elution cycle,
but with the aim of adsorbing only unwanted material. A batchwise or
column-type system may adsorb a variety of undesirable proteins, other
macromolecules, and low-molecular-weight compounds, but leave the
desired protein in solution, immediately ready for the next step in purification. An example, using both zinc hydroxide and bentonite as batchwise adsorbents, is shown in Table 5.3, describing the purification of
yeast enzyme phosphoglucose isomerase [86]. Zinc hydroxide adsorbent
is created in situ by adding zinc acetate plus alkali to maintain the pH
around 8. Many proteins are adsorbed, but the isomerase remains in
solution. After removing the hydroxide gel by centrifugation, bentonite is
added, which adsorbs more proteins, but still the isomerase remains
behind. Bentonite contains very fine gelatinous particles which require
relatively high centrifugal force to sediment. These two steps achieved a
20-fold purification from a crude extract which had not been processed by
any prior treatment.

These techniques are regarded as crude and old-fashioned for modern
practitioners of protein purification, but they can be extremely useful and
economical when scaling up for commercial production.

Hydroxyapatite and Calcium Phosphate Gels

The usefulness of hydrated calcium phosphate gels for selectively adsorbing proteins, allowing their release at higher salt concentrations,

has been recognized and applied for decades. The original and still very useful adsorbent was prepared by mixing calcium chloride with tribasic sodium phosphate, washing the gelatinous precipitate with distilled water, then allowing it to "age" for several months before it acquired optimum, consistent characteristics [117]. Being gelatinous, this material is not of any use directly in columns because of poor flow characteristics, but is ideal for batch adsorption. It has a high capacity because of its large surface area per unit weight and, until the development of ion exchangers, was widely used in protein purification. As with other inorganic adsorbents, calcium phosphate gel is now regarded as old-fashioned, but it still has uses in rapid batchwise adsorption.

To prepare hydroxyapatite, the calcium phosphate that is crystalline and suitable for use in columns, $0.5\,M$ solutions of $CaCl_2$ and Na_2HPO_4 are slowly mixed together with a solution of $1\,M$ NaCl. This makes brushite, $CaHPO_4 \cdot 2H_2O$. The brushite is boiled with NaOH, which converts it to hydroxyapatite, $Ca_{10}(PO_4)_6(OH)_2$ [118]. These days most laboratories buy ready-made hydroxyapatite.

The crystalline particles in the column adsorb proteins on their surface; unlike biologically based matrices, the proteins cannot penetrate the particles. Consequently, the adsorptive capacity is limited; for example, a monolayer of a typical protein over the surface of spheres 0.1 mm in diameter amounts to no more than $0.1\,\text{mg cm}^{-3}$. Capacities with hydroxyapatite are higher than this, probably due to crevices and cracks within the particles, giving a much higher effective surface area. But the limited capacity of crystalline calcium phosphate makes its use in enzyme purification mainly a late-stage procedure, often the final one after other techniques have failed to result in homogeneity.

Bernardi [118], in extensive studies of the behavior of proteins on hydroxyapatite, has concluded that the adsorption of basic proteins, positively charged at the operating pH, is rather different from that of neutral or acidic proteins. Adsorbed at low K-phosphate concentrations, all proteins could be eluted by increasing phosphate concentration, with basic proteins tending to require stronger buffers than acidic ones. Other salts such as KCl and NaCl had no effect in eluting acidic proteins, nor did $CaCl_2$ at up to $3\,M$ concentrations. Basic proteins, on the other hand, were eluted by chloride salts, and very low concentrations of $CaCl_2$ $(1-10\,\text{m}M)$ were sufficient to displace such basic proteins as lysozyme and ribonuclease from hydroxyapatite. The conclusions drawn from these experiments were that acidic proteins were more adsorbed by Ca^{2+} sites on the crystal surfaces, and anions such as Cl^- were not effective in displacing them as they have little affinity for Ca^{2+}. On the other hand, basic proteins, binding more through PO_4^{n-} sites, can be displaced by monovalent cations reasonably effectively, and very effectively by Ca^{2+} ions which have high affinity for the PO_4^{n-} sites. Although these theoretical and model experiments have demonstrated some new possibilities for obtaining separations on hydroxyapatite columns, actual examples of using selective salt elutions are few. A gradient of phosphate buffer remains the normal procedure adopted, perhaps through ignorance of the alternatives. Further discussion of these interactions can be found in [92].

6.4 Hydrophobic Adsorbents

During the development of affinity chromatographic support mediums, a number of "control" experiments were carried out, in particular, looking at the behavior of matrices containing spacer arms but no ligand. In most cases these behaved as expected, with no adsorption noted. But in a few cases, proteins were found to bind strongly to hexamethylene arms, even without charged ends (amino or carboxylate) which might have been acting as ion exchangers [119]. These observations were put to good use in developing the techniques of hydrophobic chromatography, for it was the interaction between aliphatic chains on the adsorbent and corresponding hydrophobic regions on the surface of the proteins that was causing binding [120,121]. Indeed, it is this "nonspecific" contribution to affinity adsorption that is most helpful in achieving tight binding (Chapter 7).

Whereas only the more hydrophobic proteins bind to short immobilized aliphatic chains at low salt concentration, hydrophobic chromatography can be extended to cover all proteins, since the hydrophobic interactions *increase* in strength with increasing salt concentration. In particular, the salts which are most effective in salting out precipitation, such as ammonium sulfate, are the ones that strengthen the hydrophobic interaction most. The reasons are the same; in salting out, a main cause of aggregation is the strengthening of hydrophobic interaction between proteins. In hydrophobic chromatography, the aggregation is between hydrophobic patches on the proteins and the immobilized ligand on the adsorbent.

Typical hydrophobic adsorbents available commercially include C_4-, C_6-, C_8- and C_{10}- linear aliphatic chains (Figure 5.19), and the same chains containing a terminal amino group. The latter were designed for affinity ligand attachment, but in themselves are interesting hydrophobic adsorbents with a hydrogen- or electrostatic-bonding group as additional potentials for interaction with proteins. It is also worth noting that products that are based on cyanogen-bromide-activated agarose have a positively charged group at the other end of the aliphatic chain (see Figure 7.4). This also applies to many other chemistries in which coupling of amines is involved, and is discussed in greater depth in Chapter 7. These adsorbents are more correctly described as multifunctional (see Section 6.6). Phenylagaroses are also commonly used; a benzene ring is attached via an ether linkage to the matrix (Figure 5.19). As with dye ligand adsorbents (Section 7.3), the potential for varying the nature of the substituent is infinite, but the behavior of each is similar. The most commonly used (and available) hydrophobic adsorbents are octyl- (C_8-) and phenyl-substituted matrices.

Hydrophobic adsorbents function as a result of interactions different from other adsorbents; thus, in theory, they have a good chance of effecting a separation of components that cannot be easily separated by, for example, ion-exchange chromatography. Hydrophobic chromatography has not been used as extensively as ion-exchange chromatography, not just because it is a newer development, but also because it does not usually achieve such sharp separations. This is partly because the α value changes only slowly with altering conditions; there is also a relatively slow association–dissociation process, meaning that columns are not run in equilibrium mode. But the main obstacle to sharp separations is protein–protein interactions: similar proteins will interact with each other as well as with the adsorbent (whereas in ion-exchange chromatography, similar proteins repel each other because of their like charges). Consequently, there tends to be a large degree of overlap between the eluting components. Also, since the principles are very similar to salting out fractionation, if the latter has already been carried out there is less chance of getting a good separation on a hydrophobic column. Weighed against these pessimistic views are several positive points. First, the capacity for proteins is very high, in the same range as ion exchangers, namely from 10 to 100 mg cm^{-3}. Second, because adsorption is carried out at high salt concentration, it is not necessary to change the buffer of a sample before application; one needs only to add sufficient salt to ensure binding of the component required. And because of the stabilizing influence of salts, recoveries are often excellent.

Application of Sample to a Hydrophobic Column

The sample will in most cases need to be in a salt that increases hydrophobic interactions in order for the protein you are purifying to bind. Most commonly used are sodium sulfate or ammonium sulfate, but sometimes chloride salts (1 to 2 M) are used. The sulfates are commonly used at 0.5 M; if the protein does not bind to a C_8- or phenyl- adsorbent in these conditions, then it is very hydrophilic, and the flow-through may be taken on to the next step, having removed most hydrophobic proteins. But if you really want to adsorb your protein in these circumstances, the salt concentration may be increased until it eventually does bind. If enough ammonium sulfate is added, the protein will begin to precipitate anyway; just below this point it is certain to adsorb to any hydrophobic column. If the protein binds strongly (as determined by elution procedures, see below), then it is worth considering less hydrophobic adsorbents. We use varieties of C_4-, C_6-, C_8-, and other more complex hydrophobics in a screening process to determine the optimum adsorbent. In some cases it may be desirable to choose one that binds only in the presence of salt, if the sample is already in that state, and elution can occur simply using low salt buffer. In other cases, it may be preferable to choose an adsorbent that binds the protein even at low ionic strength, which can be more selective. Lowering of pH usually strengthens interactions between proteins and aromatic groups and other structures having π-electron orbitals; the negative electron cloud presumably associates with positive areas on the protein adjacent to the hydrophobic amino acid residues. Positive charges on the protein increase (and repulsive negative charges decrease) as the pH is lowered. For less clear reasons, hydrophobic interactions with proteins increase with lower pH even with aliphatic adsorbents. With simple adsorbents, attached through a linkage which creates a positive charge (see above), the opposite effect might be observed, due to ionic repulsion. But the pH is not critical. It is probably best to operate at a pH you know is most stabilizing for your protein. As indicated above, loading levels can be high because capacities are high.

Elution of Protein from Hydrophobic Columns

Proteins that are strongly adsorbed even at low salt concentration are generally those with low water solubility: globulins, membrane-associated proteins, and others that precipitate in a low range of ammonium sulfate saturation (up to 40% saturation). These may not be readily removed from the column without an eluant that specifically reduces the hydrophobic interactions. These can be weakened by: (1) lowering of temperature, (2) inclusion of organic solvent, (3) inclusion of polyols, especially ethylene glycol, (4) inclusion of detergent, or (5) by increasing the pH.

If the sample was applied in high salt, then elution can be achieved by lowering this salt concentration. Either a downwards gradient of concentration, or stepwise lowering are suitable. As resolution is not high, gradients may not achieve any better result than stepwise processes. However, use of alternative salts that have less water-structuring activity than sulfate, such as chlorides or even bromides, may give a better result. Binary and ternary salt gradients have been used at least experimentally in investigated hydrophobic adsorbents [122], and there may be some benefit in considering the nature of the salts and buffers being used. At low salt concentration, proteins still bound may be eluted by buffers containing an increasing concentration of such solutes as ethylene glycol, alcohols, detergents, or combinations of these. Ethylene glycol concentrations up to 50% v/v have been used, but removing this solute can be a nuisance. Nonionic detergents such as Triton X-100 are used up to about 1% w/v, and probably the most convenient is the use of *i*-propanol at up to 30% v/v. Acetonitrile, commonly used in reverse phase chromatography (below), has also been used with these adsorbents [123].

Reverse Phase Chromatography

The hydrophobic adsorbents described above have been developed in "low-performance" systems, later adapted to high-performance chromatographic equipment. They are usually operated in aqueous media at neutral pH, although as indicated above, elution with organic solvents is not uncommon. Much more commonly used in biochemical separations is the other type of hydrophobic chromatography, known as "reverse phase" adsorbents, as used in HPLC. In this context the "normal phase" relates to chromatography used in organic chemistry in which the adsorbent is hydrophilic, and the liquid in the column is an organic solvent. Reverse phase HPLC involves the use of hydrophobic adsorbents, typically aliphatic chains between C_8- and C_{18}-, attached to silica beads. The samples are applied in an aqueous solvent, often a dilute acid, and are eluted by a gradient of miscible organic solvent such as methanol or acetonitrile. In general, smaller proteins are resolved earlier, and proteins above about 40 kDa may not be readily removed from the column. In many cases, the eluted protein is not in its fully native state; in fact, it has been proposed that reverse-phase chromatography of proteins takes place by denaturation on adsorption and renaturation on elution. For successful reverse phase purification of a protein, either the protein must be sturdy enough to withstand the rigors of the environment during chromatography, or, if not, an inactive protein must be adequate for the purpose.

Reverse phase HPLC has one major positive feature, that of very high resolution. That is why it is so widely used, and in particular for *peptide* separations, such as obtained by enzymic digestion of proteins. Peptides are small, sometimes with no real secondary structure to preserve, and so cannot denature in the conventional sense. Many of the proteins that have been of interest for clinical and therapeutic use recently do come under these categories of small size and sturdiness—indeed, some can be better described as peptides rather than proteins. They are relatively small, acid-stable, and resistant to denaturation (or easily renatured). Consequently, reverse-phase HPLC has been very important in the development of purification strategies for many of these proteins. But it is not suitable in general, especially for the many thousands of more sensitive intracellular proteins and enzymes.

Other Hydrophobic Techniques

Hydrophobic chromatography and salting out precipitation are closely related; each is based on strengthening hydrophobic interactions at high salt concentrations. Adding a slurry of ammonium-sulfate-precipitated protein to a column preequilibrated with ammonium sulfate, and then developing with a decreasing salt gradient had been successfully achieved before hydrophobic adsorbents were synthesized [124]. The columns consisted of cellulose or dextran, and later agarose [125]—neutral, "inert" materials. However, the "inertness" of these materials at high salt concentration is not strictly true; in fact, they behave very much like hydrophobic adsorbents. Cellulose is insoluble in water largely because of its hydrophobic character. The hydrogens on glucose residues are on the crystal surfaces, with the hydroxyls in the interior forming hydrogen bonds with adjacent molecules. But specially treated cellulose, exposing more hydroxyl groups and opening up its structure, has been shown to act very successfully as a "salting out adsorbent" [126]. Moreover, substituted ion-exchange celluloses appeared to adsorb proteins even more effectively, despite the fact that at high salt concentration the ion-exchange character is expected to be lost. Proteins adsorbed to these cellulose matrices could be eluted *at the same high salt concentration* as used for their adsorption, using hydrogen-bond-weakening solutes such as glycerol, sucrose, urea, or ethanol [126]. Consequently, it was surmised that hydrogen bonding in these conditions was at least as important an attractive force as the hydrophobic interactions.

Agarose contains many methylene residues as well as hydroxyls, since it contains anhydrogalactan residues. The amphiphilic nature of these matrices (partly hydrophobic, partly hydrophilic) may be even more suited to protein adsorption at high salt concentrations than truly hydrophobic materials [127]. Indeed, agarose does appear to be a most suitable matrix for salting-out chromatography. Proteins adsorb to the agarose when the salt concentration is somewhat less than the concentration required to precipitate them in free solution; thus, it is not necessary, nor desirable, to preprecipitate the protein and apply it as a slurry. Two techniques are possible: (1) add ammonium sulfate to a concentration just below that needed to precipitate the protein of interest, centrifuge off any precipitate, and apply supernatant to column preequilibrated at the same ammonium sulfate concentration, or (2) add agarose to sample, approximately $1 \, cm^3$ per 20 mg protein, and slowly dissolve in ammonium sulfate until the concentration (based on the whole volume of agarose + sample) is about that required to precipitate the protein from free solution. Then transfer agarose to top of partly poured column, as above. In either case, the proteins, rather than self-aggregate, adsorb to the agarose and coat both the surface and the interior of the beads with protein, up to $30 \, \text{mg cm}^{-3}$. After washing with starting buffer, the gradient or stepwise lowering of salt concentration can begin. This is, in principle, the same as hydrophobic chromatography, but most proteins will come off the column well before the salt concentration has decreased to $0.5 \, M$, which is the normal concentration for *applying* samples on true hydrophobic adsorbents. The poor hydrophobic character of agarose is compensated for at the adsorption stage by using a higher salt concentration. This is, in effect, a C_0- hydrophobic column!

A further extension to this procedure involves an "affinity" technique in which the presence of a ligand in the buffer alters the adsorption characteristics [128]. Either with hydrophobic adsorbents or on agarose, the principles of affinity methods remain the same as with more specific adsorbents (Chapter 7) or ion exchangers (Section 6.1). First one establishes conditions without the ligand, which are not quite sufficient to cause the effect, e.g., in this case, elution. Then addition of the ligand just tips the balance and causes the effect. It is essential that the ligand makes a detectable difference to the property concerned, in this case the hydrophobic interaction. It must be able to bind to the enzyme in the conditions on the column, which usually involves a high salt concentration. One example of "hydrophobic affinity" is the purification of some amino tRNA transferases [128]; binding of the large tRNA molecule causes a sufficient decrease in adsorption properties to allow elution. Another example is affinity elution of alcohol dehydrogenases in the presence of salt using alcohol substrates [129].

6.5   Immobilized Metal Affinity Chromatography (IMAC)

General Principles

This technique has been included in the present chapter despite the inclusion of the word "affinity" in its title. Chapter 7 is reserved for affinity methods that operate principally through a *biospecific* interaction, namely through a natural ligand binding site. Immobilized metal affinity chromatography (IMAC) does not in general act biospecifically, and so we will deal with it here. The mode of action is quite different from the other adsorbents described in this chapter. It relies on the formation of weak coordinate bonds between metal ions immobilized on a column, and basic groups on proteins, mainly histidine residues. The technique was introduced in 1975 [130], and has slowly but steadily increased in usage. A recent summary by Porath, the originator of the method, summarizes the current status of IMAC [131].

The adsorbent is formed by attaching to the matrix a suitable spacer arm plus a simple metal chelator, usually based on imino diacetate struc-

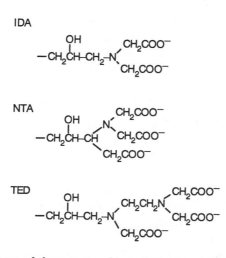

Figure 6.16. Structure of three types of immobilized metal ion chelating ligands. IDA, iminodiacetic acid, is tridentate; NTA, nitrilo triacetic acid is tetradentate, and TED, tris(carboxymethyl) ethylene diamine is pentadentate. In each case the coupling to the matrix shown is through epoxy activation. (From [131].)

tures, as commonly used in biochemistry in the form of EDTA (Figure 6.16). The most commonly used are imino diacetate (IDA) and tris(carboxymethyl) ethylene diamine (TED). These chelating ligands will bind tightly to metals ions, in particular to the divalent ions of the transition metals Fe, Co, Ni, Cu, and Zn, but also trivalent metal ions (Fe, Al). The structure of the chelating ligand is such that a metal ion, once bound, does not have all its coordination sphere occupied; there are no soluble ligand molecules that can attach to the half-complexed metal. These spare coordination sites are weakly occupied by water or buffer molecules, which can be displaced by more strongly complexing sites on proteins.

It has been found that histidine residues on proteins are most attracted to immobilized metals, but other potential electron donating side chains include tryptophan and cysteine (at neutral pH). Since all proteins contain these amino acids, it might be expected that all proteins are capable of binding to metal chelate columns. But the residues must be located on the surface of the protein (this is rarely so for tryptophan), and the strength of interaction will depend on the number of such linkages for the size of the protein. It has been shown that a single histidine in a small protein is sufficient to cause it to bind, with a K_p value of around $10^{-6} M$ [132]. In practice, the bulk of proteins in a crude extract do not bind strongly to IMAC columns, but certain specific proteins do. Thus the application is very much like an affinity technique, even though there is no connection between a natural ligand and the chelated metal. In a tandem column scheme for fractionating serum proteins, the first column is Zn-TED, which specifically bind α_2-macroglobulins, while almost all other serum proteins pass through [133,134]. Similarly, when a muscle extract was passed through Fe^{3+}-IDA, the only major proteins to bind were phosphorylase (probably by hydrophobic interactions in the high salt conditions) and lactate dehydrogenase [131]. But small proteins and large peptides do bind more strongly, probably due to the fact that only one strong linkage is needed to attach a small protein to the column.

Operating Conditions for IMAC

Metal chelate adsorbents are provided without the metal. The first step is to load up the column with the metal. A solution, typically $50\,mM$ of the metal salt, e.g., $CuSO_4$, Zn acetate, is passed through the column until it is saturated with the metal. Excess metal ions are washed out, but some more loosely bound ions should also be actively removed by a wash with a weak complexing agent such as $1-10\,mM$ imidazole or $0.5\,M$ glycine. Since it is undesirable for the metal ions to leach out into the final preparation, it may be appropriate to give a wash at this stage with the buffer that is going to be used for elution. If this causes a steady slow leakage of the metal ions, a trap of "uncharged" adsorbent (i.e., not loaded with metal) following the main column can be used. Metal chelate columns are likely to possess an overall charge; the chelate is negative, the metals are positive, and they do not always cancel out. Consequently, to avoid ion exchange effects, IMAC is operated at high ionic strength, for example in $1\,M$ NaCl. This seems to increase capacities and binding strengths also, probably by adding hydrophobic interactions to the metal liganding. Application buffers typically are at pH 6–8, not containing complexing agents (and not using histidine or imidazole as buffer!), with $1\,M$ NaCl. Elution is achieved by one or both of two principles. The proteins can be displaced by a stronger complexing agent, such as imidazole or EDTA. Alternatively, by lowering the pH of the buffer, the histidine residues on the protein become protonated, and so are unable to coordinate with the metal ions. A typical elution procedure would be to form a gradient with the starting buffer and a lower pH buffer containing complexing agent. Alternatively, stepwise elution changing one or both parameters successively may be better. A final cleanup of the column to remove strongly bound proteins may require stripping the metal ions out with strong EDTA solution.

The choice of metal ion is critical in obtaining suitable binding and selectivity. The most commonly used metals have been Zn^{2+}, Ni^{2+}, Cu^{2+}, and, more recently, Fe^{3+}. The latter has been found selective for phosphorylated proteins, and so may be used to pull out important enzymes that are regulated by phosphorylation, and to separate phospho- and unphosphorylated forms of the same protein [135,136]. With a new protein purification project there are few rules to go by; it will be necessary to do trials on a number of different metals, and possibly different chelating matrices, in order to find the ideal one. It should also be remembered that many metals inactivate proteins, especially Cu^{2+}, which reacts with cysteines. Finally, it should be stated that immobilized-metal-ion chromatography *is not necessarily* useful for purification of metalloproteins. This is a common misconception; the metalloprotein already has the metal ion attached to it, and so will not have any specific affinity for IMAC columns. On the other hand, the apo-protein may have a strong affinity. So if the metal ion can first be removed by treating with a complexing agent, IMAC may be an ideal (and more truly affinity) method. After removing excess complexing agent, the fraction is applied to the IMAC column charged with the appropriate metal. The apo-protein *may* bind, if there are sufficient coordination sites available around the metal. It may just pull the metal off the chelating matrix, and not be retained on the column. If it binds, it may be necessary to strip all metal off the column with EDTA to get the protein off. These approaches are worth considering but, in general, IMAC is not used specifically for metalloproteins.

6.6   Miscellaneous Adsorbents

Cationic Polymer–Nucleic Acid Complexes as Batch Adsorbents

The first stage in cleaning up extracts of microorganisms and other fast-proliferating tissues which contain large quantities of nucleic acids, can be to clarify the extract by precipitation with a cationic polymer such as polyethylene imine, or natural histones such as protamine sulfate (see Section 2.3). The process removes much undesirable material from the extract, including all the nucleic acids. The resultant precipitate usually contains some proteins which aggregate readily, but also many nucleic acid-binding proteins, which bind to the nucleic acid that is precipitated by the cationic polymer. So, sometimes, this clarification step results in a considerable decrease in a particular enzyme activity; perhaps all is lost in the precipitate. As long as the enzyme (or other nonenzymic protein) remains bioactive, it can be recovered from the precipitate, and so a substantial purification is achieved compared with the original untreated extract. For instance, *Escherichia coli* α-ketoglutarate dehydrogenase adsorbs to the protamine–nucleic acid precipitate, and can be extracted from it using $0.1\,M$ phosphate, pH 7 [137].

The most extensively used example of this technique is for the purification of nucleic acid-binding proteins such as RNA polymerases, DNA ligases, and restriction enzymes, as well as many other enzymes that do not specifically bind to DNA or RNA. Very acidic proteins might be expected to precipitate by forming complexes with the polyethylene imine in much the same way as the nucleic acids do. A number of examples have been described [138]. The precipitate formed is highly dependent on ionic strength, since it is the ionic attraction between the cationic polymer and the nucleic acids that results in the aggregation. Thus, for reproducible results, the salt concentration and pH in the starting extract must be carefully defined. The amount of polyethylene imine used is up to 0.4% final concentration; more than this will remain in the supernatant and possibly affect any further fractionations of the supernatant that may be carried out. The amount of precipitate decreases with ionic strength, and at 0.5 M NaCl, not all the nucleic acid can be precipitated. The precipitate is solubilized by successive extractions with increasing salt concentrations, which release different enzymes. For instance, many RNA polymerases may be retained in the precipitate at 0.2 M NaCl, but are eluted by 0.5–1 M NaCl [139]. The solubilized enzyme contains nucleic acid as well, and, on dilution of the salt, the complex reprecipitates. So the next step must be carried out at high salt concentration; hydrophobic chromatography is appropriate, as the protein can bind and allow the nucleic acids to pass through.

Thiophilic Adsorbents

Of the many methods of activating matrices for attachment of hydrophobic chains or of affinity ligands, divinyl sulfone has been used (see Chapter 7). One of the first steps in preparing any such adsorbent is to block any unreacted groups with either ethanolamine or β-mercaptoethanol. In addition, it is desirable to have a "control" adsorbent to which the ligand has not been attached. Thus, the control consists of vinyl sulfone-linked mercaptoethanol. It was found that this was not a neutral control; in the normal conditions of salt-promoted hydrophobic buffers containing 0.5 M Na$_2$SO$_4$, a few proteins bound to it strongly; principal among these are γ-globulins [140,141,142]. The structure is not apparently hydrophobic; in fact there is another type of interaction coming into play, which has been termed "thiophilic." The requirements for a thiophilic adsorbent are that it has a nucleophilic atom spaced three away from the sulfone group. The most effective is sulfur, as in the mercaptoethanol derivative illustrated in Figure 5.19, but unprotonated nitrogen (at high pH) and, weakly, oxygen also show thiophilic effects. The sulfone group is the key to the interaction in these ligands, but other adsorbents have been reported to exhibit thiophilic effects, including a pyridyl–disulfide linkage, in which the disulfide seems to take the place of the sulfone [143]. The original mercaptoethanol-linked adsorbent was named

"T-gel," and has been used for isolation of γ-globulins from serum; it is also useful for isolation of monoclonal antibodies (Section 9.3).

Since the discovery of T-gel, which was both interesting from a theoretical point of view, and useful commercially, there have been a number of further developments. Alternative thiophilic adsorbents which are still more specific for γ-globulins have been investigated [144,145]. In addition, the buffer conditions for binding have been researched, and it was found that the requirement for salt was not absolute. In fact, although the presence of water-structuring salts such as sulfates did increase the binding of γ-globulins, chlorides decreased the binding. Adsorption in the presence of $0.5 M$ sulfate salt, followed by elution in the presence of $0.5 M$ chloride was the best protocol. Thus, thiophilic adsorption is quite different from true hydrophobic adsorption, although it can be classed along with hydrophobics as "salt-promoted" adsorption.

Mixed-Function Adsorbents

A variety of adsorbents which bind proteins by more than one predominant mechanism have been described. Some of these act in a pseudo-biospecific fashion, e.g., dyes, and will be described in the next chapter. Others do not obviously interact in any biospecific mode, but selectively bind proteins, often in an unpredictable way. Some of the first such materials were the (intended) dipolar adsorbents, which selectively bound some blood plasma fractions [146]. Sulfanilic acid coupled to epichlohydrin-activated agarose produced the adsorbent illustrated in Figure 6.17, which is zwitterionic-dipolar. However, the low basicity of the aromatic nitrogen is such that it would have been uncharged at the pH used in the chromatography. Other dipolar/hydrophobic adsorbents have been used for specific purifications [147,148].

There is an infinite range of possible ligand structures that can be built on to an adsorbent, which potentially operate in several modes, such as ion exchange, hydrophobic, hydrogen bonding, and charge transfer interactions. We have investigated many such adsorbents, and because of

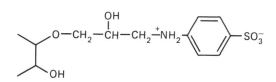

Figure 6.17. Dipolar "ion exchange" adsorbent made by coupling sulfanilic acid to epoxy-activated agarose. As the aromatic nitrogen is a very weak base, it is likely that at pHs much above 5 it would not in fact be protonated. (From [144].)

the variety, it is difficult to generalize on their behavior ([129] and unpublished work). Completely hydrophilic ligands mostly only bind proteins by ion exchange—if they have a net charge. But introduction of hydrophobicity results in many unexpected effects. Some ligands will bind proteins only in the salt-promoted mode (i.e., at $0.5\,M$ Na_2SO_4). Others bind some proteins at low salt concentrations, releasing them at higher ones, and bind a different batch of proteins at high salt concentrations, releasing them at low salt! It has been very useful to have a wide range of such adsorbents to screen for optimum selective binding of a particular protein, usually in the salt-promoted mode. Using such a screening process, we discovered the completely unexpected interaction between a thiamine ligand and an FAD-containing NADH oxidase enzyme, enabling a high degree of purification in one step [149]. Further description of screening multifunctional adsorbents is given in Section 7.3, dealing with dye ligands.

Chapter 7
Separation by Adsorption— Affinity Techniques

7.1 Principles of Affinity Chromatography

Affinity chromatography, developed during the 1960s and 1970s, referred originally to the use of an immobilized natural ligand, which specifically interacts with the desired protein. The ligand is immobilized on suitable particles which can be packed into a column. Then a sample containing the protein is passed into the column, and the specific interaction holds back the desired protein, while others pass through (Figure 7.1). "Affinity" techniques in general make use of a biospecific interaction that results in a change in properties of the protein such that it can be separated from other proteins. Not only can it apply to column chromatography, but also to precipitation, electrophoresis, liquid-phase partitioning, and other separation methods. It should be noted that many workers have used the words "affinity chromatography" to describe processes that do *not* involve biospecific interactions. "Affinity" is then being used in its more general sense of "attraction"; so we have affinity chromatography on adsorbents such as dyes, immobilized metals, and mixed-function ligands, which do not necessarily interact at natural ligands' binding sites. Used in this way, the word "affinity" becomes a tautology, since the word "chromatography" is sufficient to imply that an attraction is occurring. For the purposes of this book, we will reserve the use of the word "affinity" to interactions that are biologically significant, i.e., between a protein's natural ligand binding site and the actual ligand itself, or with a competitive binding inhibitor that may or may not be natural. There are other terminologies within this definition, such as "pseudo-affinity," and "biomimetic," which describe the interaction at the protein's binding site with an unnatural compound, and "bioaffinity" which is the natural interaction.

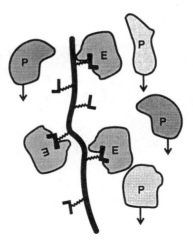

Figure 7.1. Basic principle of affinity adsorption chromatography. A ligand L is covalently attached to the backbone matrix. Only enzymes E with a specific affinity for L bind to the adsorbent. Proteins P pass through unaffected.

Most affinity procedures are chromatographic in the sense that an affinity adsorbent is packed in a column and the separation takes place in the column. Other affinity techniques include precipitation (Chapter 4) and methods in which the proteins remain in solution (Chapter 8). The present chapter will deal with column chromatographic affinity methods.

Synthesis of Affinity Adsorbents

Few people these days synthesize their own affinity adsorbents from scratch. Either they use standard affinity ligands fully prepared, or they purchase activated matrix to which their own ligand can be coupled directly. Nevertheless, there are times when the right materials are not immediately available, and a knowledge of the basic chemistry of synthesizing an adsorbent can save both time and money. First, let us consider the matrix itself. As with all protein adsorbents, it must be porous to maximize effective surface area, and rigid enough to allow good flow rates. The latter point is not so important when dealing with ligands that are proteins themselves (especially antibody–antigen interactions), since the relatively slow protein-protein interaction requires slow flow rates to allow adsorption and elution processes to be established. Most affinity chromatography matrices are agarose-based, but a number of synthetic organic and inorganic porous bead matrices are also available [150].

The next consideration is the chemistry of attachment for the ligand. There are many specific chemistries which will only briefly be touched on here; publications exclusively concerned with affinity techniques cover the subject in greater detail [151,152,153]. The main requirements for a successful affinity adsorbent are:

1. The ligand should be attached to the matrix in such a way that the ligand's binding to the protein concerned is not seriously disturbed;
2. A "spacer arm" setting the ligand away from the matrix should be used to make it more accessible to the protein (or for other reasons; see below);
3. Nonspecific interactions should not be so great that many other proteins are adsorbed in addition to the one required;
4. The linkages should be stable to the likely conditions to be used during the chromatography, including "cleaning-up" procedures before reuse.

Methods for synthesizing an affinity adsorbent are varied and dependent on the chemistry of the ligand itself, and whether a spacer arm is required. The matrix must be "activated" in such a way that quite gentle chemistry allows covalent attachment of spacer arm and/or ligand. (There are some processes in which the ligand is reactive enough to attach itself without matrix activation, e.g., dyes, Section 7.3.) The activated matrix is, in most examples, an electrophile that will react readily with amines. Some activation processes introduce a sufficiently long group to act as spacer arm; otherwise, it will be the spacer that reacts directly with the activated matrix. In the latter case, there may be the problem that the end of the spacer arm is not reactive enough to link up with the ligand. In that case, it may be necessary to synthesize a ligand–spacer arm complex first, which is reacted with the matrix (Figure 7.2).

The chemistry of the activation process involves the reaction of hydroxyl groups on the (usually) agarose matrix, and the most commonly used reaction has been with cyanogen bromide. The original method [154] has been superseded by a simpler technique in which the reaction is complete in a few minutes [155]. Cyanogen bromide-activated agarose is available commercially and has been widely used. The reactions of cyanogen bromide are complex; both (i) cyclic (Figure 7.3) and (ii) acyclic imidocarbonates are formed, with (iii) some carbamates and (iv) carbonates as side products, but the principal reactive component has been demonstrated to be (v) the cyanate ester [156]. Indeed, the imidocarbonates, being unstable in acid, can be destroyed by an acid treatment before the material is used. Cyanogen bromide-activated agarose reacts swiftly in weakly alkaline conditions (pH 9–10) with primary amines to give principally the isourea derivative (Figure 7.4). Some simultaneous aqueous hydrolysis gives carbamates (iii) and carbonates (iv), as in Figure 7.3.

The isourea substituent is positively charged at neutral pH, which can influence the behavior of affinity adsorbents by introducing an element of

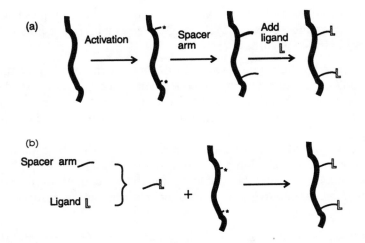

Figure 7.2. Alternative approaches to synthesizing an affinity adsorbent. In (a), the spacer arm is attached to the matrix, then the ligand is covalently attached. In (b) the spacer arm is attached to the ligand first, and this is then reacted to link up with the matrix.

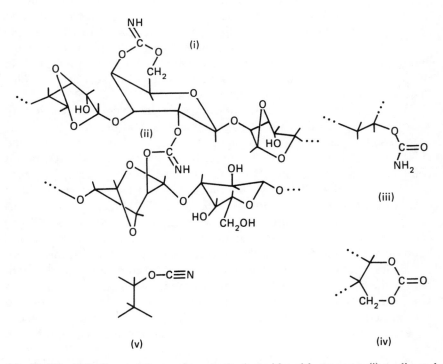

Figure 7.3. Reaction products of cyanogen bromide with agarose: (i) cyclic and (ii) acyclic imidocarbonates, (iii) carbamate, (iv) carbonate, (v) cyanate ester.

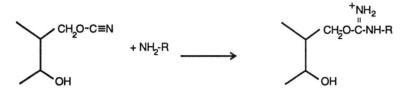

Figure 7.4. Reaction of cyanate ester (Figure 5.19) with amino compound to give isourea derivatives.

anion exchange character. On the other hand, many of the ligands that are attached are negatively charged, so the isourea derivative may cancel out possible cation exchange effects.

Although cyanogen bromide activation is still widely used, especially for attachment of proteins through ε-lysine groups, other more convenient and efficient methods are now available and should eventually supplant the cyanogen bromide procedure. I shall describe a selection of methods, but this is not comprehensive. The first of these is "epoxy activation," in which an epoxy group is introduced as the active electrophile. There are two reagents; one is epichlorohydrin which results in a three-carbon propane 2-ol spacer arm after reaction with the ligand [157]. The other is 1,4-butanediol diglycidyl ether (a bisoxirane) which results in a 12-atom spacer arm (Figure 7.5) [158]. Although the latter result is preferable, the reagent must be used in much greater amounts as it is less reactive; consequently, the bisoxirane activation is a more expensive process.

Reaction of a primary amine ligand with epoxy-activated matrices forms a secondary amine linkage which is very stable, but being protonated at pHs below 8, will introduce a positive charge into the linkage. It is often necessary to synthesize a natural ligand analog containing an amine for coupling, in which case this positive charge is an unnatural feature. Epoxy groups also react rapidly with sulfhydryls (as is the case with all the activation methods), but also slowly with hydroxyls. Consequently, it is possible, at high pH, to couple a hydroxyl-containing ligand with an epoxy-activated matrix, resulting in a stable, uncharged ether linkage (Figure 7.6). An oxygen linkage is the most desirable, but is the

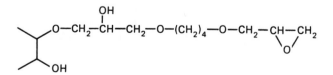

Figure 7.5. Bisoxirane-activated (epoxy-activated) agarose structures.

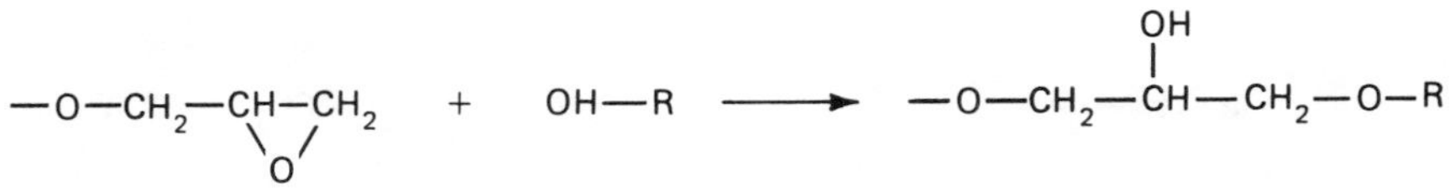

$$\text{—O—CH}_2\text{—CH—CH}_2 \quad + \quad \text{OH—R} \quad \longrightarrow \quad \text{—O—CH}_2\text{—CH—CH}_2\text{—O—R}$$

Figure 7.6. Reaction of epoxy-activated agarose with hydroxyl residue.

slowest and most difficult to "direct" properly, and requires more extreme conditions. A sulfur (thioether) linkage is the fastest, but requires a suitable ligand with a thiol, and this linkage is unstable in alkali. The amine linkage through nitrogen is the most common and easiest approach, but results in the introduction of a positive charge.

The next method involves activation with 1,1'-carbonyldiimidazole, which gives a product more reactive than epoxy, about on a par with cyanogen bromide activation. Coupling (again to a primary amine) does not leave a positive charge, but a urethane linkage with minimal spacing from the matrix ([159]; Figure 7.7). The reagent is also less noxious than either cyanogen bromide or epoxides.

A method that does not introduce any spacer at all uses toluene sulfonyl chloride (tosyl chloride) [160], or the more reactive 3,3,3-trifluoroethanesulfonyl chloride (tresyl chloride) [161]. The tosyl/tresyl group introduced on activation reacts smoothly and rapidly to give the very stable secondary amine from the primary amine ligand (Figure 7.8). At a neutral pH, the secondary amine group is charged, so the ultimate behavior is very similar to that of cyanogen bromide activation, which also barely adds any spacer and introduces a positive charge. But the

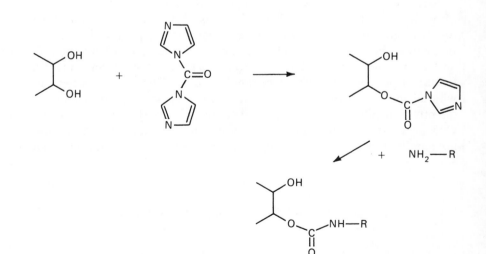

Figure 7.7. Activation and reaction of carbonyldiimidazole-activated agarose.

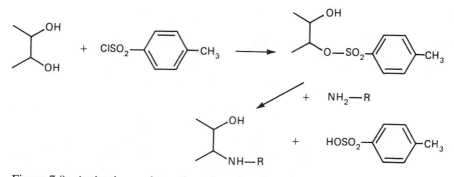

Figure 7.8. Activation and reaction of tosyl-activated agarose.

secondary amine linkage is much more stable than the isourea produced by cyanogen bromide.

Another very useful chemistry involves the reaction of divinyl sulfone (vinyl sulfone) with the matrix to yield a vinyl sulfone-activated form. This also reacts with amino and thiol groups and, slowly, with hydroxyls [162,140]; (Figure 7.9). Vinyl sulfone-activated gels are useful for the synthesis of thiophilic adsorbents (Section 6.6), and for coupling of proteins, especially immunoglobulins, to the matrix.

Other chemistries include the introduction of an N-hydroxysuccinimide ester as a leaving group to react with primary amines forming an amide linkage, and the simpler introduction of a carboxylate which can be coupled to the amine in the ligand using a carbodiimide reagent. This also forms an amide linkage, and can be reversed so that the carboxylate is in the ligand, and the amine on the matrix. Amides are alkaline unstable, but otherwise the linkage is satisfactory; it is uncharged.

In the examples given above, there are two types of activators, the monofunctional (cyanogen bromide, tosyl/tresyl chlorides), and the bifunctional which react with the agarose at one end, the other remaining for reaction with the ligand. With the bifunctional reagents epichlorohydrin and divinyl sulfone, the first reaction is faster; once one end of the reagent has reacted, the other end becomes less reactive. Nevertheless, there is opportunity for cross-linking of the agarose, and in fact, epichlorohydrin is purposely used for this. The amount of coupled active groups increases with time to a maximum, then declines as cross-linking exceeds new substitution. Preparation of these materials must take this into account.

With a few exceptions, the activation methods require a nucleophilic group in the ligand. Mostly, this is a primary amine (some secondary amines can also react); after coupling, some chemistries leave a protonatable nitrogen, some do not. Some chemistries leave a sufficiently long spacer such that no further addition is needed, whereas with others a spacer arm inclusion is desirable. A summary of some of the methods is given in Table 7.1.

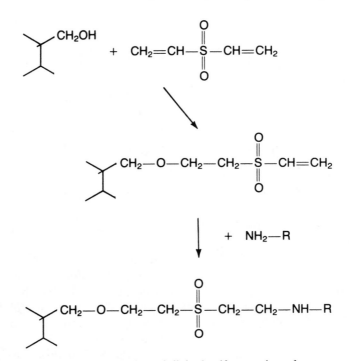

Figure 7.9. Activation and reaction of divinyl sulfone-activated agarose.

If a spacer arm additional to that provided by the activation procedure is required, then the final chemistry of linking the ligand must be considered; it is not possible to couple an amine spacer and expect it to be highly reactive at the other end. As indicated above (Figure 7.2), it may be best to construct a ligand analog including a spacer ending with an amine which can attach directly to the activated matrix. Alternatively, the spacer arm can be ε-amino caproate, leaving a carboxylate which can be linked to an amine in the ligand with carbodiimide. Carbodiimides are very reactive compounds which can react with other portions of complex ligands.

Most of the early work on affinity chromatography was concerned with the purification of enzymes using their substrates, or competitive inhibitors as affinity ligands. More recently, protein ligands have become commonplace; some people are unaware that an affinity chromatography column does not *always* have attached proteins. Some of the commonest uses currently are with the adsorbents Proteins A and G (see Section 9.3) for purifying antibodies, or with antibodies themselves attached for purifying the antigens, and lectins such as concanavalin A for purifying glycoproteins. All these are true biospecific interactions between two different proteins.

Table 7.1. Summary of Common Methods for Activation of Matrices

Method	Length of spacer introduced (atoms)	Alkali lability[a]	Protonated nitrogen[a]
Cyanogen bromide	1	Yes	Yes
Epichlorohydrin	3	No	Yes
Bisoxirane	11	No	Yes
Divinyl sulfone	5	Yes	Yes
Carbonyldiimidazole	1	Yes	No
N-hydroxysuccinimide	8	Yes	No
Tosyl/tresyl chloride	0	No	Yes

[a] lability and protonation of a coupled amine–ligand

When coupling proteins to activated matrices, it is important that the conditions do not cause denaturation or other inactivation of the protein. It is normally assumed that coupling occurs through externally exposed lysine residues; as with any other amine, for reaction to take place it must be nucleophilic, i.e., unprotonated. Thus, the pH must be high enough for a significant number of the lysines to be at least partly unprotonated, but not so high that the protein is inactivated. Generally, a pH around 9 is most suitable, but reaction at this pH is slow with some activated gels, e.g., epoxy-activated. The latter prefer heating to 50–60°C, but this is likely to denature the protein. Cyanogen bromide activation has been widely used, as it is among the more reactive forms, and will couple proteins in a few hours at pH 8–9. But leakage of protein from the final product has always been a problem. Vinyl sulfone-activated gels are fast acting in mild conditions, as are the tosyl- and tresyl-activated gels, and these are recommended for protein coupling. Never use a buffer that contains a primary amine, or else that is all that will couple! So Tris buffer must be avoided; normally, a $0.1–0.2\,M$ bicarbonate/carbonate buffer is used, or borate, as neither of these can react with the activated groups (this applies to *all* coupling of ligands, not just proteins).

Many affinity adsorbent matrices, activated, with spacer arms or complete, are available commercially. The complete adsorbents mostly contain ligands that are not specific for one protein, but are general ligands that bind to many different proteins. For example, nucleotide ligands have been extensively investigated since so many enzymes use nucleotides as cofactors or substrates. Thus, although the specificity of these adsorbents is far from absolute, they are generally useful, being applicable to a range of different enzymes. It has been stated that about 30% of all known enzymes utilize one of four nucleotide cofactors [163]; if one includes nucleic acids as substrates, and enzymes that bind nucleotides as activators or inhibitors, the total exceeds 50%. This does not necessarily mean that 50% of the protein in a given extract consists of these enzymes, nor that many of them will necessarily bind to a particular nucleotide affinity adsorbent in specified conditions.

Application of Chromatographic Theory to Affinity Adsorbents

The theory developed in Section 5.1 can easily be applied to affinity chromatography, with some interesting conclusions. Since the value of K_p, the dissociation constant between adsorbent and protein, should be similar or related to the dissociation constant between protein and free ligand, it is possible to calculate directly what the partition coefficient ought to be. But one of the drawbacks of the affinity adsorbents is that, although a fairly high substitution of ligand can be achieved ($1-10\,\mu\text{mol cm}^{-3}$ for small molecules), only a small proportion of the immobilized ligand seems to be oriented correctly even when spacer arms are used. Typically, only a few per cent of potential sites may actually bind the protein. Capacities of $2-4\,\text{mg cm}^{-3}$ of the desired protein are regarded as good, which for a protein of MW 100,000 corresponds to an m_t value of 0.02 to $0.04\,\text{m}M$. If we let m_t be 0.04 and p_t be 0.02 in Eq. (5.11), then α values can be calculated for various K_p values, as shown in Table 7.2.

The rather surprising result is that, for good adsorption ($\alpha > 0.9$), K_p has to be less than $0.003\,\text{m}M$, or $3 \times 10^{-6}\,M$, smaller than most protein–ligand dissociation constants. Considering that the specific interaction with an immobilized ligand on a column is likely to be weaker than with the free ligand, one may wonder how affinity adsorption chromatography ever works.

The answer to this dilemma is found in the extensive search for suitable methods of attaching the ligands to the supporting matrix. Direct attachment was rarely satisfactory, but use of a spacer arm, usually hexamethylene, gave good adsorption. A continuing problem was nonspecific hydrophobic interaction with the hexamethylene chain causing unwanted

Table 7.2. α Values Using $m_t = 0.04\,\text{m}M$, $p_t = 0.02\,\text{m}M$

K_p (mM)	α
1.0	0.04
0.3	0.11
0.1	0.26
0.03	0.50
0.01	0.72
0.003	0.88
0.001	0.95
0.0003	0.985
0.0001	0.995

Table 7.3. α Values in the Presence and the Absence
of Free Ligand

K_p (mM)	K_L (mM)	K_N (mM)	α in absence of free ligand	α in presence of free ligand
0.001	0.1	10	0.95	0.004
0.001	1.0	1.0	0.95	0.04
0.001	10	0.1	0.95	0.26
0.0001	0.1	10	0.995	0.04
0.0001	1.0	1.0	0.995	0.26
0.0001	10	0.1	0.995	0.72

protein to adsorb to this spacer arm. Since the adsorption is hydrophobic in character, introduction of hydrophilic groups into the spacer arm (carbonyls, amides) should lessen the interactions and so make the adsorbent more specific for the protein for which it was designed. Unfortunately, these chemically highly sophisticated adsorbents often proved to have little attraction for the desired protein or anything else [164,165]. As it turned out, the hydrophobic interactions were a necessary part of the binding to the adsorbent. Indeed, some of the original attempts at making affinity adsorbents without spacer arms may have failed not because the closeness of attachment of the ligand to the agarose backbone caused steric hindrance, but because some additional binding forces, provided by a hydrophobic spacer arm, were needed. Some thermodynamic calculations show the magnitude of additional forces that are required.

The energy of interaction $\Delta G^0 = -\text{RT}\ln K_p$ can be considered to be made up of the specific interaction between the protein and ligand, ΔG_L^0, and nonspecific interactions ΔG_N^0.

If $K_p = 0.001\,\text{mM}$, then $\Delta G^0 = \Delta G_L^0 + \Delta G_N^0 = 34.5\,\text{kJ}\,\text{mol}^{-1}$.
If K_L, the specific dissociation constant, is $0.1\,\text{mM}$, then
$\Delta G_L^0 = 23\,\text{kJ}\,\text{mol}^{-1}$ and $\Delta G_N^0 = 11.5\,\text{kJ}\,\text{mol}^{-1}$.

If in this case $m_t = 0.04\,\text{mM}$, and $p_t = 0.02\,\text{mM}$, α can be calculated as 0.95.

Suppose now that a free ligand is introduced into the buffer, and this completely displaces all biospecific interactions, as the protein binds to it rather than the immobilized ligand. Then ΔG_L^0 becomes zero, and only the nonspecific forces remain. These amount to only $11.5\,\text{kJ}\,\text{mol}^{-1}$, which is far too low to cause any significant retention on the column; α becomes effectively zero.

If we repeat these calculations with alternative parameters, we get the results shown in Table 7.3.

It can be seen that even weak nonspecific interactions are sufficient to add to the specific ones to create quite strong binding overall. More significantly, displacement by soluble ligand reduces the interactions so that the protein is rapidly eluted (affinity elution; see Section 7.4).

As an example, the ligand 5'-AMP is commonly used to purify NAD-linked dehydrogenases. The ligand is attached through a hexamethylene spacer arm which provides interactions to add to the interaction between the AMP and the NAD-binding site, making a sufficient total for tight binding. The K_i value for 5'-AMP for various dehydrogenases is in the range 1–10 mM, which is far too weak in itself. Similarly, the nonspecific interactions are too weak in themselves, so displacement with NAD is effective.

A convenient affinity adsorbent for hexokinases can be made by coupling glucosamine to agarose through an alkyl chain [166]. N-aminoacyl glucosamines (in free solution) were found to have K_i values greater than for glucosamine itself when the number of C atoms in the aminoacyl chain was 6 or less, indicating disturbance of binding caused by the acyl chain, but when 8 C atoms were present, the K_i value was much lower, i.e., some enzyme–acyl chain interaction was occurring. Whereas yeast hexokinase would not bind to agarose-C_6-acylglucosamine, it was retarded on a column of agarose-C_8-acylglucosamine, and could be eluted by glucose. The nonspecific interaction was needed for binding to the adsorbent. Mammalian hexokinase isoenzymes showed selective binding to these adsorbents; types II and III could be chromatographed on adsorbents with C_6-spacer arms, whereas the kidney type I isoenzyme needed C_8- for retardation.

The calculations made in Tables 7.2 and 7.3 are based on two assumptions that may not be justified. The first concerns the possible heterogeneity of distribution of attached ligands; the m_t value may be much higher in restricted portions of the adsorbent, so that much more protein can bind even with weak binding constants (e.g., $K_p \sim 10^{-4} M$). An investigation of this possibility utilized spin-labeled model ligands to determine distribution [167]. It was found that inhomogeneity was not extensive. However, it was revealed (as can be calculated from substitution levels) that the closeness of the attached ligands permitted multiple attachment to an average-sized protein, if the protein had more than one binding site. This leads to the second assumption above—single-site attachment may not be the whole story; the low occupancy may in some cases be due to the requirement for just the correct spacing between adjacent ligands to allow two-site attachment (Figure 7.10). This could not of course apply to monovalent (single active site) enzymes. For the same reasons, single weak binding, as for example the case of lactate dehydrogenase to 5'-AMP above, may be greatly strengthened by multipoint attachment, without the need to invoke nonspecific interactions (but see [165]).

In conclusion, it can be seen that affinity chromatography needs a total energy of interaction between protein and matrix of at least 30 kJ mol^{-1}, which can rarely be supplied by a single protein–ligand interaction. Nonspecific interactions may amount to half or more of this value, due to

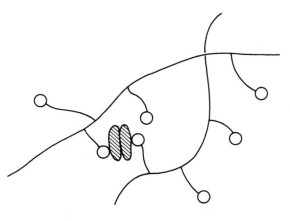

Figure 7.10. Exact spatial requirement for two-point specific attachment to dimeric enzyme on an affinity adsorbent (enzyme *shaded*).

mainly hydrophobic forces between surface residues on the protein and the spacer arm. Multipoint specific attachments can also assist. Affinity elution, in which free ligand is added to the buffer to displace the protein (see Section 7.4) can be very effective, and quite weak specific interactions are adequate. The low-percentage occupancy of affinity adsorbents may in some cases be explained by the requirement for two-point biospecific attachment (Figure 7.10), but is more likely to be due to a requirement for nonspecific interactions in addition to biospecific binding. If these two attachments have rigorous spatial requirements, relatively few ligands will be correctly oriented to provide a protein binding site (Figure 7.11).

In view of the requirement for nonspecific interactions, it is interesting to speculate on the possibility of purposely introducing them, by either using a matrix which is more hydrophobic or including a high density of

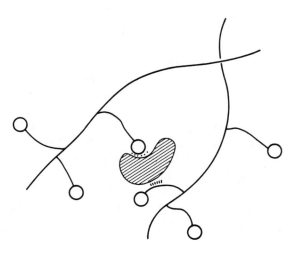

Figure 7.11. Exact spatial requirement for both specific and nonspecific attachment on an affinity adsorbent (enzyme *shaded*).

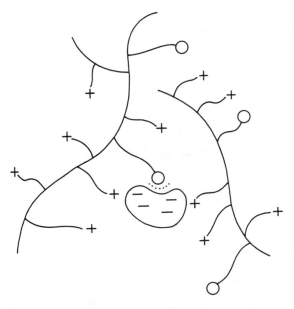

Figure 7.12. Combined affinity and ion-exchange adsorption for negatively charged proteins.

other attracting groups, e.g., DEAE- (Figure 7.12). The latter type of adsorbent would be operated in conditions where the enzyme would be hardly bound at all on a simple DEAE-ion-exchange column (α about 0.1–0.4), but the ion exchange interaction would be a sufficient addition to the specific interaction to achieve α 0.99 or higher on the combined adsorbent. Affinity elution would then work well, and the capacity for the enzyme in question would be high. This sort of effect is undoubtedly occurring with many dye ligand adsorbents, which can also act as cation exchangers (see Section 7.3).

General Techniques and Procedures in Affinity Adsorption Chromatography

In the early 1970s when affinity chromatography was proving highly successful in many instances, the feelings among protein purifiers, and particularly those dealing with enzymes were, "Why did we not think of it before?" and "This will make all other methods obsolete." The general euphoria dissipated after many people had spent months, sometimes years, trying to develop a suitable adsorbent for their particular enzyme and often not succeeding. For some applications, especially when the enzyme has a unique substrate, and it makes up only a fraction of the total protein present, once a suitable adsorbent has been found, there could never be any questions of not using the technique. One-step purifications of 1,000-fold with nearly 100% recovery have been reported; the ease and speed of the method has revolutionized, indeed made possible, the isolation of a large number of important proteins occurring in small amounts. But for every success story there is at least one complete failure and a lot of wasted effort. The main problems have been the low capacity of the adsorbents (though newer methods of attaching ligands are improving this factor) and nonspecific adsorption of other proteins. But as we have seen, some nonspecific interactions are a necessary evil if the desired enzyme is to adsorb, so very careful trial-and-error designs of adsorbent and buffer conditions are needed to reach the ideal conditions.

To design an adsorbent we must first consider the ligands and their likely mode of interaction with the protein. If possible, the attachment should be at a point that is not involved in the protein/ligand interaction; some knowledge of substrate analogues and the K_i values can help here. If there is a choice of two ligands, a large one and a small one, choose the large one, as there is more opportunity for varying the mode of attachment to the matrix. A small molecule such as alanine is unlikely to be attached successfully, since an enzyme recognizing alanine would almost certainly make use of both its charged groups and would recognize the methyl as being distinct from other amino acids. Thus, the chances of the methods of attachment illustrated in Figure 7.13 *not* destroying the enzyme's ability to recognize the molecule are slim. On the other hand, adenine nucleotides have been successfully attached to agarose through a number of positions (Figure 7.14).

The starting material can be unactivated matrix, such as agarose, or preactivated (e.g., CNBr, epoxy, carboxyldiimidazole, *N*-hydroxysuccinimidyl, etc.), or with a hexyl side-chain terminating in a carboxyl or amino group. Reactions with excess of ligand or analog may take place in organic or aqueous solvents, depending on the chemistry involved. If 100% organic solvent is needed, then a careful dehydration and rehydration of biopolymers such as agarose is required. After carrying

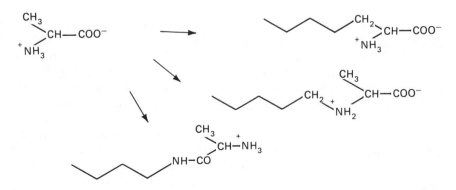

Figure 7.13. Possible ways of immobilizing alanine as a ligand for an affinity adsorbent.

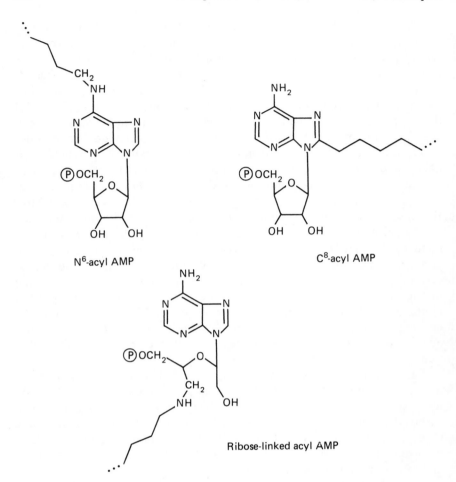

Figure 7.14. Three ways of attaching AMP as an affinity adsorbent.

out the chemistry, an estimation should be made to determine the quantity of ligand that has been immobilized.

Trials of the effectiveness of the adsorbent can be carried out in a Pasteur pipette or a small 1- to 2-ml disposable column; a small sample of the protein fraction in a suitable buffer is applied, and the column washed. If the desired protein sticks under these conditions, one can assume that its α value is greater than 0.9 and the adsorption has been achieved rapidly: a Pasteur pipette or small open column packed with affinity adsorbent has a flow rate rather higher than optimal. The buffer used can be important. Many affinity ligands are charged, and at low ionic strength will act as weak ion exchangers. To avoid binding unwanted proteins, the ionic strength should be reasonably high. But it is important that the biospecific interaction being exploited is not itself weakened by higher ionic strength or other features of the buffer conditions. It is unlikely that you would have much idea of how the K_L value changes with salt concentration or with pH, but these are points to consider if the protein does not seem to bind in some conditions. And if the buffer chosen is one that promotes hydrophobic interactions, which might be needed to strengthen the nonspecific forces, then other unwanted hydrophobic proteins might bind also.

Trials for elution can now be carried out. For elution there are several alternative methods. Affinity elution, by inclusion of free ligand in the buffer, is the ideal method, but can sometimes prove costly. Often, quite high concentrations of ligand have been used; this has been because the interaction with the column was so strong ($\alpha > 0.999$) in these particular buffer conditions that a large ligand concentration was needed. If the buffer is changed after application of the sample so as to reduce the α value, a much lower concentration of ligand can be used. The principles of affinity elution are described in Section 7.4 in more detail; they apply equally to affinity adsorbents, ion exchangers, or any other adsorbent. To achieve this lowering of α before affinity elution, some knowledge of the interactions involved is needed. Increasing salt concentration may well decrease the biospecific component of binding, so it may also decrease the subsequent interaction with free ligand in affinity elution (increase the value of K_L). Increased salt would minimize any nonspecific ionic inter-actions, but it would also increase hydrophobic interactions, especially with the spacer arm. If the latter are very important, *decreasing* salt concentration may be more effective. Alternatively, introduction of a surface tension-reducing agent to lessen hydrophobic interactions may be used. This could be a nonionic detergent such as Triton X-100, ethylene glycol, or isopropanol (see Section 6.4). These are unlikely to affect the value of K_L with charged ligands. After a prewash with this α-lowering buffer, ligand is introduced in the same buffer for the affinity elution process.

A second method for eluting the protein is to use a high salt concen-tration, which is expected to disrupt all nonhydrophobic interactions between immobilized ligand and protein, including any biospecific ionic or other polar forces. But as increasing salt concentration increases hydrophobic forces, some of the nonspecific interactions may become stronger. Nevertheless, in most examples salt displaces the protein. In some cases, very tight binding occurs and neither free ligand nor salt will dislodge the protein (this is particularly true for immunoadsorbents; see Section 7.2). Chaotropic salts, such as thiocyanate at high concentration, or use of subdenaturing concentrations of urea can be successful in these cases. Alternatively, an extreme pH (but not extreme enough to denature the protein) may be best.

Other methods for elution include temperature changes, dilution of the buffer, even washing with water (decreasing temperature and ionic strength decrease hydrophobic interactions, see Section 6.4), and electrophoresis to remove charged proteins from the column [168].

Many ready-made adsorbents are available commercially, and are suitable for a large range of proteins. Because of the low capacity and expense of the adsorbents, the method is most useful for purifying proteins that make up only a small proportion of the total protein present. In some cases it can be used as the first (and sometimes only) step in purification. On the other hand, for dealing with large quantities of the required protein, other methods may be more suitable as preliminary steps in the fractionation procedure, with an affinity procedure placed only near the end. The adsorbents can be reused many times, provided that they are cleaned immediately after use (e.g., high salt concentrations, denaturing agents, detergents, and alkali if sufficiently stable), and stored in the presence of bacteriocidal agents (azide, merthiolate, chlorbutol). Leakage of ligand due to the inherent instability of the isourea linkage formed from cyanogen bromide activation has been a problem. This can be overcome by using one of the other activation processes which gives a more stable linkage.

In view of the many excellent reviews and monographs on (biospecific) affinity chromatography [151,152,153], I have not attempted a more detailed coverage of this topic, nor to list comprehensibly the many different sorts of adsorbent that have been developed. However, the particular example of immunoadsorbents, in which the "ligand" is an antibody to the protein being purified, is dealt with separately below, as are a selection of commonly used affinity materials in Section 7.5, and the pseudo-affinity dyes in more detail in Section 7.3.

7.2   Immunoadsorbents

The ultimate adsorbent for any enzyme or protein is an immobilized antibody raised against that protein, and specifically interacting with a single surface feature of the protein, not necessarily a ligand binding site. Antibodies are not only very specific for the exact amino acid sequence and three-dimensional organization, but also have very high binding constants (low value of K_L) for their antigens. Indeed, too tight binding to immunoadsorbents is often a major problem. Either polyclonal or monoclonal antibodies may be used; each has advantages and disadvantages compared with the other. A considerable number of protein isolations involving immunoadsorbents have now been reported; reproducibility of the procedure from one laboratory to another depends on the availability of the same antibody preparation.

Basic Principles

1(a). The protein is purified by other techniques, from a source evolutionarily distinct from the immunization animal. For monoclonal antibodies the animal is normally a mouse; for polyclonal, a rabbit or goat. Or see 1(b).

1(b). The sequence of part of the protein is known, probably from gene isolation and sequencing. A peptide corresponding to about 15 amino acids from this sequence is synthesized chemically; it is convenient to add a cysteine to this sequence, a reactive residue that makes conjugation to bovine serum albumin or keyhole limpet hemocyanin easier. The conjugate is used as the antigen.

2. Raise antibodies to the antigen. For polyclonal antibodies, this is simply a matter of injecting suitably prepared sample into the animal at intervals, and testing its serum for the presence of antibodies (for details, see [169]). But it is essential that the antigen (i.e., the protein of interest) be as pure as possible; trace amounts of other proteins may be much more antigenic, resulting in a mixed specificity antibody product. For monoclonal antibodies, the purity of the antigen is relatively unimportant if the screening procedure to detect suitable clones uses a bioassay. Indeed, monoclonal antibodies have been isolated using antigens containing only about 1% of the desired protein, since the screening procedure disregards antibody clones specific to the other 99% of protein [170].

3. Partially purify antibody from rabbit serum (polyclonal) by precipitation (e.g., 33% saturation of ammonium sulfate or 5% wt/vol polyethylene glycol, or on a column such as T-Gel or Protein A [171], see Section 9.3). Better still, purify by "reverse" immunoaffinity chromatography, in which the antigen itself is immobilized on a column, and the serum is passed through; only the specific antibodies should bind.

4. Couple purified antibody to activated affinity adsorbent (cyanogen bromide-, tosyl-, vinylsulfone-activated, etc.) at pH 8 to 9 in bicarbonate/carbonate buffer.

5. Apply crude (or preferably partly purified) sample containing protein being isolated to the column. Wash extensively with suitable buffer to remove unbound material.

6. Elute desired protein from column.

The difficult steps are 1, 2 and 6. We will assume that step 1 has been achieved by alternative techniques in the first place.

Methods Using Polyclonal Antibodies

Polyclonal antibodies can be raised in a rabbit with less than 1 mg of pure protein in many cases; in larger animals such as a goat, not much more is required. Larger animals of course yield more antibody. Some proteins are far more antigenic than others; so much so that even traces of an impurity in the protein preparation, undetectable by normal analytical techniques (see Chapter 11), may also raise quite strong antibodies. Subsequently, these antibodies would then combine with their impurity antigen, which in a crude preparation would be present in larger proportions than in the purified protein. Consequently, the specificity of the antibody can be far poorer than expected. Such problems are particularly acute when the desired protein is a relatively poor antigen, as a result, for example, of being strongly conserved through evolution.

One quite successful technique that can overcome this problem is to carry out a gel-electrophoretic separation of the desired protein from other proteins [172]. The protein sample is run into a fairly large cross-section gel and electrophoresis continued until it is clearly separated from other impurities. After locating the protein, the gel is sliced and the section containing the protein is mashed to give a fine suspension of polyacrylamide gel plus protein. This is then used for the first injection; a few hundred micrograms of protein is sufficient. More detailed discussion of raising antibodies can be found in [173,169].

As much as 10% of the immunoglobulins raised in this way can consist of antibodies to the protein required, the remaining 90% being non-specific immunoglobulins that happened to be present in the animal at the time. Tens of milligrams of immunoglobulin can be attached per cm^3 of agarose [174]; up to 1 mg of specific antibody per cm^3 may be present. However, since the attachment of the antibody molecules to the adsorbent will in many cases be in a way that prevents it binding to its antigen, perhaps only about $0.1\,mg\,cm^{-3}$ is effective. If one of these antibodies binds one molecule of antigen of a similar molecular weight, then the capacity of such a column would be of the order of $0.1\,mg\,cm^{-3}$—not great, but nevertheless very useful for purifying small amounts of protein from a crude mixture. One rabbit may provide as much as 1–2 g of immunoglobulins. This might make several hundred cm^3 of adsorbent, sufficient to purify at least 10 mg of protein *each time it is used*. Thus the potential problem of not regaining the amount of protein used in making the antibody is not serious, especially if the column can be used more than once.

The final step, elution from the immunoadsorbent column, has often proved to be the stumbling block in the whole process. Antigen–antibody interactions characteristically have dissociation constants of the order 10^{-8}–$10^{-12}\,M$. Putting this value as K_p in Eq. (5.11) gives α values which are effectively unity. To reduce the specific interactions without denaturing either the antibody (so that the column can be used again) or the protein being isolated, may not be easy. A pH of 2–3 (typically a glycine-HCl buffer) is commonly used to displace the antigen without destroying the antibody, but if the antigen is an enzyme it will probably be inactivated

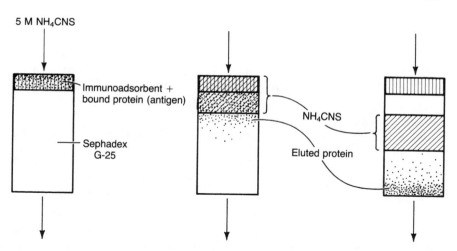

Figure 7.15. Removal of antigen from immunoadsorbent using chaotropic salt and rapid separation of salt from antigen (protein being purified).

by this low pH. Alternatively, partial disruption of H-bonding and hydrophobic interactions using urea or guanidine HCl, or the use of organic solvents, detergent, or ethylene glycol may be useful with sufficiently stable antigens. Any procedure can be used if denaturation of the protein being isolated does not matter. High salt concentration may be used if the interaction happens to be largely electrostatic, but conversely, it will strengthen any hydrophobic forces. Chaotropic salts such as thiocyanate or iodide, lithium bromide, etc., have been used with some success, but again, the concentration needed can cause denaturation of the displaced protein. A rapid elution procedure for obtaining active antigens after elution with a denaturing agent has been reported [175]. In this technique the adsorbent is placed on top of a desalting column, so that as soon as the antigen emerges from the adsorbent it separates from the eluting agent by gel filtration action (Figure 7.15). On a small scale, exposure to the denaturing eluting agent may be kept to only a few minutes.

There have been reports of successful elutions using *cold distilled water* [176,177]. The mode of action is a weakening of hydrophobic interactions, which are minimized at low ionic strength and low temperature. Many reported techniques may be suited only to the particular protein that they describe, but it is worth trying a number of different eluants that are known to work in specific cases. Quite mild eluant examples include $2\,M$ guanidine HCl, $0.6\,M$ KCNS, and dilute ammonia solutions (pH 10–11).

Another technique for displacing the antigen is electrophoresis [168]. At a pH where the antigen is sufficiently charged, electrophoretic mobilization of the small fraction of antigen in solution slowly moves it away from the adsorbent without the excessive dilution that simply passing buffer through would cause. A novel technique involves the use of a cleavable spacer arm [178]. The antibody is attached through a phenyl ester linkage to the agarose matrix. This is chemically cleaved with an imidazole-glycine buffer at pH 7.4 to release the antibody–antigen complex. Unfortunately this does not separate the antibody from antigens, and the adsorbent can be used only once.

Ideally, removal of adsorbed antigens from the column would occur by weakening of the dissociation constant from a value of about $10^{-8} M$ to around $10^{-5} M$, about a 1,000-fold decrease in affinity, without causing denaturation of the antigen. These values represent α values of about 0.98 and 0.15 respectively, assuming the low total capacity of $0.1 \, \mathrm{mg \, cm^{-3}}$. In order to start with a value around $10^{-8} M$ (rather than the more usual affinity of several orders of magnitude stronger), it may be practical to vary the species from which the original enzyme sample was purified. But more commonly, one will use monoclonal antibody techniques to obtain a suitable binding strength.

Methods Using Monoclonal Antibodies

Most of the problems encountered using polyclonal antibodies can be avoided with the newer technologies of monoclonal antibody production [179]. The facilities needed and the techniques used, although simple enough, are more demanding than those required for ordinary antibody production. The advantages over polyclonal antibodies for making immunoadsorbents are:

1. The protein preparation used does not have to be pure; the screening process for detecting suitable clones rejects antibodies directed against other proteins. Here 1% "purity" can be sufficient [180].
2. As mice are usually used, the amount of antigen required is very small; a total of $50 \, \mu g$ may be sufficient.
3. Many different clones of antibody types against the desired protein may be found, directed against different antigenic determinants. The dissociation constants for the antibody–enzyme complex will vary, and it is possible to select a clone giving K_d in the desirable range of $10^{-6} - 10^{-8} M$, so lessening problems when eluting columns.
4. Once a suitable clone has been found, storage in liquid nitrogen enables an indefinite supply of that antibody to be obtained from many individual animals; the supply does not terminate with one animal's death.
5. As the monoclonal antibody is virtually pure, *all* the coupled immunoglobulin on the adsorbent is specific for the desired enzyme, so capacities of the immunoadsorbent are generally at least tenfold higher

than with unpurified polyclonals. This in itself will lead to the need for a tenfold decrease in strength of binding.

Most of the methods for preparing a monoclonal antibody adsorbent column and for its operation are identical to those described above for polyclonal antibodies. The purification of the monoclonal antibody itself may differ, as it does not occur as a component of blood serum, but as a culture supernatant. Methods for purification are described in Section 9.3.

Relative Advantages of Polyclonal and Monoclonal Antibodies

The choice of type of antibody must be made according to the following considerations:

1. The availability of facilities and expertise for monoclonal work;
2. The ultimate aim of the exercise (e.g., is this a one-off purification for research purposes, or is there a foreseeable continuous requirement for the product?);
3. The quantities of end product needed;
4. The overall costs of the procedure (particularly important for a long-term commercial operation);
5. Does the simplest method give the best product?

The relative merits are listed in Table 7.4.

Finally, a note on operating procedures. Antibody–antigen interactions are slow in comparison with, say, an enzyme and its low-molecular-weight substrate. The formation of a tight antigen–antibody complex is generally considered to occur in more than one step, a rapid complex formation determined by collision rates between the macromolecules (typically about $10^6 M^{-1} s^{-1}$), and a much slower formation of "secondary bonds" [181] which are responsible for the tight binding. The K_d for the first step may be quite high, e.g., $10^{-5} M$, with a further similar K_{eq} for

Table 7.4. Merits and Demerits of Monoclonal and Polyclonal Antibodies

	Monoclonal	Polyclonal
Equipment and expertise	Considerable	Simple
Man-hours of work	Considerable	Few
Quantity of antibody	Large	Moderate
Permanancy of supply	Permanent	Transient
Purity of antibody	High	Low
Specificity and affinity	Can be either high or low in each case	A range of many products, mostly high specificity
Suitability as immunoadsorbents	Excellent operating parameters	Problems due to range of affinities

the second step giving an overall antigen–antibody dissociation constant of $10^{-10} M$.

$$\text{Antigen} + \text{Antibody} \xrightarrow[\text{rapid}]{} \underset{\substack{\text{Primary} \\ \text{complex}}}{(\text{Ag–Ab})} \xrightarrow[\text{slow}]{} \underset{\substack{\text{Secondary} \\ \text{complex}}}{\text{Ag–Ab}}$$

Because the rate constants for the second, slow step are in the range of 10^{-3} down to $10^{-6} s^{-1}$, the rate of formation of the secondary complex may be measured in minutes to hours. So flow rate during application should be slow. Batchwise treatment, and allowing the sample to be in contact overnight may be a good plan. Elution changes these parameters principally to give a faster off-rate. But mild eluants may still be very slow in acting, so low flow rates should be used in elution also, unless the eluant is denaturing and liable to destroy the antigen.

Finally, mention must be made of the latest technologies involving generation of antibody molecules in *Escherichia coli* by recombinant technology [182]. It is likely that future immunoaffinity chromatography will make use of antibodies that have never been in an animal, but isolated from antibody recombinant libraries with all the possibilities of selection, mutation, and reselection to create a semiartifical, but ideal antibody for the proposed task.

7.3   Dye Ligand Chromatography

Developmental History

In the 1960s, Pharmacia Fine Chemicals, Uppsala, Sweden, attached a blue dye known as Cibacron Blue F3G-A to a high-molecular-weight soluble dextran for use as a visible void volume marker for gel filtration. Had they selected a different dye, it is possible that the rapid expansion in the use of colored adsorbents would not yet have happened. Workers were puzzled when yeast pyruvate kinase appeared in the void volume during gel filtration, indicating a much higher molecular weight than expected. But when the blue marker was omitted, the enzyme came out after the void volume [183]. By carrying out gel filtration first in the absence of marker, and then in its presence, purification was achieved— although the marker had to be removed afterwards. The pyruvate kinase was binding to the blue dye. By coupling the blue dextran to agarose by standard affinity chromatographic methods, a new type of adsorbent was produced. Similar observations were made almost simultaneously concerning yeast phosphofructokinase [184]. Soon afterwards other enzymes, especially dehydrogenases, were found to have affinity for this blue dye, and it is now generally accepted that most enzymes that bind a purine nucleotide will show an affinity for Cibacron Blue F3G-A. Although many other related dyes also bind to various enzymes, the original blue one has been used in the majority of cases to date, and not just for nucleotide enzymes.

The specificity of dye adsorbents is such that they can qualify as affinity adsorbents, even though the dye used bears little obvious resemblance to the true ligand; "pseudo-affinity adsorbents" is perhaps a better description. Molecular models do show a rough resemblance between Blue F3G-A and NAD [185], but the most important correspondences are with the planar ring structure and the negative charged groups. It has been shown by X-ray crystallography that Blue F3G-A binds to liver alcohol dehydrogenase in an NAD site, with correspondences of the adenine and ribose rings but not the nicotinamide [186]. Thus the dye behaves as an analogue of ADP-ribose, and it binds to the "nucleotide fold" found in AMP, IMP, ATP, NAD, NADP, and CTP binding sites. The binding is commonly tighter than the true substrate, with K_L values in the micromolar range. As a result, further nonspecific interactions resulting from spacer arms are not always needed to enable a protein to bind to an immobilized dye column. Moreover, although the original adsorbent used was "Blue Dextran," the dextran component acted only as something to couple to the agarose, a sort of giant globular spacer arm, so large that much of the inside of the agarose beads is not accessible by it. Even so, the binding capacity of Blue Dextran agarose is quite high, with a much higher percentage of the attached dye ligand being active in binding proteins than is the case with a normal affinity adsorbent. Modern Blue adsorbents dispense with the dextran, and either link the dye directly to the unactivated matrix, or to a short spacer.

The dye (Cibacron Blue) itself is sufficiently reactive to couple directly to nonactivated agarose. Indeed, this dye and the others described below were developed for the purpose of reactive coupling to biological polymers (wool, cotton) and do so through the triazinyl chloride group, in slightly alkaline conditions (Figure 7.16).

Agarose provides a great number of hydroxyls for the reaction, and direct coupling can be manipulated to provide a range of substitution percentages, up to about $5\,\mu\mathrm{mol\,cm}^{-3}$ (swollen gel). Capacity for proteins is some 10–20 times that for a comparable true affinity adsorbent; up to $30\,\mathrm{mg\,cm}^{-3}$ of protein will bind, although with gel exclusion effects the capacity decreases as the protein size increases.

Note that direct attachment to the matrix backbone does not detract from the dye's ability to bind enzymes: a spacer arm does not seem to be so necessary, although the triazinyl group may be acting as a spacer itself. As was mentioned in the previous section, the reason why spacer arms are needed when weak enzyme–ligand interactions are involved may be not so much to position the ligand farther away from the matrix as to provide nonspecific interaction forces to strengthen the binding. With typical K_i (K_p) around $10^{-6}\,M$, and a higher value for m_t, it is seen [from

Eq. (5.11)] that strong adsorption does not need extra interactions for dye ligand columns, e.g., $m_t = 0.1\,mM$, $p_t = 0.05\,mM$, $K_p = 0.001\,mM$, $\alpha = 0.98$.

From this simple description, it would seem that Cibacron Blue should be ideal for nucleotide-binding proteins, especially in combination with affinity elution. In practice, there are many complications, for the Blue dye is not nearly as specific for just nucleotide-binding proteins as was first thought. Moreover, there are a great many nucleotide-binding proteins—some estimates are that as many as 50% of enzymes are in that category—so even if the dye were so specific, there would still be a range of proteins that would bind from a complex mixture applied. But the lack of specificity has been put to good use, in that a large number of other proteins (including human serum albumin, interferon, and many enzymes) have been purified with the aid of Cibacron Blue columns.

It was nearly 10 years before any serious consideration was given to looking at other dyes' abilities to bind proteins. Dyes in the same category (the reactive triazinyl dyes) were soon found to behave in a fashion similar to Cibacron Blue; moreover, the interactions were very variable in strength, and certain dyes showed selectivity for particular proteins. One of the first new specificities reported was that a bright red dye, Procion

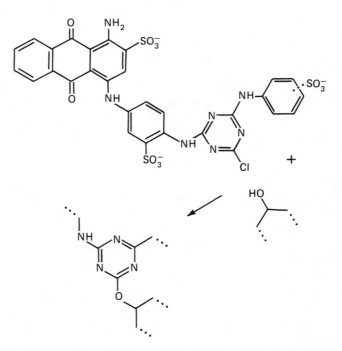

Figure 7.16. The structure of Cibacron Blue F3GA and its reaction with hydroxyl groups.

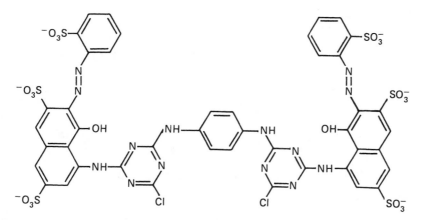

Figure 7.17. The structure of Procion Red H-E3B (reactive red 120). This is a bifunctional dye; one of the triazinyl chlorides attaches to the agarose when synthesizing the adsorbent, whereas the other may be hydrolyzed.

Red H-E3B, which has a large dimeric structure, showed selective specificity toward NADP-binding proteins (Figure 7.17) [187]. In retrospect, it now appears that this was a generalization based on too few examples; both the Red and Blue adsorbents have since been used successfully in purifying dehydrogenases both NAD- and NADP-specific, as well as many other proteins. Other tests with ranges of dyes for particular enzymes have been tried [188,189,190,191]. At least 90 different dyes of the triazinyl chloride type have been used in columns for screening for protein binding, and many other types such as the vinyl sulfone reactive dyes (Remazol: Hoechst, Frankfurt, Germany) are also used. Some of the chemistries of reaction with hydroxyl groups are shown in Figure 7.18. Amicon Corporation, Beverly, MA, introduced the first multiple-dye screening kit (5 dyes), which led to a number of successful purifications using one or more of these five dyes (Cibacron Blue F3G-A, Procion Turquoise H-A, Procion Green H-E4BD, Procion Red H-E3B, and Procion Yellow H-A). Our group has also distributed a 40-dye column kit in Australasia and Japan (Rainbow-Sep Wako Pure Chemicals, Osaka, Japan), and a 10-dye kit with greatly improved properties is available (Mimetic ligands: Affinity Chromatography Ltd., Isle of Man, U.K.).

There are some spectacular examples of successful applications of dye adsorbents. In most cases, it has been used as just a step in a purification protocol, preceded and followed by others, but occasionally it has been used as the only step in a true affinity process (e.g., [191,192,193]). Proper applications of the principles of affinity elution, together with the differential chromatographic procedures and more extensive screening to

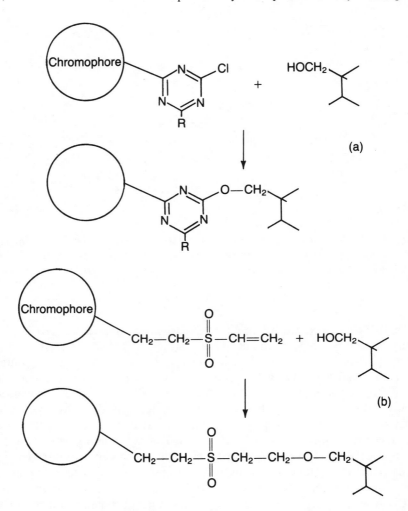

Figure 7.18. Reaction of (a) monochlorotriazinyl dye (Cibacron, Procion H-type) with hydroxyl groups on, e.g., agarose. The R group is generally aniline or p-sulfanilic acid. (b) Reaction of vinyl sulfone dye (Remazol). See Figure 7.21 for dichlorotriazinyl dye reaction.

find the most suitable dye (see below), would undoubtedly improve many of the reported methods.

Preparation of Dye Ligand Adsorbents

Three factors define a dye ligand adsorbent, as indeed they define most other adsorbents: (1) the nature of the matrix, (2) the structure of the dye, and (3) the degree of substitution of dye.

The matrix used most has been agarose, which generally gives adsorbents with the highest protein-binding capacity, though not necessarily with optimum selectivity. But other synthetic and natural polymers have been used. The reactive group is the primary hydroxyl in agarose; most other matrix polymers used in protein chromatography also utilize hydroxyls for coupling of the ligand. We generally use cross-linked agarose (for stability and flow rate) with porosity capable of taking in proteins of several million Da (ca. 4% agarose). More control over dye substitution can be obtained using amino-substituted matrices, described below.

The structure (and purity) of the dye used is very important in terms of selectivity of protein binding, but knowledge of it does not in itself help choose which dye to use [194,195]. The structures of some of the dyes that are commercially available are not public knowledge. The dyes used are reactive dyes known by a variety of names. The most extensively used dyes originate from, though are not necessarily made by, ICI (Manchester, U.K.) with their reactive dye series known as Procion, whereas Ciba-Geigy, Basel, Switzerland, supplies the overlapping Cibacron series. The Procion and Cibacron reactive dyes are based on triazinyl rings. Procion H-type dyes and the main Cibacron series (including Cibacron F3G-A, but not the F-range) have monochlorotriazinyl reactive groups. The Cibacron F-range has a monofluorotriazinyl reactive group. The Procion MX- dyes are much more reactive, being based on the dichlorotriazinyl group, but these leave one chlorine unreacted after coupling. This MX-series, representing the original reactive dyes, is being phased out.

Hoechst provides Remazol dyes, which have a quite different reactive group, intermediate in reactivity between the Procion H- and MX-series (Figure 7.18) and there are many more types from chemical companies such as Sandoz, Basel, Switzerland (Drimarine dyes), and less specified consumer dyes obtainable in art suppliers, haberdasheries, and drug stores. Although in some cases the same chromophore may be used by different companies, different reactive groups result in different end products which may not behave the same way when interacting with proteins. Some of these dyes have been used with such success in the textile industry that they continue to be manufactured in multiton batches. Others that find less demand are withdrawn from production, and new ones are introduced all the time. Many are not very pure, containing a variety of minor components, sometimes of quite different color, and often there are mixtures of isomers of the main component present [196]. Thus, a column made from the crude dye may contain a number of different compounds attached to it, in different proportions, and they may vary from one batch of dye to another. For accurate, controlled work this raises a dilemma: should any attempt be made to purify the dyes first, which would involve a great amount of time and expense? The alternative is to ignore possible variations, hoping that they will not have a great effect, and where possible, to always use the same batch of dye, and store it in good conditions. One positive advantage of the latter procedure is that scaling up does not involve significant expense, since the crude dye costs little compared with the matrix. But to purify large quantities of the dye may be economically unjustifiable. On the other hand, for a truly reproducible procedure, the adsorbent must be accurately defined—and obtainable. Several of the dyes, including Cibacron Blue F3G-A, are no longer manufactured by the original producers, although new alternatives may be very similar (e.g., we have successfully replaced Procion Blue H-EG with Procion Blue H-EGN, the latter having the same chromophore, but a slightly different reactive chemistry).

The dyes also have color index numbers and are known as "reactive (color) (number)." Thus, the most widely used "Cibacron Blue F3G-A" has also been manufactured by ICI, and known as "Procion Blue H-B" (no longer available), "Color Index 61211," and "Reactive Blue 2" (structure in Figure 7.16). The next most widely used dye is Procion Red H-E3B (Reactive Red 120; Figure 7.17).

To couple the dyes directly to agarose, the gel is treated with an approximately 0.2% dye solution at pH 10–11 (Na_2CO_3 or NaOH) in the presence of salt (2% NaCl). At 20–30°C, coupling is complete in 2–3 days for the monochlorotriazine dyes (Cibacron and ICI Procion H series), but the dichlorotriazine dyes (ICI Procion MX series) being more reactive need only 1–2 h, and in that time generally give much higher substitution levels that the monochlorotriazyl dyes do in 3 days. This is because there is competition between the hydroxyls on the matrix (which must be ionized to react) and hydroxyl ions in the buffer, which causes most excess dye to hydrolyze to an inactive form within 3 days [197]. Using cross-linked agarose and other heat-stable matrices, the solution can be heated to higher temperatures to speed up the reaction. It has been found that 2 h at 80–90°C using a 0.2% dye solution, gives a similar substitution of Procion H- dyes as 2 days at 20°C. The mixture is allowed to cool overnight, and excess dye washed off using a Buchner funnel with large amounts of water.

The dye adsorbent should be stored in a dilute buffer solution at pH 8–9, with a bacteriostatic agent such as azide; before and after use it should be treated with a solution of $6M$ urea–$0.5M$ NaOH, which removes any noncovalently interacting dye (before use) and all strongly bound protein (after use).

The degree of substitution is an important feature in dye ligand chromatography, as it is in other affinity methods. If more than one dye substituent can be involved in the binding, then much tighter adsorption is obtained (see Figure 7.10). Moreover, the higher the value of m_l, the higher is α [Eq. (5.11)]. It may be that a highly substituted gel binds the required protein so tightly that subsequent recovery is low, in which case more weakly substituted gels can be made by shortening the treatment with dye during their preparation. For consistent results, the degree of substitution should be recorded [198], but this can only be done properly if the molar absorptivity is known. With many dyes this detail is not available, so it can be expressed only in terms of milligrams of dye per milliliter of adsorbent, assuming that all the original material is in fact dye. In practice, there are stabilizers and humectants added, which means that dye content is always less than 100%. Because of the widely different dye structures, the (molar) degree of substitution using the standard coupling procedures is very variable, ranging from as little as $0.3\,\mu$mol cm^{-3} to $5\,\mu$mol cm^{-3} or even higher with dichlorotriazinyl dyes. Higher and/or more consistent degrees of substitution can be obtained by using an amino- (or sulfhydryl-) substituted matrix which reacts with even mono-chlorotriazinyl dyes in minutes (sulfhydryl-) to hours (amino-). We have created a new range of "high-dyes" with between 5 and 10 times as much substitution as normal ($10-20\,\mu$mol per ml wet weight). This uses an ammonia-treated epichlohydrin-activated agarose to create an amino-substituted matrix, with which the dyes are reacted (N-linked dye). Using sodium sulfide instead of ammonia, a thiol matrix is created. Dyes react almost instantaneously with this, but the resultant S-linked dye is alkaline labile and not suited for NaOH cleaning. The N-linked dyes are more stable than the conventional type (O-linked dye). These high-dyes behave qualitatively similarly to the lower substituted analogues, but have 2–3 times higher capacity and somewhat less selectivity. It is often more difficult to get the proteins off these adsorbents.

Dye–Protein Interactions

The interactions between the dyes and proteins are complex and not well understood. We may classify them for affinity purposes as "specific" and "nonspecific" according to whether the interaction is at the ligand-binding site or not; the interactions could include practically all types of forces described between molecules. The dyes generally contain aromatic rings, sometimes heterocyclic, often fused, and by definition of a dye, a long series of conjugated double bonds must be present to give a strong adsorption of light in the visible spectrum. Additionally, sulfonic acid groups are included to confer aqueous solubility; these groups are negatively charged at all pH values above zero. Some dyes contain carboxylates, amino, chloride, or metal complexing groups; most contain nitrogen both in and out of aromatic rings. Thus, interactions with proteins as a result of hydrophobic, electrostatic, and hydrogen bonding are all likely. Electrostatic interactions would be of the cation exchange type, leading to the expectation that stronger binding would occur at lower pH. This is indeed the case; an increasing pH gradient is often an effective mode of

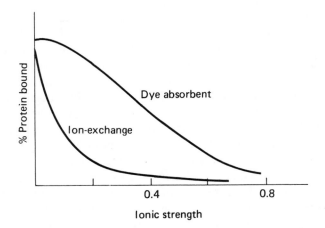

Figure 7.19. Effect of ionic strength on amount of protein (from a crude mixture) bound to dye adsorbent, compared with simple cation exchanger.

elution. However, if cation exchange were the predominant mode of interaction, we would expect a very sharp effect of ionic strength in the range $0-0.1\,M$. In fact, there is relatively little change in protein binding in this range (Figure 7.19), and it generally takes a much higher salt concentration to elute a protein. Moreover, the protein does not have to be basic (positively charged) to bind to dyes. High salt concentrations do remove most bound proteins, which suggests that hydrophobic interactions are not predominant either. Not only cation exchange forces get stronger at lower pH; as the proteins gain more positive charges they can interact more strongly with the π-electrons in the conjugated structures of the dyes. The dyes are not very hydrophobic overall (otherwise they would not be water-soluble), but they do have planar aromatic surfaces that prefer to interact with hydrophobic residues in proteins. Finally, several dyes interact by charge transfer processes in which polarization of both dye and protein occurs when they come in close proximity.

More specific forces are involved when considering the roles of phosphate and divalent metal ions in the interactions between dyes and proteins. Many enzymes interact with phosphorylated substrates. Phosphate ions bind weakly (and sometimes strongly) to the sites that would normally be binding the phosphate ester on the substrate; these often include arginine residues. These positively charged sites can also interact with the sulfonate groups on the dyes, which resemble phosphates. Thus the presence of phosphate ions in the buffer usually weakens the overall interaction, especially at pH values above 6.5 when much of the phosphate is doubly charged, by occupying the site that would bind to the dye sulfonate.

The presence of divalent metal ions has been shown to make a substantial difference to the binding strength of many proteins to dye columns [199,200]. Generally, binding is tighter, and the transition ions Cu, Ni, Co, Mn, and Zn in particular are effective; Mg, although needing higher concentrations to get a similar effect, has the advantage of not forming a precipitate with most buffer salts. Schemes have been developed in which the sample is applied to the column in the presence of a metal ion, and the enzymes eluted simply by omitting the metal ion from the buffer [201,202].

Thus, there are many variables, pH, ionic strength, metal ions, and buffer composition (as well as temperature and the presence of nonionic detergents or organic solvents), that can influence the behavior of proteins on dye columns; and there are many different dye columns. The problem comes in deciding where to start.

Screening to Obtain a Suitable Adsorbent

This section will assume that large varieties of dye adsorbents are available, and is concerned with the selection of the optimum dye, which (as well as being readily available) should bind the desired protein totally in mild conditions, and release it totally on a buffer change. A further desirable characteristic is for affinity elution to be applicable. The selected dye should also bind as little as possible of the contaminating proteins in the sample being applied. The latter point can be optimized by operating in what I have described as a differential mode, in which opposite, differential selectivities of two dye adsorbents are made use of in a tandem column procedure (Figure 7.20) [191,203].

When screening the range of dyes, two measurements on each *non-adsorbed* fraction are made: total protein content and protein being isolated. For differential operation, the adsorbent for the first column (the "negative" column) is chosen as that dye that binds the most protein without binding the desired one. The second, "positive" column is that dye that binds the desired protein, but as little as possible of total protein. We have occasionally found a "negative" dye that binds more of the total protein from a crude mixture than the "positive" dye does, despite the selectivity toward the protein we want being the other way about. This has two great advantages. First, it means that very little protein overall will be binding to the second column, as most that would have bound has been removed by the first column. Second, the process *by definition* has selected a positive column that has some specificity for the protein concerned, which will as likely as not mean that interactions are through the ligand-binding site, so affinity elution techniques are likely to be successful.

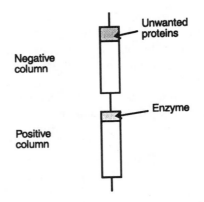

Figure 7.20. Principle of differential column chromatography. The first, "negative" column binds many proteins that would otherwise have bound to the second, "positive" column, but allows all the enzyme through. The positive column binds all the enzyme, but little more protein.

So use of differential chromatography, after screening as large a range of dyes as possible, results in a high degree of purification *at the adsorption stage*. If affinity elution is also successful, a single-step isolation procedure may be achieved (e.g., [191,192,193]). This differential chromatography principle is not restricted to dye adsorbents. But dyes are particularly suitable because of the range and subtleties of the interactions between different dyes and proteins.

A number of processes for screening dye adsorbents have been reported. Initial screening can be done in a batch process, remembering that requirements for binding are more stringent (i.e., higher α value) than in columns. Using a 96-well ELISA plate, small amounts of dyed adsorbent were trapped in glass wool, so that extract samples could be mixed in, and the unbound proteins in the liquid could be sampled without taking up any adsorbent particles [204]. Alternatively, extract is added to weighed samples of adsorbent in microfuge tubes, shaken for a period, and either allowed to settle, or given a brief centrifuging before sampling the supernatant [205]. Using prepacked columns, more extract sample is needed. The columns (usually 1–2 ml) can be lined up on test-tubes, and extract applied to all simultaneously. The sample is washed in with starting buffer, and as with all these procedures, the amount of the protein that you want (usually the activity of an enzyme) is measured in the flow-through (or supernatant for the batch processes). As indicated above, the *total* amount of protein in these fractions should also be measured, so as to find the dye that is the most specific for your protein. If you do not have much extract, or the assay is tedious, the total number of columns to be screened can be reduced by an approach which relies on knowing something about the general protein-binding ability of each dye [203]:

Table 7.5. Grouping of Dyes

Dye adsorbents were made by a standard technique, and the resulting materials were analyzed as to what proportion of the protein in typical crude extracts bound to them under standardized buffer conditions. Group 1 dyes bind the least protein; group 5 dyes the most. The groupings are indicative only; the amount of protein binding to a dye adsorbent increases with the dye content, and the dye contents are very variable when using the standard coupling technique. Thus, a group 2 dye, made by the standard technique, may become a group 4 dye if a nonstandard coupling method is used that results in much more bound dye. A more meaningful grouping would compare dye adsorbents with identical dye contents. Nevertheless, *selectivity* for specific proteins remains, independent of dye content and apparent grouping. Many of the dyes listed here are different from those listed in the earlier editions of this book, as many dyes are no longer available, and new dyes have appeared.

Group 1	Group 2	Group 3	Group 4	Group 5
P^a Blue MX-7RX	C^a Brown 3-GRA	C Blue F3-GA	P Blue MX-G	P Blue H-ERD
C Blue 2-RA	P Red MX-5B	P Blue H-4R	P Blue MX-4GD	P Green H-4G
P Turquoise H-A	P Scarlet MX-G	P Red H-E3G	P Orange MX-2R	P Green H-E4BD
P Yellow MX-6G	P Yellow MX-4R	P Yellow H-5G	P Red MX-8B	P Navy H-ER
P Yellow MX-8G	R^a Black GF	R Blue R	C Red 3-BA	P Red H-3B
P Yellow MX-3R	R Orange FR	R Brill Blue B	P Violet H-3R	P Yellow H-E3G
R Violet R	C Turquoise GFP	P Blue H-EGN	P Black H-EXL	P Yellow MX-GR
R Orange 3R	C Yellow R-A	P Yellow H-E6R	P Red MX-7B	C Blue F-R

^{a}P = Procion (I.C.I.); C = Cibacron (Ciba-Geigy); R = Remazol (Hoechst).

Each dye is classified into a group that relates to its ability to bind protein in general. We have used extracts of mammalian tissue, yeasts, plants, and bacteria, and found that (with a few exceptions) dyes that bind a lot of protein from one source will bind a lot from another, and similarly for weak-binding dyes. So each dye has been allotted a group number between 1 and 5, according to its average protein-binding ability (Table 7.5). To approach a new isolation problem, one column from each group is selected, and screening carried out, measuring for the moment just the behavior of the desired protein, which we will assume is an enzyme. There are many possible outcomes; we will discuss three of them:

1. The enzyme binds to all five dye columns.

In this case (assuming that the enzyme is recoverable and has not been inactivated) we are dealing with a "sticky" enzyme, which will bind to most columns, and to those in groups 4 and 5 very firmly. The simplest approach is now to screen all the group 1 columns, and no others (though there could be a suitable "negative" found among the group 2). It may not be necessary to use a negative column since relatively small amounts of protein bind to the group 1 columns. Alternatively, the buffer composition could be changed (e.g., increase pH, remove divalent metal ions, include phosphate) so that the binding of the enzyme and proteins in general is weakened. The five columns are screened again with the new buffer, in the expectation of a different outcome.

Another approach which can be highly successful in this situation is to increase the loading on the column; eventually the required protein will pass through, but you may discover a column that not only has a selectivity towards your protein, but also can be used at high loading, so that a small column can deal with a large amount of sample (see also Section 5.6)

2. The enzyme binds only to the group 5 columns, or to none at all.

In this case we have a "nonsticky" enzyme, and probably only group 5 columns need to be looked at further. A "negative" column chosen from group 5 or 4 will be very useful as it will remove a lot of unwanted protein. Alternatively, buffer conditions may be changed to increase the binding (lower pH, include divalent metal ions), thereby shifting the binding range into lower dye groups.

3. The enzyme binds to the dyes in groups 4 and 5, and is partly retained on the group 3 dye.

This more complicated intermediate situation is perhaps the best, as it leaves more flexibility in obtaining optimum dye selection. We will assume that we do not wish to alter the buffer conditions. Based on the results found, it is probable that the enzyme will bind strongly to all group 5 columns, and is unlikely to bind to any group 1 column. Initially, we should simply screen all the group 3 columns to see how it behaves. At this stage, we measure the amount of protein binding as well as enzyme binding. The results will probably be that the enzyme will bind totally to some group 3 column, not at all to others, and be partly bound on the remainder (Table 7.6). From these results we may be able to select an excellent pair of dyes. However, there is always the possibility that we have missed a better "positive" dye in a lower group, so if there is time and enough material, it would be worth screening all the group 2 columns as well.

Table 7.6. Possible Results of Screening of Group 3
Dye Columns

Column number	Is enzyme bound?	Percent protein bound
3–1	Yes	32
3–2	No	26
3–3	Partly	33
3–4	No	34
3–5	Yes	29
3–6	Yes	38
3–7	No	28
3–8	Partly	30
3–9	No	31
3–10	No	27
Best negative column is 3–4; Best positive column is 3–5.		

See text for details.

The above screening procedures usually work well, and less than half of the complement of columns need to be screened. But the possibility remains that a column that shows high selectivity to our enzyme has been missed because the buffer conditions were such that either the interactions were too weak despite the selectivity, or too strong. In the latter case, that dye was not investigated because it bound too much protein overall under those conditions.

The screening procedure is carried out on small columns containing 1–2 ml of adsorbent, using 5–30 mg of protein mixture (we use the dye ligand chromatography as the first step in an isolation, so the protein mixture is the initial crude extract). Alternatively, about 100 mg wet weight of adsorbent is placed in a microfuge tube with 0.5 ml of extract. A suitable buffer is chosen, meeting any requirement of the enzyme for stability, preferably at an ionic strength of 0.05–0.15 to minimize cation exchange behavior. To start with, pH values in the range 6–7 are best, but higher pH values may be suitable if you find you are dealing with a "sticky" enzyme, lower pH values if not and your enzyme is stable enough at low pH. Mes buffers (see Chapter 12) are particularly suitable for pH 6–7, Mops, triethanolamine, Tes, and Tris for higher pH's. Phosphate is excellent (especially if planning very large-scale operations, as it is cheap) provided that it does not reduce the enzyme's binding too much and/or interfere with possible affinity elution procedures (be wary if the ligand you intend to use in affinity elution is a phosphate ester). We usually start with a pH 6 Mes buffer containing magnesium ions to see what happens; if the enzyme does not stick to any columns in this buffer then we are dealing with a minority case.

Elution of Proteins and Enzymes from Dye Columns

Enzymes and other proteins bound to dye columns may be eluted either nonspecifically or by affinity elution techniques. In almost every case, the latter will give a better degree of purification, but careful attention to the principles of affinity elution (Section 7.4) is needed to get a good recovery of activity in a reasonably small volume of eluant. The "negative/positive" differential adsorption technique described above is as likely as not to have selected out a dye with some biospecific affinity for the protein being isolated, so the addition of buffer containing an appropriate ligand has a good chance of eluting it. Moreover, the screening technique selects columns that do not bind the protein too strongly, so sometimes no changes to the buffer composition need be made before adding the ligand. For further details, refer to Section 7.4.

Buffer changes that result in weakening of the protein–dye interaction, and eventually elution of the protein, resemble those used in cation-exchange chromatography. However, there are significant differences, the most notable being the response to changes in ionic strength in the range $0-0.2\,M$ (Figure 7.19). Whereas cation exchangers bind protein with a strength that resembles an exponential decay as salt concentration increases, dye adsorbents are relatively little affected in the low salt range, releasing the proteins gradually as the salt increases to higher concentration. Not all proteins are released from the group 4 and 5 dyes even at $1-2\,M$ NaCl, probably reflecting greater importance of hydrophobic interactions between these dyes and some proteins. Increases in pH, including the use of a pH gradient, will elute some proteins and this can be a better procedure than a gradient in salt concentration. As the dyes in general do not buffer in the pH range used, and quite strong buffering mixtures can be used because of the low importance of ionic strength, smooth pH gradients can be generated much more successfully than on ion exchangers (cf. Figure 6.6). Elution conditions are, of course, inversely related to adsorption conditions, so the effects of changes in buffer composition, in particular the involvement of phosphate ions or divalent metals in the protein–dye interactions, are important. For simple trials to see if an adsorbed protein can be recovered quantitatively, elution with $1\,M$ NaCl in starting buffer is normally adequate. Exploitation of hydrophobic interactions could be tried by inclusion of solvents (e.g., 20% ethanol or i-propanol, up to 50% ethylene glycol), nonionic detergents in the buffers, or changes in temperature, but this is not a normal procedure. Chaotropic salts such as LiBr or KCNS have also been used to help elute proteins, especially in the procedure to use dye adsorption in the purification of human serum albumin [206].

Cleaning and Storage of Dye Adsorbents

After use, some proteins may remain on columns, and a treatment with alkali will normally remove them. Better still, inclusion of urea (e.g., $6\,M$ urea in $0.5\,M$ NaOH) helps to strip off even denatured protein. Approximately one-half volume of this cleaning solution should be passed through, followed by copious amounts of dilute NaCl (e.g., $10\,mM$), then the equilibrating buffer for the next use. If the column is to be stored for long periods, it should be in a slightly alkaline phosphate buffer containing $10\,mM$ sodium azide. Before use after long storage (and preferably before every use), a wash with dilute alkali is beneficial in helping to remove any dye that has leaked from the column. Many dyes will leak from agarose columns, but this is a very variable phenomenon due to the precise chemistry close to the reactive group on the dye. The leakage of Cibacron Blue from both agarose (Sepharose) and Trisacryl at different pHs has recently been quantitated [207]; minimum leakage from agarose was in the pH range 4–11, with the least around 10. Trisacryl had almost no leakage from pH 1 to 9; exposure to high pH for long periods led to

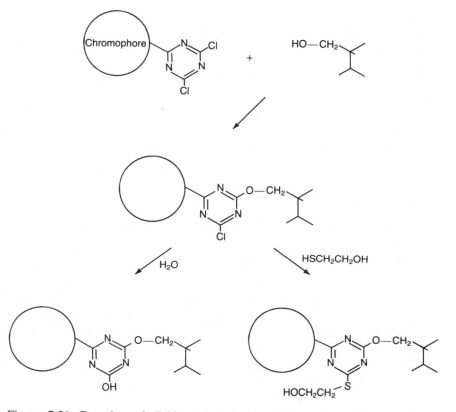

Figure 7.21. Reaction of dichlorotriazinyl dye with agarose, and subsequent blocking of the remaining chlorine with mercaptoethanol, as an alternative to slow hydrolysis to a hydroxyl group.

losses in both examples. But the leakage is mainly a property of the dyes themselves; some of the earlier Procion MX- red dyes were very bad in this respect. Presumably, the dye chromophore broke away from the linkage. Acid treatment can cause hydrolysis of the agarose matrix, and is not recommended.

Dichlorotriazinyl dyes after reacting retain one chlorine atom on the triazine ring (Figure 7.21). This may remain for weeks, and be potentially hazardous in that it could react a susceptible nucleophile in an adsorbing protein. Otherwise, it hydrolyzes, creating a less stable hydroxyl-substituted triazine ring, which can break down and release the dye. This process can be ameliorated by taking a freshly prepared MX- dye adsorbent, and treating it in dilute alkali with excess β-mercaptoethanol overnight, resulting in a thioethanol substitution on the triazine ring. This is more stable, and does not seem to change the protein-binding specificities.

Cleaning operations should be carried out as soon as possible after use to avoid bacterial action. But strong alkali-urea solutions should not be allowed to remain in the columns for long, or the agarose will degrade. By attention to the care of the dye adsorbents, they can be continually used over periods of years, with only an occasional need to repack the column.

Most dye columns can be reused many times, but there are occasions when the nature of a crude extract sample renders the column inoperative after a few uses, despite cleaning with urea and alkaline solution. In such a case, the cheapness of manufacture of the dye columns is somewhat negated by the need to replace them frequently; an alternative first step, e.g., salt precipitation, before application to the column may solve this problem.

Finally, some words of warning: As has been mentioned above, the dyes used are not pure compounds, they vary from batch to batch, and variations could have substantial quantitative effects on protein binding, and in some cases, qualitative effects also. Leakage of dyes from columns has concerned workers wishing to purify proteins for clinical use. Although the dyes are not suspected of being carcinogenic (they are, after all, used in the clothes we wear), such foreign compounds cannot be allowed in products to be taken by or injected into humans. Because all the commonly used dyes have multiple negative charges, they bind tightly to anion exchangers. So trace amounts of dye can easily be removed by passage through DEAE-cellulose, even in the presence of $0.5 M$ salt. Nevertheless, there has been a lot of attention to leakage, with antibodies used for quantitating amounts of dye that cannot be detected spectrophotometrically [207,208]. Dye adsorbents for use in therapeutic protein isolation have been developed that have negligible leakage and highly consistent quality (Mimetic ligands; see above). In addition, "designer dyes" have been synthesized to specifically interact with a particular enzyme, on the basis of modeling the substrate-binding site from 3-dimensional X-ray crystallographic structural knowledge [209].

7.4   Affinity Elution from Ion Exchangers and Other Adsorbents

To most people, affinity chromatography implies the use of an immobilized ligand on an adsorbent that specifically selects out proteins binding to that ligand. After adsorption, the enzyme required can either be nonspecifically eluted by, for instance, an increase in salt concentration, or specifically eluted by displacement with ligand free in solution. Thus, affinity chromatography can involve two specific stages, "affinity adsorption" and "affinity elution." The mechanism by which affinity elution from a biospecific adsorbent operates is clear. In practical terms, the presence of a ligand reduces the partition coefficient from near 1.0 to a significantly lower value. However, the latter could occur with other adsorbents; the term *affinity elution* does not require any particular property of the adsorbent itself.

Affinity Elution from Ion Exchangers

Affinity elution from ion exchangers has been found to be a very convenient and specific method for purifying many enzymes, where they have a certain combination of properties [210,211,212,213,214], and may be applied to other proteins with suitable ligands.

The basic principles of affinity elution from ion exchangers are illustrated in Figures 7.22, 7.23, and 7.24. An enzyme is adsorbed at a certain pH because the electrostatic interactions between the matrix of the adsorbent and the charges on the protein are sufficiently strong to hold it, with a partition coefficient close to 1.0. If a ligand of the enzyme is introduced which is charged and of opposite sign to the net charge on the enzyme, then binding of this ligand decreases the net charge on the enzyme. In the right conditions, the reduction of the net charge on the enzyme weakens its electrostatic interaction with the adsorbent sufficiently to reduce the partition coefficient substantially. This might result in a reduction in α from 0.95 (when the enzyme band moves slowly down the column at a rate of one-twentieth that of the buffer flow) to 0.5 (the enzyme moves down at one-half the rate of the buffer). This causes specific elution of that enzyme and no other—except those which might also bind that ligand *and* have similar adsorption characteristics to the ion exchanger.

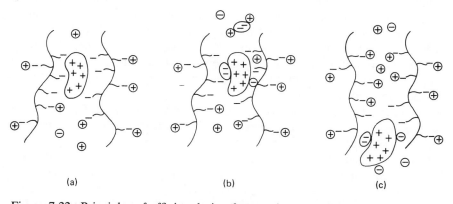

(a) (b) (c)

Figure 7.22. Principles of affinity elution from cation exchanger. In (a) the positively charged protein is adsorbed to the exchanger. In (b) a negatively charged ligand that binds to the protein is introduced, lessening the total charge on the protein. In (c) the protein is eluted, since α has decreased due to the lower total charge on the protein.

Apply sample   Increase pH   Add ligand

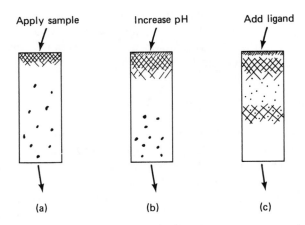

         (a)   (b)   (c)

Figure 7.23. Principles of affinity elution from cation exchanger. In (a) the protein mixture is applied; the enzyme required adsorbs, some other proteins pass through the column. In (b) the pH is increased so that the enzyme required begins to move down the column ($0.8 < \alpha < 0.95$)—some other proteins may be totally eluted ($\alpha < 0.8$). In (c) the addition of specific ligand decreases α ($0 < \alpha < 0.5$) so that the enzyme required is eluted.

The first proteins for which this method was successfully used were fructose 1,6-bisphosphate aldolase and fructose 1,6-bisphosphatase [210,211]. In both cases, the tetrameric enzymes (of about 160,000 MW) bound 4 substrate molecules having 4 negative charges each. Consequently, a reduction in total (positive) charge of 16 units was able to cause a large diminution in adsorption (Figure 7.25). Calculations based on experimental values [82] show that for rabbit aldolase on CM-cellulose at pH 7.1, the charge addition should reduce α from 0.95 to 0.01; even if adsorbed much more strongly initially, at $\alpha = 0.995$ (at pH 6.8), the addition of substrate would reduce the value to 0.35. However, many enzymes do not bind so many molecules of substrate, nor are their substrates as charged, so the total effect in general is expected to be weaker than for aldolase.

A detailed study on the adsorption of various enzymes to CM-cellulose in the presence and absence of substrates demonstrated the magnitude of effect. The α/pH curves were always shifted to lower pH in the presence of negatively charged ligand [82]. At a given α value the total charge on the enzyme and on the enzyme–ligand complex could be calculated, using an enzyme titration curve to relate the two pH values. With some enzymes it was found that the shift corresponded to no more than the change in charge expected by titrating the enzyme through that pH range. For example, the α/pH curve for rabbit muscle pyruvate kinase shifted only 0.25 pH unit in the presence of 0.1 mM phosphoenol pyruvate. This tetrameric enzyme is expected to acquire 12 negative charges if it

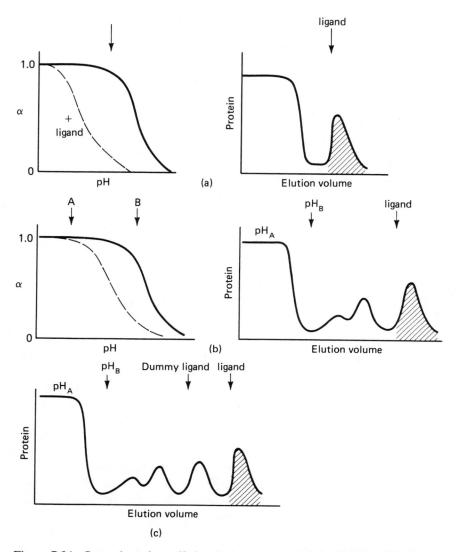

Figure 7.24. Operation of an affinity elution column. (a) Application at the same pH as used for elution; the effect of the ligand is so large that α decreases from >0.95 to a very low value. (b) Application at pH_A where $\alpha \simeq 1.0$, and increase of pH to B to make $\alpha \simeq 0.9$, so that addition of ligand has sufficient effect (as in Figure 7.23). (c) As (b), but inclusion of "dummy ligand." In practice the "dummy ligand" can be used at the commencement of pH_B wash. *Shaded area* represents the enzyme being eluted.

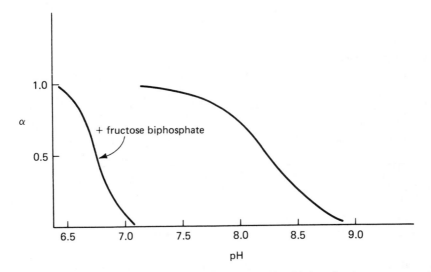

Figure 7.25. Shift in α against pH curve for rabbit aldolase in the presence of 0.1 mM fructose 1,6-bisphosphate (FBP). (From Scopes [82].)

becomes saturated with its substrate. However, it would lose only about 8 positive charges by titrating through 0.25 pH unit. In this instance, 0.1 mM phosphoenol pyruvate was probably insufficient to saturate the enzyme. On the other hand, lactate dehydrogenase, on binding 4 molecules of NADH, would acquire 8 negative charges, corresponding to a pH shift of about 0.8 pH unit (in the range pH 8.5–7.0). But in fact, the addition of 0.1 mM NADH shifted the curve by up to 1.5 pH units— nearly twice the effect expected. Similarly with yeast phosphoglycerate kinase [108], the shifts on binding 3-phosphoglycerate or ATP were rather more than anticipated, and with 1,3-bisphosphoglycerate, substantially more. In each of these examples, a conformational change occurs on binding the substrate, which itself lessens the strength of binding to the ion exchanger.

Thus several explanations for the affinity elution effect can be proposed. Besides simply canceling out total charge, the ligand binds at a site that probably has a cluster of charges opposite in sign to those of the ligand. This binding may mask a rather specific, strong interaction between these charges and the adsorbent. Also, since the active site is likely to be close to the surface of the protein molecule, charged residues at the active site would be having more influence on the total interaction than other charges further from the surface. The masking of these charges with ligand would weaken the interaction more than would the loss of a similar number of charges randomly distributed through the protein molecule. In addition, a protein conformational change usually occurs when the ligand binds, and this can alter the interaction with adsorbent—in either direction.

In general, it can be said that an efficient affinity elution procedure for an enzyme from an ion exchanger can be found if the following characteristics apply to the enzyme. For nonenzyme proteins, the same applies; the ligand must be small and charged:

1. The enzyme can be adsorbed, without inactivation, on an ion exchanger which has the same sign of immobilized charge as the charge on the ligand to be used in affinity elution. For anion exchangers (DEAE-, Q-, etc.), the ligand must be positively charged; for cation exchangers (CM-, S-, phospho-, etc.), the ligand must be negatively charged.
2. The ligand must be capable of binding to the enzyme at a pH where the enzyme otherwise remains on the column, i.e., has an α value of at least 0.9.
3. The number of charges added by the ligand per 100 kDa of enzyme should be at least 4, unless an additional conformational change assists the elution.
4. The dissociation constant K_L for the ligand should preferably be below $10^{-3}\,M$; the smaller the better.

In view of (1) above, since practically all charged ligands (other than metal ions) are negative, affinity elution from ion exchangers is virtually restricted to proteins of a neutral or basic nature that will adsorb to a cation exchanger at neutral pH (6–8). Polycationic ligands are rarely encountered. Many animal enzymes have isoelectric points in the appropriate range for cation exchange affinity elution (>pH 6), but corresponding enzymes in plants or microorganisms tend to have lower isoelectric points. Since the methods depend quite critically on isoelectric points, they are species-specific; changing species may require a re-examination of the appropriate buffer pH values for optimum procedure.

Affinity Elution from Other Adsorbents

Phosphocellulose is an ion exchanger, and several published affinity elution procedures use this adsorbent. However, there are specific effects involved with phosphocellulose, in particular with enzymes that bind phosphorylated substrates. Nucleic acid enzymes have often been purified by adsorption to P-cellulose, and several tRNA synthetases have been successfully eluted using RNA substrates [215]. Glucose 6-phosphate dehydrogenase [92], hexokinase [93], and phosphoglucose isomerase [212] also can bind to and can be specifically eluted from phosphocellulose; in these cases one may consider that phosphorylated nonreducing ends of the cellulose are acting in a biospecific fashion (Figure 7.26). But in most other cases, phosphocellulose should be considered simply as an ion exchanger, and discussion of its behavior belongs to the previous section.

Dye ligands represent another group of adsorbents that affinity elution is applicable to. As indicated in the previous section, dyes interact with many proteins, and especially enzymes, at natural ligand-binding sites. Consequently, displacement from the adsorbent using the ligand in the buffer is widely used. Very often, dyes are interacting with NAD(P)- or ATP-binding sites on the protein, so addition of the relevant nucleotide should cause elution. But other similar enzymes with the same nucleotide substrate may be eluted also, if the dye adsorbent has not been chosen with specificity in mind.

The concentration of ligand needed for affinity elution is generally about 10 times the K_L, but that value may not be known. Usually $1\,\mathrm{m}M$ ATP or NAD^+ is adequate, and $0.2\,\mathrm{m}M$ NADH or $NADP^+$ (since these in general bind more tightly). But the buffer conditions for elution must be such that the desired protein is already beginning to move down the column (see Figure 7.24). If the protein bound to a dye is not a nucleotide-binding protein, then its relevant ligand can be tried. Note that there are relatively few examples in which an uncharged ligand has successfully eluted a protein biospecifically from a dye adsorbent. The use of negatively charged ligands may operate partly in a similar fashion to affinity elution from cation exchangers, by canceling out electrostatic interactions.

Other adsorbents for which affinity elution has been used successfully include multifunctional [148] and hydrophobic materials [129]. Hydrophobic adsorbents are suitable for purification of lipophilic proteins, and affinity elution of lipid-binding proteins using the relevant lipid is possible. It is notable that successful elution of alcohol dehydrogenases with alcohols from hydrophobic adsorbents, in the presence of salt, has been achieved, though it is difficult to know whether this was biospecific, or a general reduction in hydrophobic interactions caused by the alcohol [129].

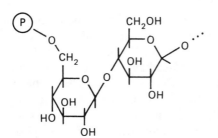

Figure 7.26. Phosphorylated nonreducing end of phosphocellulose as an affinity ligand for glucose 6-phosphate dehydrogenase and phosphoglucose isomerase.

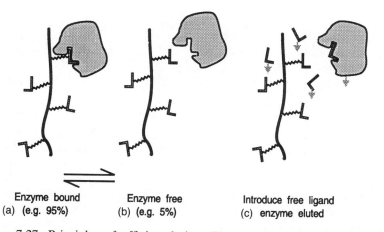

Enzyme bound	Enzyme free	Introduce free ligand
(a) (e.g. 95%)	(b) (e.g. 5%)	(c) enzyme eluted

Figure 7.27. Principles of affinity elution: The bound enzyme cannot interact with its natural free ligand in the buffer (a), but if the enzyme comes away from the immobilized ligand (b), when $\alpha < 1.0$, it can interact with free ligand, and not go back on to the immobilized ligand (c), so is displaced into the buffer stream.

Practice and Theory of Affinity Elution

Affinity elution, whether from an ion exchanger, a biospecific affinity adsorbent, dye, or other material, is successful when the interaction with the adsorbent is through the ligand binding site, and it is possible to replace this interaction by free ligand in solution. With ion exchangers, the interaction with the adsorbent is not so sharply defined, being more an overall electrostatic attraction. Nevertheless, the basic principles remain the same. The loss of energy of interaction should be sufficient to reduce the α value substantially; some representative calculations have already been presented (see Table 7.3). But if the binding to the adsorbent is too strong, then it becomes almost impossible for the ligand to insert itself. There must be a significant proportion of protein that is *not* binding at any given moment, which can pick up the ligand and not go back on to the column. This is another way of saying that the α value must not be too close to unity; probably the range 0.90 to 0.95 is ideal, as illustrated in Figure 7.27. In order to reach this situation, it may be necessary to change the buffer composition before the elution step. Thus, application of the sample is best made under conditions of high α, >0.95, to ensure that all the desired protein sticks even with a large application volume. Then the buffer is changed, by for example increasing pH with cation exchangers (and dyes), increasing ionic strength for many sorts of adsorbent, but perhaps decreasing ionic strength for hydrophobics. Then, when the protein is slowly moving down the column, the ligand is added,

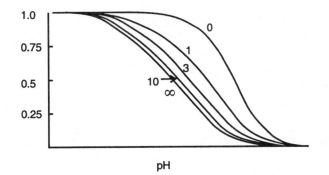

Figure 7.28. Theoretical estimates of the value of α changing with pH and with the concentration of eluting ligand in the buffer, for affinity elution from a cation exchanger. The concentration values represent multiples of the dissociation constant of the enzyme–ligand complex.

generally at 5–10 times the K_L, if known, and the desired protein may well come off with the ligand front (Figure 7.24).

The efficiency of affinity elution will depend on many things. Non-specific proteins may come off if adding the ligand significantly alters the basic parameters of the buffer, e.g., ionic strength, surface tension, pH (the latter can be readjusted). To overcome this problem, one can use what is termed a "dummy ligand," which has similar properties to the true one which is going to be used, and so is able to remove these nonspecific proteins first. For example, 1 mM of a trivalent anion has a significant ionic effect when operating at low buffer concentration, so a prewash with 1 mM EDTA or sodium citrate, with similar ionic properties would act as a dummy ligand.

Another major factor affecting the efficiency of elution is the number of binding sites for the ligand on the protein. A polymeric protein such as the enzyme lactate dehydrogenase (4 subunits) has 4 binding sites for its substrate, and more than one of these may be involved in the interaction with the column. Even if not, the occupancy of these sites by free ligand is important in not allowing the freed protein to recombine with the adsorbent. The concentration of ligand needed depends on the dissociation constant between protein and free ligand, K_L. If the concentration of free ligand is 10 times K_L, then, in theory, only one-eleventh of the protein will not have bound ligand at any given moment: if the α value is 1.0 for this fraction, and zero for the protein–ligand complex, then the overall α value, assuming equilibrium behavior at all times, would be 0.1. In the previous edition of this book, I presented theoretical calculations showing how the value of α should vary in ideal circumstances, with examples of affinity elution from a cation exchanger. Without repeating these details, it is worth showing again the simplest example (but recalculated on a different basis), the response to various concentrations of ligand with pH as variable, Figure 7.28.

When designing an affinity system for a multisubstrate enzyme, it is helpful to know the sequence of binding of the substrates, since it is obviously necessary that a substrate to be used for affinity elution must bind to the free enzyme. The example of lactate dehydrogenase illustrates several points here. Lactate dehydrogenase will bind to 5'-AMP-affinity adsorbents and a large number of different dyes, and can be affinity eluted from them. It does not bind lactate or pyruvate unless NAD is already bound (either NAD^+ or NADH). Consequently, no amount of pyruvate will ever elute lactate dehydrogenase specifically, though at high enough concentration it might, simply by increasing the ionic strength. But NAD^+ at 1 mM or NADH at 0.1 mM are usually very effective in eluting the enzyme. Because NADH binds about 10 times more tightly, it can be used at 10 times lower concentration. But even more effective and specific for lactate dehydrogenase is a combination of NAD^+ at low concentration, ca. 0.1 mM, plus sulfite ions, 1 mM [216]. Sulfite acts as a substrate analog by fitting to the pyruvate-binding site, reinforcing the conformational change and effectiveness of affinity elution by NAD^+ alone.

As an alternative affinity adsorbent for lactate dehydrogenase, the inhibitor oxamate has been coupled to agarose; this, like sulfite, is an analog of pyruvate. In the absence of NAD^+ or NADH, the enzyme does not bind to oxamate, but does so in the presence of 1 mM NAD^+. The sample containing the enzyme is applied to an oxamate column in the presence of NAD^+, and the enzyme sticks on; all other proteins are expected to pass through. Then the NAD^+ is removed from the buffer, and the interaction with the oxamate column is lost: lactate dehydrogenase is eluted [217]. This is a form of negative affinity elution (see below).

Most examples of affinity elution from ion exchangers involve cation exchangers, for reasons mentioned before. Typically, the sample is applied at a lower pH than that being used for elution. For anion exchangers, the sample would be applied at a higher pH, and eluted after reducing the pH. But remember that when using anion exchangers, the ligands must be *positively* charged. If a negatively charged ligand is added, complex effects can occur; possibly the ligand itself will displace proteins unspecifically by a conventional ion-exchange procedure—the ligand could bind to the exchanger more tightly than the proteins. It will also bind specifically to its protein, but (at least theoretically) this should make the protein adsorb even more tightly, since it acquires more negative charges. Negative affinity elution has been attempted in which the enzyme is bound, washed at a critical pH in the presence of its ligand, and then washed off in the absence of the ligand. If the ionic conditions are such that the ligand is not substantially adsorbed to the exchanger, this can work.

7.5   Commonly Used Affinity and Pseudo-Affinity Adsorbents

Whereas this chapter so far has concentrated on synthesis and use of affinity adsorbents, most people are not going to make their own, but use a commercially available material. This short section is put in to mention some of the types available and their uses.

Small Ligands

There are many "small" ligands, i.e., those that are not biopolymers, many of them used for very specific purposes. For instance, benzamidine for isolating trypsin: benzamidine is a potent inhibitor of the enzyme. But more generally, the nucleotides as a group are important for isolating many enzymes. Dehydrogenases utilizing either NAD or NADP can be adsorbed on columns of immobilized NAD^+ or $NADP^+$. But more commonly, the ligands 5'-AMP and 3',5'-ABP (adenosine bis-phosphate—commonly, though incorrectly, known as 3',5'-ADP, adenosine diphosphate) are used. 5'-AMP represents half of the NAD molecule, and has affinity for NAD-linked dehydrogenases as well as kinases using ATP, and some other AMP-binding proteins. 3',5'-ABP represents half of the NADP molecule, and so specifically interacts with NADP-linked dehydrogenases. 2',5'-ABP is also available for purifying coenzyme A-binding enzymes. ATP columns are available; these must be treated with care, as trace amounts of ATPase will quickly convert them to ADP columns!

For purification of recombinant proteins with tags (flags), maltose and glutathione adsorbents are used (see Chapter 9) to bind the fused portions of the maltose-binding protein and the glutathione S-transferase, respectively, and there are many more similar materials coming on the market.

Biopolymer Ligands

In this category come proteins, polysaccharides, and nucleic acids. Of the many proteins immobilized as adsorbents, the most important are the lectins, Proteins A and G, and antibodies in general. Lectins are natural carbohydrate-binding proteins, and their value in protein purification is that they bind to many glycoproteins. Glycoproteins are widespread in plants and animals as extracellular proteins; carbohydrate moieties are covalently attached by posttranslational modification during excretion from the cell.

By far the most widely used lectin is concanavalin A, a mannose- and glucose-binding protein. Most glycoproteins have exposed mannose residues, which interact strongly with concanavalin A. Elution is by displacement with a buffer containing mannose or glucose. Other lectins are available with specificities towards galactose, N-acetyl galactosamine, etc.

Protein A and Protein G have specificity towards γ-globulins, and their use is described in more detail in Section 9.3. Antibodies, in particular monoclonal antibodies, have uses; most of those commercially available are either antibodies to other antibodies, e.g., anti-rabbit IgG raised in goats, or mixed preparations of γ-globulins. Use of antibodies in protein purification was described earlier in this chapter (Section 7.2).

Heparin is a sulfated polysaccharide, highly negatively charged, which mimics nucleic acids and so is useful in purifying nucleic acid-binding proteins and enzymes. Immobilized heparin adsorbents are widely used for this purpose, as a pseudo-affinity column, which can otherwise be treated much as an ion exchanger for purposes of binding and elution, since the interaction is principally electrostatic.

Finally, and also for purifying DNA-binding proteins, is DNA itself. A general DNA preparation from calf thymus is the normal material, and this is expected to bind the DNA-binding proteins in an applied sample. Control of ionic strength is needed to ensure that nonspecific cation-exchange effects are prevented, but that there is still sufficient interaction to hold on the true DNA-binding proteins. Calf thymus DNA is not likely to pull out proteins that interact with specific sequences of DNA; for that purpose, the specific DNA itself must be isolated or synthesized for immobilization to a column [218]. DNA columns must be handled and cleaned with care, to minimize contamination with DNases.

Chapter 8
Separation in Solution

The methods described in Chapters 4–7 involve phase changes. The proteins pass from liquid phase (dissolved) to solid (precipitated or adsorbed) and back again. Not all proteins withstand the stresses occurring in these methods; gentler methods in which the proteins remain in solution at all times are available. One of these, gel filtration, is one of the principal techniques used in purifying enzymes and other proteins, and it will be considered in detail. Other methods, grouped together as electrophoretic techniques, are less widely used, for reasons that will be outlined in Section 8.2. A third method of separation in solution involves phase partitioning where proteins may move from one liquid phase into another (Section 8.3), and a fourth method, ultrafiltration, is described in Section 8.4. As a general technique, separation in solution is an important procedure both in research and industrial protein purification.

8.1   Gel Filtration

Several other names for this method have been put forward, including *gel permeation*, *gel exclusion* and *molecular sieving*. However, the use of the term *gel filtration* is now so widespread and generally recognized that it will be used here. This name is rather unfortunate, since the procedure does not depend on the material used having a gelatinous nature, and it is not really filtration. Originally introduced by Porath and Flodin [219], the dextran-based materials of Pharmacia chemicals, Sephadex™, quickly acquired a reputation as a gel filtration medium for (what was then) rapid separation of macromolecules based on size. These cross-linked dextran beads were used both for protein separation and for protein molecular weight determination [220,221]. Since then the method has been devel-

238

Figure 8.1. Two-dimensional representation of
pores of gel filtration material and accessibility of
various areas by molecules of different sizes.

oped and extended by improving the gels to give faster processing and to
cover a larger range of molecular sizes. The basic principles are simple.
The gel consists of an open, cross-linked, three-dimensional molecular
network, cast in bead form for easy column packing, and optimum flow
characteristics. The pores within the beads are of such sizes that some are
not accessible by large molecules, but smaller molecules can penetrate all
pores.

The inaccessibility may be due to the fact that a particular pore is too
narrow for the molecules to pass down, or that, even if large enough,
there is no channel from the surface by which it can be reached. This is
illustrated in Figure 8.1 in two dimensions. The two-dimensional representation is misleading, since a cross-link apparently preventing passage
may be bypassed by motion out of the plane of the paper. Under optimum flow conditions all accessible pores can be filled as the proteins pass
by. Therefore, equilibrium rather than kinetic effects are involved. Any
particular grade of material has a certain proportion of pores accessible to
each size macromolecule, ranging from 0 to 100%. It is not possible to
make a material with just one size of pore which would give a very sharp
cutoff. The best gel filtration beads have pore sizes that result in 90%
exclusion of macromolecules that are 5–6 times the size (volume) of
those macromolecules that are excluded from only 10% of the bead volume. This means that 80% of the pores (at best) cover a twofold range of
linear dimensions for spherical particles, and this gives optimum separation of similar-sized protein molecules.

Diagrams to demonstrate the theory of gel-filtration chromatography
are usually clear enough, but often give a false impression of the dimen-

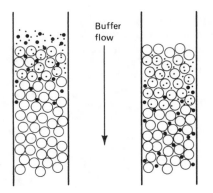

Figure 8.2. Simple diagram illustrating principles of gel filtration to separate different sized molecules in a column. Large molecules are excluded from most of the available column volume, and so move rapidly through ahead of the "solvent front."

sions of things. Thus, Figure 8.2 illustrates the principles of the technique, but the true dimensions are approximated better by Figure 8.3. Beads are in the range 10^3 to 10^4 times the linear dimensions of the molecules, and their surfaces on a molecular scale are of course "fuzzy."

In a column of gel-filtration medium, the behavior of a particular size molecule can be expressed in a number of ways. It can be related to the total column volume V_t and the void (outside the bead) volume V_0 by expressions such as $K_{av} = (V_e - V_0)/(V_t - V_0)$, where V_e is the elution volume of the molecule being considered, and K_{av} is a coefficient which defines the proportion of pores that can be occupied by that molecule. This expression is independent of column size, and depends only on the protein behavior and the nature of the gel beads. Alternatively, the elution volume can be related to the elution volumes of a number of other molecules of known size, from which the molecular size of the unknown can be estimated by simple extrapolation. The elution volume relates, theoretically, to the Stokes radius, not to the size or the molecular weight. The Stokes radius describes a sphere with a hydrodynamic behavior equivalent to that of a particular irregularly shaped particle. Consequent-

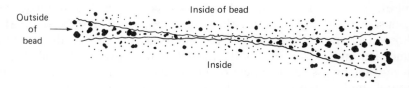

Figure 8.3. Scale diagram to illustrate relative sizes of molecules and beads used in gel filtration.

ly, provided that the shapes of both the unknown and the calibrating proteins are similar, the molecular weights can be used in calibrations.

For protein purification, the exact position of elution is not critical, provided that the unwanted proteins are well separated. Usually it is desirable to arrange for the protein required to elute between one-third and two-thirds of the way down the separation range, since here the resolution is greatest; however, for particular problems, other choices could be made (see below). Sephadex (cross-linked dextran) was used for over 20 years, and will continue to find uses, particularly where speed is not important. Cross-linked polyacrylamide beads (Biogels) have also been used widely, and cover a similar molecular size range. For very large molecules, from proteins up to virus-sized aggregates, agarose gels have been more suitable. Attempts to introduce controlled pore glass media into protein chemistry have not been so successful because of adsorption effects; these can be overcome, but the advantages of the rigidity of glass beads have largely been surpassed by recent developments. The major problem with dextran, agarose, and polyacrylamide beads has been their softness; very gentle pressures, including osmotic pressures during chromatography, cause distortion, irregular packing, and poor flow characteristics.

The original types of gel-filtration media are now largely replaced by more rigid cross-linked gels, though many are still based on agarose. Some of the best known of the modern media are the Pharmacia ranges (Sephacryl^{TM}, Superose^{TM}, Superdex^{TM}), Trisacryls^{TM}, the TSK range (Merck) and Toyopearl (Toyo-Soda) (see Appendix D).

It should be noted at this point that not all gel-filtration materials behave completely inertly. Adsorption of proteins under certain buffer conditions has been noted [222] and, indeed, can be made use of as a separation procedure. Partial adsorption delays the elution so that a protein emerges as though it has a smaller molecular size than is actually the case. Adsorption may be of an ion exchange character, in which case it can be avoided by using a high ionic strength buffer, but is more likely to be hydrophobic, in which case high ionic strength should be avoided.

With this great range of possible media, the beginner needs some advice as to what to purchase. Unless it is known that the protein of interest is of unusually low ($<15,000$) or high ($>10^6$) molecular weight, two or three materials should suffice, for instance, Sephacryl S-300, S-200, and Superdex 75 will suit most purposes. Most of the high performance materials cover a very large range of molecular sizes, which makes them poor at resolving large amounts of closely similar proteins, but ideally suited to small-scale separations and analytical work.

Whatever grade of material is used, certain values are constant. With spherical beads it is inevitable that V_0 equals between 25 and 30% of V_t, depending on how tightly they pack in the column. The useful range for resolving proteins lies within about 80% of the remaining volume

($V_t - V_0$), i.e., about 55% of V_t. The total volume accessible to liquid is slightly less than the total column volume, because the solid material of the gel matrix plus tightly bound water occupies a finite volume. For a column of 100 ml total volume, one can expect resolution of proteins within the range of elution volumes 35–90 ml, with the sharpest resolution between peaks of different molecular weights eluting at around 60 ml. The actual resolution depends not only on the above features, but also on the extent of diffusion and nonideal behavior in the column, Proteins have relatively low diffusion coefficients, so zone-spreading due to chromatographic partitioning (see Section 5.1) is more important than simple diffusion. But in additon to these fundamental factors, there are nonideal behaviors. First, gravitational instabilities can cause the more dense solution behind the protein front to break ahead of the theoretical position (Figure 8.4). Second, turbulent flow, as liquid squeezes through the narrow spaces between the beads into the wider voids beyond, leads to additional spreading. Figure 8.5 illustrates the phenomenon of turbulent flow between the bead particles. The gravitational instability at the *front* edge of the protein band could be solved by running a column upward, but then the trailing edge, normally gravitationally stable, would break up. Only by running in zero gravity could this be solved. The turbulent flow situation cannot be countered except by reduction in flow rate, which will alleviate but not eliminate the problem. And if the flow rate is cut back a lot, not only does it make the run inconveniently slow, but molecular diffusion effects increase.

Both of these nonideal behaviors are decreased by using smaller bead particles, resulting in the disturbances operating over shorter distances. In addition, smaller beads allow for a larger plate number, at least when the

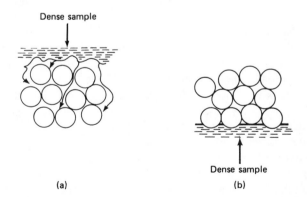

(a)   (b)

Figure 8.4. Gravitational effects at buffer-sample boundaries during application of sample to a gel filtration column. In (a), the dense sample runs ahead between the beads, causing a diffuse leading boundary. In (b), the dense sample applied in an upward flow maintains a sharp leading boundary. (From Scopes [66].)

Figure 8.5. Turbulent flow in gel filtration causing greater spreading of protein bands than diffusion alone.

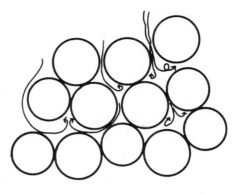

volume of sample applied is small relative to the total column volume. Thus a finer grade of bead will give a better resolution. Moreover, smaller beads mean that equilibrium between outside and inside is reached more quickly because the diffusion distances are smaller, so a fairly fast flow rate can be used without losing equilibrium conditions. Unfortunately, the smaller the beads, the higher the pressure needed to force the liquid down the column, and with nonrigid beads, this results in distortion and eventual blockage of flow, hence, the recent trends toward developing more rigid beads that can withstand higher pressures.

Practical Procedures

The range of possibilities in gel filtration is much smaller than with ion-exchange chromatography. The buffer used, within reasonable limits, *should* make no difference in the resolution achieved, and the only real variable is the grade of material, that is, the effective fractionation range that the material is designed for. Assuming that one has two or three of the more easily handled rigid gels, it is just a matter of choosing which to use and what size column is needed. The sample is applied in a small volume, because the volume containing each component protein will keep increasing due to the diffusion and nonideal flow effects discussed above. Since the effective fractionation volume is approximately half the volume of the column, sharpness of resolution relies on application of the sample in a small volume. This should not exceed 3% of the total column volume; smaller volumes than this will give slightly better results, down to about 1% of the column volume. Below this value, diffusion, natural chromatographic spreading, and nonideal flow will result in much the same spreading, however small the sample may have been (Figure 8.6).

On elution, 90% of each component is typically contained in a volume representing about 8–10% of the column volume, the spread depending somewhat on the elution position. The first fractions to be eluted, the largest-size molecules, have spent less time in the column, and have been

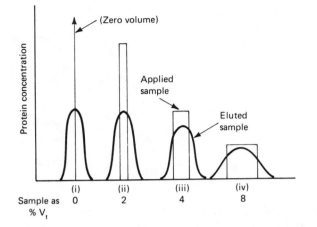

Figure 8.6. Effect of sample volume on elution pattern from a gel filtration column. Despite the theoretically zero volume in (i), diffusion spreads the peak to a volume similar to (ii). In (iii) the sample volume is larger than ideal, resulting in a larger total elution volume. In (iv) the sample volume is much too big except for separations of molecules differing in molecular size by a factor of at least 4. (From Scopes [66].)

subjected to little chromatographic processing. Consequently, the spreading is less than for fractions eluted later. Moreover, the smaller molecules have higher diffusion coefficients, so that they spread out still further. An idealized representation of these situations is shown in Figure 8.7. If the desired protein is in the lower size range, it will spread more and will have more chance of being contaminated by similar-size proteins. If it elutes early, the band will be sharper, but if a lot of larger-size unwanted protein is present, this has more chance of overlapping with it (Figure

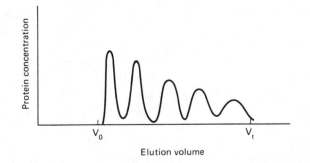

Figure 8.7. Ideal separation of five proteins, molecular weights differing each by a factor of 2, on a gel filtration column. The same amount of each protein was applied to the column; the peak heights decrease because of greater diffusion of the smaller molecules. (From Scopes [66].)

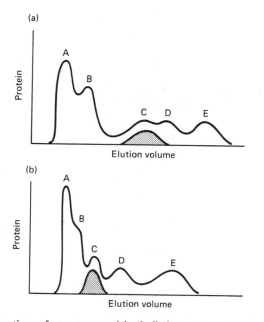

Figure 8.8. Separation of an enzyme (shaded) from other proteins on gel filtration column, using different-pore-size materials: (a) large pores—the enzyme is eluted late, and is well separated from proteins A and B, but not from D; (b) small pores—the enzyme is eluted close to the void volume, and overlaps with protein B; it is well resolved from protein D.

8.8). Thus, the choice of fractionation range must be made with consideration of the sizes of the main contaminants as well as the desired protein.

The column dimensions are important, shape as well as total volume. In ideal conditions the shape would not matter much; indeed, a squat column would have the advantage of a faster flow rate and less turbulence, leading to less spreading (see Figure 1.7). On the other hand, the resolution might be less because the number of beads passed, which determines the plate number, is smaller. This will not be so if the plate number is determined mainly by the *sample* size rather than bead size. But the main disadvantage of a squat column is the difficulty of designing and operating it so that completely homogeneous sample application and buffer flow are achieved. Very minor deviations from the perfect disc containing each component would lead to poor resolution (Figure 8.9).

A long, thin column is unlikely to suffer so much from horizontal plane distortion, and such distortion would matter less anyway, since the sample would occupy a greater column depth (Figure 8.9). In practice, most routine gel-filtration chromatography is carried out in columns 20–40 times longer than their diameter. But on scaling up for commercial operation, it is the column width that is increased, until diameter may actually

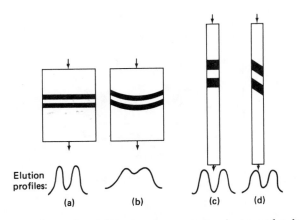

Figure 8.9. Comparison of gel filtration in a squat column and a long column. Ideal behavior (a) and (c) should give identical results. Minor deviations in flow in squat column (b) gives poor resolution, but in long column (d) they do not matter so much.

exceed height in columns of many liters' capacity (see Chapter 10). It is relatively easy to control even application on a large column, and bead size is not an important determinant of plate number in such columns, provided that the *linear* flow rate is not increased.

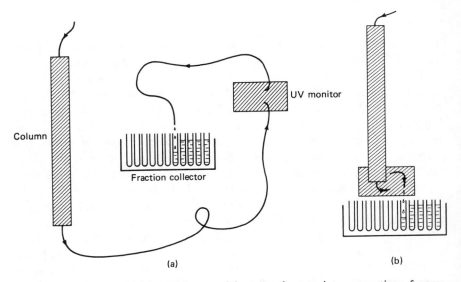

Figure 8.10. Incorrect (a) and correct (b) ways of arranging connections from a gel filtration (or any other) column to protein monitor and fraction collector. The length of tubing should be kept to a minimum, and tubing should be as narrow as possible without restricting the flow.

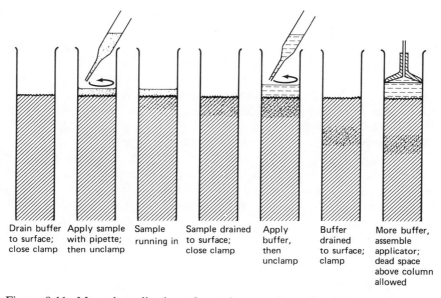

Figure 8.11. Manual application of sample to surface of gel filtration column. Care must be taken not to disturb the flat surface of the gel.

The compactness of a protein band can depend on the mode of sample application, discussed below. Equally important is attention to the fate of the protein band as it emerges from the column. Longitudinal mixing in tubing that is too wide is a common mistake, spoiling many excellent separations. Similarly, any flow cell for monitoring the protein must have small dimensions to minimize such mixing, and must be placed as close as possible to the outlet from the column (Figure 8.10).

The column should be poured in a single step, if possible, using a suspension of no more than 2 times the settled volume, so that there can be little separation of large and small beads during the settling (see Figure 1.11). This packing should be as fast as possible without compression. A good way of creating a higher flow rate is to attach a long piece of wide tubing to the bottom of the column, so as to increase the total liquid height; the pressure drop across the column itself is then greater.

Sample application is a most important consideration. The more evenly it is applied, the better the final result. Well-designed sample applicators, creating an even flow across the whole surface of the gel medium, can be used successfully (see Figure 1.8). Do not allow any substantial dead space between the top surface and the applicator, as this will create mixing (see Figure 1.10). Avoid applying a sample that is less dense (e.g., containing acetone) than the buffer, since the follow-up buffer will tend to run below the last bit of sample, and this last bit then trickles in slowly, long after the main part is in the gel. Test an applicator by using a colored sample; if uneven sample bands are observed, manual application is advisable. Remove the applicator, allow the buffer to drain to the gel surface, close a clamp on the outlet tubing, then apply the sample with a Pasteur pipette, gently, allowing it to run 5–10 mm down the side of the column, and at the same time move the pipette around the column (Figure 8.11). When all the sample is on, open the clamp and let the sample run into the surface. Then repeat the process with buffer (about the same volume as the sample), allow it to all run in, add more buffer, then fit in the applicator for a continuous buffer supply.

Remember that the buffer that the protein will be eluted in is the buffer present in the column before starting; it does not matter what the sample is washed with or eluted with. So if the desired protein needs to be eluted in a special, rather expensive buffer, don't waste it. Wash in 1 column volume of the special buffer prior to application of the sample. Then follow the sample through the column using a cheap convenient solution.

Gravitational instabilities during sample application (Figure 8.4) are particularly marked when a very dense solution, such as a redissolved ammonium sulfate precipitate, is applied to the column. The problem is greatest when using soft gels such as Sephadex G-200. With the modern rigid bead types, the firmness of the surface helps to prevent excessive mixing as the dense solution layers onto the less dense buffer. Nevertheless, better resolution, maintaining a tight band of sample, can be obtained by applying the sample upward at the bottom of the column, then when it has all run in, inverting the column so that less dense buffer layers onto the surface (Figure 8.12). It is not always convenient to do this, but it would be worth trying if resolution is not as good as expected.

Finally, here is a summary of the practical aspects of operating laboratory-scale gel filtration. Column size should be between 30 and 100 times the volume of the sample. The starting protein concentration should ideally be in the range of $10–20\,mg\,ml^{-1}$, $30\,mg\,ml^{-1}$ at the most. Thus, 100 mg of protein should be in a volume of not less than 3.5 ml, preferably about 5–6 ml, and a column of volume $200–250\,cm^3$ would be ideal. The length of the column should be about 20–40 times its diameter, so in this case, a column of 2.5 cm diameter and 50 cm in length would be suitable, or one 1.8 cm by 100 cm. One can determine an ideal column size as follows: diameter $= \sqrt[3]{m}/10$ cm, where m = amount of protein in mg and the length $= 30 \times$ diameter.

Flow rates depend on the type of bead and, in particular, their size (see Section 5.4). Modern rigid materials can sustain a flow rate that is too fast for equilibrium conditions to be attained, so some cognizance of the manufacturer's recommendations is needed. However, it has been common over the past few years to put speed ahead of resolution as a desirable trait. Yet, for long purification protocols, it may matter little whether the gel filtration step takes 2 hours or 4, but resolution is likely

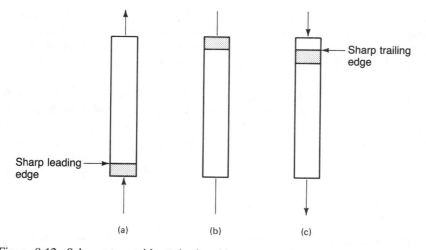

Figure 8.12. Scheme to avoid gravitational instabilities as illustrated in Figure 8.5: (a) the dense sample is applied to the bottom of the column with upward flow; (b) when the sample has all been applied, the column is inverted; (c) less dense buffer follows the sample—the column is run with downward flow.

to be better at the slower speed. So it is best to use the lower end of the manufacturer's recommendation. With bead sizes in the $50-100\,\mu$m range, flow rates of $10-20\,\mathrm{cm\,h^{-1}}$ at ambient temperature are optimal, and this can usually be achieved by a buffer head of $1-2\,$m, eliminating the requirement for a pump.

The time of separation can be calculated from the column dimensions. For example, in a column 2.5 cm in diameter and 50 cm long, if the protein was eluted in the middle of the separation range, it would emerge about 3 h after application (at a flow rate of $20\,\mathrm{cm\,h^{-1}}$). Finer, more rigid materials used in high-performance columns allow faster separation under pressure. Preparative gel filtration is not suited to very small beads at high pressures; moderate-pressure techniques are becoming common, using beads of around $30\,\mu$m diameter on HPLC equipment at $1-2$ atmospheres pressure, and flow rates between 50 and $100\,\mathrm{cm\,h^{-1}}$.

It should be noted here that resolution decreases as the plate number goes down, and the plate number in gel filtration is determined partly by the ratio of sample size to total column volume, but also by the bead size and homogeneity of the beads. Typical low pressure gels such as Ultrogels, Biogels, and Sephacryls, have an approximately twofold range of bead sizes around $80-100\,\mu$m. Consequently, the beads do not pack as ideally as "monodisperse" beads with a narrow size range; this affects the plate number also, and below a certain column size, resolution achievable begins to drop off sharply. For the above-mentioned materials, this occurs at about 200 ml; smaller columns than this will not give optimum separation. Similar calculations can be done for other materials; for instance, Superose at 30 μm (narrow range of sizes) will give optimum resolution down to a column size of about 20 ml. In general, there should be at least 5×10^8 beads in the column—somewhat fewer for a monodisperse material.

Gel filtration has become one of the major methods in protein purification, second only to ion-exchange chromatography, and complementary to it. Generally, it is used at a later stage in a purification protocol, often the last, and being a gentle procedure, often gives near to 100% recovery of bioactivity.

8.2   Electrophoretic Methods

Separation of protein mixtures by electrophoresis began early in the twentieth century. Until relatively recently, the technique was suited only for analytical purposes, and there is still no widely used electrophoretic preparative system. The original method involved separation of the charged protein molecules in free solution in a complex apparatus developed by Tiselius, which was widely used for 30 years or more, up to the 1950s. Development of zonal electrophoresis, especially in gels, quickly displaced the Tiselius apparatus by virtue of ease of operation, better resolution, multiple sample analysis, and much smaller quantities being needed. But these were only analytical methods. There have been many attempts to use electrophoresis to separate significant quantities of protein; a lot of them were quite successful. But the equipment required, and difficulty of using it has resulted in many commercial designs being withdrawn after unsuccessful marketing.

Simple electrophoresis describes the situation in which the protein molecules are separated in an electric field at a constant pH and current. Other electrophoretic methods include isoelectric focusing, in which separation occurs in a pH gradient, and isotachophoresis, in which the electric field varies according to the conductivities of the components being separated. Although these have been used mainly in the analytical mode, each is capable of being adapted to a preparative scale, using tens or even hundreds of milligrams of protein mixture.

Electrophoretic steps are not often employed in protein purification procedures. The reasons are many and include the complexity and expense of suitable equipment and the difficulty of operating it. A second, but not insignificant, consideration is that electrophoretic separation depends mainly on differences in charge/isoelectric points, a principle that is well exploited in ion-exchange chromatography. Only in the high-resolution methods of gel electrophoresis (exploiting size differences as well) and isoelectric focusing is electrophoresis capable of achieving separation that cannot easily be obtained in one step by any other method.

But preparative gel electrophoretic methods have a variety of problems, as will be outlined below. Thus, while a preparative electrophoresis system can be a valuable asset to the protein purification laboratory, beginners are advised not to consider it until other methods have failed.

Electrophoresis Principles

A protein molecule in solution at any pH other than its isoelectric point has a net average charge. This causes it to move in an applied electric field. The force is given by E, the electric field (volts $\cdot$ m^{-1}), times z, the net number of charges on the molecule. This force is opposed by viscous forces in the medium (just as in centrifugation, Section 1.2), proportional to the viscosity η, particle radius r (Stokes radius), and the velocity v; in a steady state:

$$Ez = 6\pi\eta rv \qquad (8.1)$$

The specific mobility $u = v/E$ is given by:

$$v = \frac{z}{6\pi\eta r} \qquad (8.2)$$

It is sometimes stated that electrophoresis in free solution separates molecules according to charge alone, independent of size. But from Eq. (8.2) it can be seen that the mobility u is inversely related to the Stokes radius. So a spherical molecule of 20,000 Da, charge -5, should have a different mobility to one of 40,000 Da, charge -5, but the same as one of 160,000 Da, charge -10. But in practice, the resolution in free solution would barely separate the first two of these proteins. This is because of the technical problems of maintaining stable boundaries and the considerable diffusion that occurs in a typical run. Free boundary (Tiselius) electrophoresis, operated as an analytical system, rarely produces more than about eight discernible components even from the most complex mixtures, because of diffusion, overlapping of components, and, sometimes, protein–protein interactions. Interactions between different proteins at low ionic strength may be of many types; electrostatic forces are the more important for the more soluble proteins, whereas hydrophobic interactions may predominate with the less soluble globulin-type proteins. The electrostatic interactions can be minimized by increasing the ionic strength (see discussions of salting in Section 4.2), but an increase in ionic strength causes a higher current in electrophoresis; consequently there is more heating, which aggravates other effects.

A major problem in designing a preparative electrophoresis apparatus is to allow for efficient heat removal, so that all parts of the system remain at a constant temperature. Thus, a compromise must be made between high salt concentration to minimize protein–protein interactions, and low salt concentrations to lessen heat output. Often all the salt is present as buffer. Low conductivity buffer mixtures can be employed, in which the ionic components are relatively large and have low electrophoretic mobility, e.g., $HTris^+$. As buffers, these ions have extra low mobility because part of the time they are uncharged, e.g., $borate^- \rightleftharpoons$ H-borate. Thus, Tris-borate for the pH range 8–9 has been a popular buffer for electrophoresis (see below). Electrophoresis is normally carried out at neutral or slightly alkaline pH, when most proteins migrate toward the anode.

Gel electrophoresis separates molecules of otherwise similar electrophoretic mobility if they have different molecular sizes. Originally, a gel of starch was employed [223], and starch gels are still very popular for analytical purposes. But starch gel has never been used successfully in a preparative way. Polyacrylamide gels [224] have become a universal feature of biochemical laboratories for protein (and oligonucleotide) analysis, and are also suitable for preparative electrophoretic techniques. Because the gel, a three-dimensional network of filaments forming pores of various sizes [225], presents a different effective viscosity depending on the size of the molecule, Eq. (8.2) for electrophoretic mobility, $u = z/6\pi\eta r$, becomes a more complex function. Within the "molecular sieving range," the mobility is linearly related to the log of the molecular weight for similarly shaped molecules [226]. In the presence of dodecyl sulfate (Section 11.1) [227], $u = u_0 (A - \log MW)/A$, where A is the log of the molecular weight of a molecule that would not move in the gel if the extrapolation extended linearly over the whole range (Figure 8.13). Since u_0, the mobility of a small molecule unaffected by the gel, equals $z/6\pi\eta r$, then:

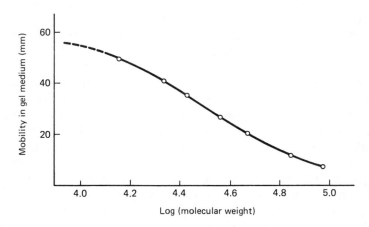

Figure 8.13. A plot of electrophoretic mobility versus log of molecular weight for similarly shaped protein molecules of the same charge density. Very small molecules all have the same mobility as the gel presents no extra viscous resistance.

$$u = \frac{z}{6\pi\eta r} \cdot \frac{A - \log MW}{A}$$

and we can regard the expression $A\eta/(A - \log MW)$ as the effective viscosity in the middle range.

Gel electrophoresis separates on the basis of both charge and size; the gel can be designed to give pore sizes suitable for the particular set of proteins being separated. A major advantage of gels is their stabilizing effect in minimizing convectional and diffusional movements of the proteins. Consequently, a sharp band remains sharp as it moves through the gel, so that complete resolution of proteins of very similar mobility can be achieved. These effects will be discussed further in Chapter 11. For analytical work, gels are excellent; for preparative work, there are major technical difficulties.

Methods for Preparative Electrophoresis—Horizontal Slabs

The requirements for successful electrophoresis are: (1) a separation channel, which may be a vertical column or a horizontal slab or block in which the proteins move; (2) a connection at each end of this channel via large buffer reservoirs to the electrodes; and (3) a cooling system to maintain the channel at a constant temperature. Because of the difficulties in maintaining stable boundaries in free solution, the buffer in the channel must be mixed with an inert powder or set in a gel to minimize gravitational and diffusional instabilities.

The simplest system is the horizontal slab. This is formed by mixing buffer with dry starch powder, Sephadex G-25 beads (the proteins being excluded), or powder of a variety of other polymeric compounds, to form a thick slurry which is poured into the apparatus; excess buffer is drained off. Alternatively, a very open-pore gel, e.g., 1–2% agar or agarose, is set in the apparatus. The ends may be supported by a wettable cloth or other material, making contact with the reservoirs of buffer which lead to the electrodes (Figure 8.14).

A vertical slot is cut in the block, and refilled with a slurry of the powder suspended in the protein sample (which has been preequilibrated in the same buffer). The current is then switched on and electrophoresis allowed to proceed. If the apparatus is in the cold room and the current kept low, natural convective cooling should be enough.

After the electrophoretic run, the block is sliced and the protein washed out from each slice with a little buffer. Resolution is not especially good and, as indicated above, would rarely be better than using ion-exchange chromatography. On the other hand, the proteins remain in solution at all times and are subjected to less stress. Capacity can be quite high (e.g., hundreds of milligrams) if the starting protein sample is sufficiently concentrated.

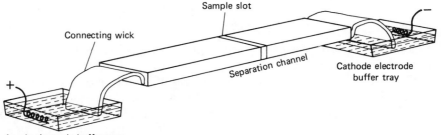

Figure 8.14. Horizontal preparative electrophoresis with powder or agar gel stabilization.

An interesting variant of the slab method is "isoelectric point electrophoresis" [228], in which the isoelectric point of the protein is known and the mixture is prepared in a buffer at that pH. The sample is put into a slot prepared in agar gel, and electrophoresis carried out. The protein, having zero mobility at its isoelectric point, remains in the slot while contaminating proteins move away in both directions (Figure 8.15). The initial requirement is to find the pH at which the protein does not move. This might not be the same as its isoelectric point as determined by isoelectric focusing (see Section 8.2). The isoelectric point may depend on the buffer used, because of binding of buffer ions. Binding of phosphate or other polyanions is a common occurrence, and this lowers the apparent isoelectric point.

Another factor affecting the real mobility, as opposed to the theoretical one, is electroendosmotic flow. This is the flow of bulk liquid solvents under the influence of electric field due to immobilized negative charge groups on the walls of the electrophoresis channel and on any stabilizing

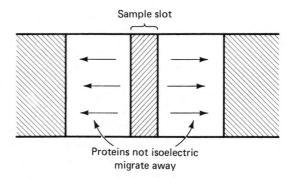

Figure 8.15. Isoelectric point electrophoresis; the enzyme required does not move, while other proteins with different isoelectric points migrate away in each direction.

Figure 8.16. The effect of electroendosmosis on the migration of proteins.

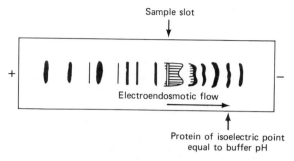

powder or gel used. Each ion has associated with it a number of water molecules which migrate with the ion. Immobilized anions, such as carboxylates on starch, cannot migrate, but the counterions associated with them do; being cations they move toward the cathode, resulting in a net movement of water in this direction. With materials such as starch and agar, electroendosmosis can be so large that many components are swept toward the cathode despite having a negative charge (Figure 8.16). In preparative electrophoresis, the electroendomosis is not a serious problem, since the *cause* of the movement of a protein band is of less concern than the separation achieved from other protein components.

Methods for Preparative Electrophoresis—Vertical Systems

Horizontal slab electrophoresis has not had a widespread application, although it does not require complicated apparatus. Vertical column electrophoresis has been used more extensively, especially using gels. Without a gel, it is possible to stabilize the buffer using a powder, as in the horizontal method described above, but much better is to use a sucrose density gradient to stabilize the buffer gravitationally. A dense sucrose solution is placed at the bottom of the column, grading to zero or near zero at the top where the sample is applied. The sucrose in the column is only to stabilize the liquid, minimizing convection currents. It affects the proteins' migration somewhat by increasing viscosity as they move down, thereby slowing them, but at the same time diffusive spreading is decreased. After electrophoresis has been completed, the separated bands are collected simply by allowing the liquid to run out, collecting the fractions with a fraction collector. This must be done slowly to avoid distorting the separate bands of protein as they move down the column. Although simple in concept, the practical problems in setting up the column, applying the sample, and avoiding convection currents have proved too great for general use.

The free solution method described above results in only crude separations; in free solution overlapping due to diffusional mixing is extensive,

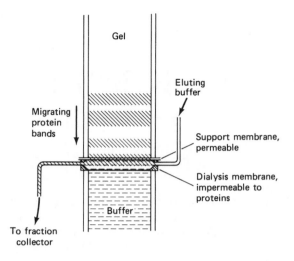

Figure 8.17. Elution of emerging protein bands from a vertical gel column as electrophoresis proceeds.

and the best results are rarely better than can be achieved by other methods. Small-pore gel methods are greatly preferable because of the high degree of resolution. However, there is one complication which results in a necessarily much more complex apparatus, and that is the problem of getting the proteins out of the gel after electrophoresis is complete. Simply applying hydrostatic pressure will not do; surface tension pressure is far too great. Chopping up the gel is possible, but even after finely mincing the particles, recovery of (active) protein is usually very low. Consequently, the procedure adopted is to allow each component to migrate right out of the gel column and collect it, as it emerges, into a perpendicular flow of buffer. This principle has been adopted on several designs of apparatus for preparative electrophoresis (Figure 8.17). The same basic concept applies to recent commercial equipment manufactured by Applied Biosystems, Foster City, CA (High Performance Electrophoresis HPEC system), and the Bio-Rad, Richmond, CA (Model 491). A difficulty is that the bands, quite concentrated in the gel, become extensively diluted as they are swept away by the buffer, unless a very low flow rate is adopted (which causes further problems). Although many home-made apparatuses for vertical gel preparative electrophoresis have been reported, e.g., [229,230] (and many more have not), a beginner is advised to try commercial equipment first.

Buffer Systems for Electrophoresis

The composition of the buffer used for pH control in electrophoresis is very important, because most of the current flowing through the separation channel is carried by the buffer ions. Very mobile ions (e.g., metal ions Na^+, K^+, Mg^{2+}, the simple anions F^-, Cl^-, Br^-, SO_4^{2-}, HPO_4^{2-})

carry much current because of their high mobility. If the electric field is to be maintained, this leads to heating, so unless there is a particular reason for the presence of such ions, they should be avoided. Bulky organic ions have much lower mobility, especially if they are buffering species of the type $n = +1$ or $n = -1$ (see Section 12.3). On the other hand, small, simple cations are more effective in preventing protein–protein interactions than these bulky ions, so if this is a problem, some simple salts may be needed.

Two systems of buffers for electrophoresis will be described: discontinuous and continuous systems. The latter are simple; the same buffer is used throughout, so that in the electrophoresis channel no change in buffer composition occurs. In order to lessen voltage losses across the connections from electrodes to the actual electrophoresis channel, the same buffer but at a higher concentration may be used (Figure 8.18a). Normally, buffer concentrations with total ionic strength in the range of 0.05–0.15 are used. This is a compromise between low conductivity, which allows application of a higher voltage but may result in protein–protein interactions, and a minimum interaction system with higher conductivity, in which case the current and voltage must be kept down to minimize heating. So many different buffer systems have been used that it would be unreasonable to list them here. For simple gel electrophoresis, the buffer most widely used is Tris borate at between pH 8 and 8.5, sometimes including a low concentration of EDTA. This is also widely used in agarose gel electrophoresis of DNA. As made up, less than half of the Tris and boric acid molecules are ionized, so the ionic strength (which determines the current flow) is far less than the concentration of the bufferring components, which can be up to $0.1\,M$ each. For somewhat

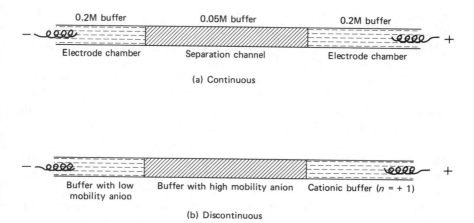

Figure 8.18. Buffer arrangements in (a) continuous and (b) discontinuous buffer system.

higher pH values, 2-amino-2-methyl-1,3-propanediol glycinate, pH 8.5–9 can be used; for pH 7.5–8, Tris-Tricine; for pH 7–7.5, imidazole-Mops (see Section 12.3). For special purposes, many low-pH buffer systems have also been described.

Discontinuous buffer systems are used in gel electrophoresis to sharpen up protein bands. Mainly used in analytical systems, they can also be used in preparative electrophoresis. The principle is that a Kohlrausch discontinuity is established between the starting buffer ions and the following ions which have a lower mobility (Figure 8.18b). The discontinuity junction is self-sharpening, and can often be observed as a result of refractive index changes due to sharp concentration changes at the discontinuity junction. The features that cause protein band sharpening are: (1) a low conductivity immediately behind the junction, which implies a local high electric field—thus, trailing protein molecules migrate more rapidly to catch up with the main band in front; and (2) a higher pH immediately behind the junction—this has the same results as (1). The principle of establishing buffer discontinuities is further exploited in isotachophoresis (see below).

Typical discontinuous buffer systems (with the discontinuity being in the *anions*) involve phosphate, citrate, EDTAate, or chloride as the mobile starting anion, to be followed by a relatively slow-moving anion which is usually a buffer of the type $BH = B^- + H^+$. The low mobility of the follow-up buffer is partly due to the fact that a proportion of it is uncharged at a given instant. At pH 8–9, borate ions or glycinate ions are most commonly used [231].

Isoelectric Focusing

Simple electrophoresis involves separation of proteins on the basis of their mobility at a particular pH. Isoelectric focusing involves setting up a pH gradient and allowing the proteins to migrate in an electric field to the point in the system where the pH equals their isoelectric point (Figure 8.19). To establish a pH gradient requires the use of polymeric buffer compounds which rather resemble proteins themselves, as they have large numbers of both positive and negative charges, and possess isoelectric points in the same pH range. Such buffers are called ampholytes; they, too, migrate to their isoelectric point, and as there are hundreds or even thousands of individual ampholyte species, they spread across the whole slab/column between the cathode contact (a weakly conducting base) and the anode contact (a weak acid). Thus, after a few hours of application of the electric field, the ampholytes have migrated and formed a pH gradient; the pH range depends on their composition (i.e., the range of isoelectric points of the ampholytes themselves). Any proteins present also move. Since the proteins are mostly larger, they move more slowly,

Figure 8.19. Principles of isoelectric focusing.

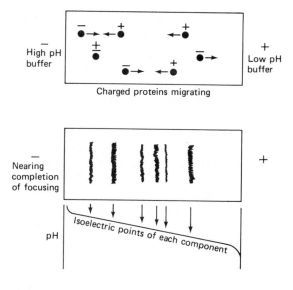

Charged proteins migrating

so the pH gradient becomes established before the proteins reach their isoelectric position.

The high resolution of this method is due to one great advantage over simple electrophoresis. This is implied in the term *focusing*. In most other protein separation methods, diffusion and zone mixing increase with time. In isoelectric focusing, diffusion is countered, because as soon as a protein molecule diffuses away from its isoelectric zone, it becomes charged and so migrates back again. When focusing is complete, theoretically nothing moves in the system, and each component is separated according to its isoelectric point. There should be almost no current (there are no other ions present). Consequently, high electric fields can be applied with little heating, and the focusing is rapid. In practice, a small current flows, and if the applied field is too great, the pH gradient can break down and become less clearly defined.

Isolectric focusing (electrofocusing) was originally designed as a preparative method, but it has found widespread use as an analytical procedure (see Chapter 11). Isoelectric focusing is more flexible in its demands on apparatus design than conventional electrophoresis because of a relative lack of heating. Use of a vertical column with sucrose density stabilization has been highly successful on occasion. A number of novel systems have been developed also. These include placing the sample plus ampholytes in a long, flexible tube and winding that tube into a horizontal coil. After separation, individual components should have separated gravitationally to the lower parts of the coil, which is cut and the material from each segment collected separately. More recently, Bio-Rad have introduced the Rotofor™, which also separates in free solution, with 15–20 fractions collected from adjacent parts of the apparatus after focusing is complete. Electrofocusing can also be carried out in a horizontal slab with a liquid stabilizer such as Sephadex powder, as is used in horizontal slab electrophoresis. The layer of powder may be only a millimeter thick, yet quite a large amount of protein (10s of mg) can be applied to a slab 10–20 cm wide. This method of preparative isoelectric focusing has largely superseded other preparative electrophoretic methods in slabs because of its much higher resolution.

But isoelectric focusing does have some problems and is by no means suitable for all tasks. The properties of the desired protein and, to a lesser extent, those of the contaminants are important. Two features are essential for clean results: (1) the protein must be stable at its isoelectric point; and (2) it must be soluble around its isoelectric point even at zero ionic strength. Neither of these properties can be taken for granted. It is often mistakenly thought that proteins are most stable at their isoelectric points. There is no real reason why this should be true, and in fact is often not; there are many examples of proteins denaturing rapidly at their isoelectric points. Most intracellular enzymes have a sharp acid-instability curve in the pH range 4–6; such a protein may not survive pH 6.0 for long. Yet if its isoelectric point is 5.5, by no means a low value, the method of isoelectric focusing creates exposure to a pH below 6.0. (Similar arguments arise for chromatofocusing, an alternative isoelectric focusing method; see Section 6.1.) Extracellular proteins may naturally exist at lower pH values (and so be more acid-stable) but also may have very low isoelectric points, so instability at the isoelectric point is also possible.

If the stability conditions are suitable, then the solubility situation must be considered. In a column, the formation of a precipitate at the isoelectric point will upset the electrical conductivity, and the precipitate may simply fall down to the bottom! Note that the protein is unlikely to be precipitated *close to* its isoelectric point. Precipitation is a solubility feature, and high enough concentrations to exceed the solubility are expected only after focusing has occurred to the isoelectric point itself (see Figure 4.3). But we must not only consider the protein that we are trying to purify. If one of the impurities has an insolubility problem, then if it precipitates it is likely to have an unsettling effect on the whole system. Precipitation can also occur by denaturation of an impurity as it gets out of its pH-stability range.

To conduct current to the slabs, at the anodal end a weak acid of the type $HA = H^+ + A^-$ is used, since any diffusion of this acid into the column causes it to ionize as it meets a higher pH, and the anions then return under electrophoresis. The protons are picked up by an ampholyte, which becomes positively charged, moves away from the acid end, and quickly neutralizes again, passing the proton on to the next ampholyte. In this way a small current passes, protons flowing from the anode to cathode. In an exactly analogous way, hydroxyl ions can flow from the cathode to the anode; the relative contribution of these two ionic flows to the total steady-state current depends on the pH range of the ampholytes and the pK_a values of the acid and base at each end. Ampholytes covering large or small pH ranges are available, and if the isoelectric point of the enzyme is known with reasonable accuracy, a narrow range (e.g., 2 pH units) can be tried at once.

Because of the focusing, the sample can be applied at any position, or even mixed in with the ampholyte buffer when preparing the column or slab system. However, the latter is not recommended, since contact with quite low and high pH values at the ends may cause denaturation and precipitation. The top of a sucrose density column is usually made the cathodic end, and the sample can conveniently be applied near the top (alkaline denaturation and precipitation are less likely than acid) and overlaid with a little more ampholyte before placing the cathode buffer on top. Care should be taken that the successive densities are correct; the sample protein will increase the density, so less sucrose should be present (Figure 8.20).

Further refinement of isoelectric focusing makes use of immobilizing the ampholytes. A slab acrylamide gel can be copolymerized with ampholytes that have acrylic groups in, so the individual ampholyte molecules are unable to move. But pH gradients can still be established, and can cover a very small pH range accurately. Although developed as an analytical tool, it has been scaled up to process gram quantities of proteins [232].

There is little doubt that isoelectric focusing, when used on protein mixtures which satisfy the criteria of stability and solubility at their iso-

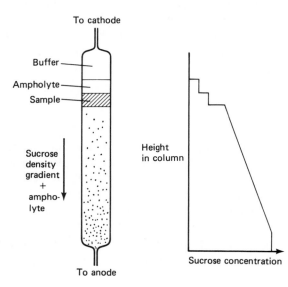

Figure 8.20. An isoelectric focusing column with sucrose density stabilization.

electric points, is the highest-resolution system, even though it does not discriminate between proteins on the basis of their molecular size. So many minor bands have been found with supposedly pure proteins that isoelectric focusing seems to be the ultimate analytical test. However, there are instances where interaction between proteins and ampholytes of very similar isoelectric points results in multiple banding as an artifact. Because ampholytes are not present as a continuum but as discrete compounds with closely spaced isoelectric points, they, too, focus into bands [233]. If an ampholyte of isoelectric point (IEP) 6.1 interacts strongly with a protein of IEP 6.0, a conjugate of intermediate IEP is formed. Because of the low charge on either component at around pH 6.0., the electric force is not sufficient to separate them. Thus, a protein of IEP 6.0 might show other conjugate bands around that value which are not due to heterogeneity of protein, but rather to interaction with discrete molecules of ampholyte. This possibility is decreased by having ampholytes with more individual components, in which case conjugates would have IEP values so close together that discrete bands would not be resolved; the only effect of conjugate formation would be a diffusion of the protein position.

Considerable work was done in the early days of isoelectric focusing on the use of low-molecular-weight, low-conductivity buffers for establishing a pH gradient as an alternative to the expensive ampholytes [234,235,236]. Although successful separations in such systems have been presented, the resolution, because of higher conductivity, has never been as impressive as when using polymeric ampholytes.

The final problem (which the low-molecular-weight buffers overcome) is the removal of the ampholytes from the protein sample after focusing. By definition, the ampholytes have the same isoelectric point, and so ion-exchange chromatography is likely to be of limited use. Gel filtration is suitable, remembering that a gel with lower size limits of at least 10 kDa should be chosen, as the ampholytes themselves are quite large molecules. For the latter reason, dialysis is not useful.

Isotachophoresis

Isotachophoresis is the third method of separating proteins in an electric field. Simple electrophoresis takes place in a constant electric field, at constant pH, and separates according to actual mobility. Isoelectric focusing is achieved with a constant electric field but a pH gradient, allowing the components to move until they become uncharged. Isotachophoresis also develops a pH gradient, but it allows all the components to move at the same speed through an electric field that is of different strength at the location of each component; whereas it separates on the basis of mobility

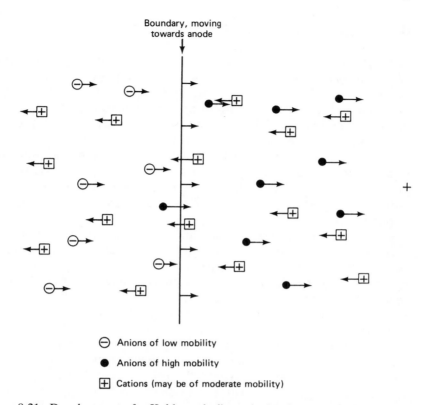

Figure 8.21. Development of a Kohlrausch discontinuity during isotachophoresis (also see Figure 8.18b).

(mobility per unit electric field), *actual* velocities of each component are the same. The buffer situation is such that the only significant anions are the proteins themselves. In the absence of ions of intermediate mobility, two protein ions of different mobility separate until all the faster ones are ahead of the slower ones, at which point a sharp boundary forms. The faster ions cannot actually move ahead of the slow ones and leave a void, since there are counterions traveling in the opposite direction across the boundary whose charge must be balanced at each location. Consequently, a sharp Kohlrausch discontinuity develops, in principle the same as when using a discontinuous buffer system in simple electrophoresis (Figure 8.21).

Isotachophoresis of proteins has not been widely used (it is more useful as an analytical tool for small ions), but it has significant potential for resolving components of very similar mobility, which may be required for removing small amounts of impurity in recombinant proteins intended for therapeutic use.

8.3   Liquid Phase Partitioning

The technique of counter-current distribution has been a popular method in organic chemistry. Organic compounds are separated by partitioning between two immiscible phases, typically an aqueous phase and an organic solvent. This has been tried with proteins, but with little success. The main problem was to find water-immiscible (or partially miscible) solvents in which proteins were soluble and stable. Few proteins are either stable or soluble in an organic solvent; the only real applications were for very small hormone-type polypeptides. But two-phase systems suitable for proteins are now available; they can be used in the counter-current distribution mode [237,238], but more convenient single step procedures are normally used.

For separation of proteins by liquid phase partitioning, it must be possible to create two immiscible liquid phases, each of which can dissolve proteins, with the proteins remaining bioactive. Addition of certain pairs of hydrophilic polymers to aqueous solutions can cause a phase separation without the presence of any hydrophobic solvent [239]. In particular, dextran and polyethylene glycol (PEG), when dissolved in water in appropriate proportions, develop two phases, a dextran-rich one which is denser and settles on the bottom, and a polyethylene glycol-rich phase on top. With more polymer components, still more phases can separate out; 18-phase systems have been described [240]. In such circumstances, a protein mixture distributes itself through the solution, and depending on a variety of solubility properties of the individual protein components, relative enrichment of a particular protein can occur in one of the phases. Separation of the phases is simple; a brief centrifugation may help to sharpen the boundary, after which a separatory funnel can be used (Figure 8.22). The partitioning of a particular protein component in this system is rarely extreme. Defining the partition coefficient as the *ratio* between the concentrations in the upper (PEG-rich) and lower (dextran-rich) phases, typical values for a simple PEG–dextran system lie between 0.1 and 1 [241]. Note that this definition of partition coefficient is different from that used in Chapter 5 for adsorption phenomena. Thus, most proteins are partitioned in the lower phase, and recovery of even the more PEG-favoring proteins in the upper phase may be only 50%. The distribution of various proteins in 3-phase systems has been reported [242,243]. The pH of the system is found to be an important factor [244].

So most proteins, as well as particulate matter and cellular debris, partition into the lower, denser phase in the typical dextran–PEG phase system. The breakthrough in the usefulness of phase partitioning came with the attachment of affinity ligands, mainly dyes, to PEG, which attract specific proteins into the upper, PEG-rich phase [245]. This is referred to as "affinity partitioning," in line with other affinity methods such as chromatography, precipitation, electrophoresis, etc. Polyethylene

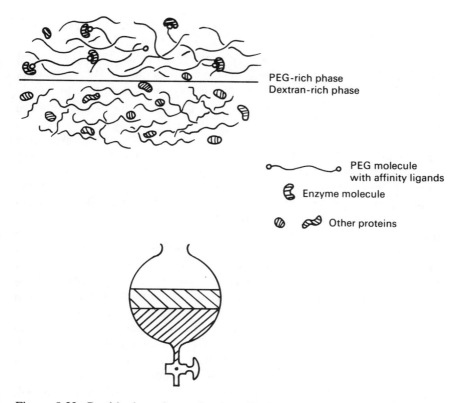

PEG-rich phase
Dextran-rich phase

PEG molecule
with affinity ligands
Enzyme molecule

Other proteins

Figure 8.22. Partitioning of proteins into liquid phases in the presence of an affinity ligand. Few protein molecules partition into the PEG-rich phase, but a specific enzyme is attracted into it by binding to the PEG-ligand complexes. After the phases have separated, the dextran-rich phase can be removed (*lower diagram*), leaving the desired enzyme with the PEG.

glycol molecules have two hydroxyl groups, one at each end of the molecule. These can be "activated" by procedures used to activate agarose (see Section 7.1), or can be reacted directly with reactive dyes such as the triazinyl variety (see Section 7.3). Small amounts of PEG-containing ligand are mixed with the bulk PEG, and the ligand–PEG complex partitions into the upper phase (Figure 8.22). Proteins with an affinity for the ligand enter the upper phase, and conditions are chosen such that most other protein stays in the lower phase. After agitation to mix the phases, the liquid is allowed to settle, and the phases separate (gentle centrifugation may be appropriate). Purification factors of 5, up to much higher values can be achieved by selecting the ideal conditions, and then repeating the distribution by replacing the unwanted lower phase by a fresh one.

The next step is to remove the desired protein from the PEG phase. This can usually be done by adding a high concentration of a phosphate buffer, which causes separation again into two phases, a PEG-depleted phosphate-rich lower phase, and PEG-rich upper phase, still containing the PEG–ligand complex. Due to the high salt concentration, the interaction between the PEG–ligand and the protein is lost, so the protein partitions preferentially into the lower phase.

Probably the greatest advantage of this system lies in large-scale and commercial protein isolations, particularly using yeasts and bacteria, for the whole cell homogenate/lysate can be dealt with without any prior centrifugation step to separate out the cell debris. The debris and unwanted proteins partition into the lower phase; the upper phase clarifies, with the desired protein extracted into it. The volumetric capacities are huge, with protein concentrations of $100\,mg\,ml^{-1}$ being handled with ease [246]. Indeed, high protein concentrations can affect and even cause phase formation. Most effort has gone into developing affinity phase partitioning for large-scale commercial enzyme production, because there are handling problems that make it rather unsuitable for laboratory scale purifications.

Being a single step process, phase partitioning does not have the resolving power of column chromatography. But multiple steps can be introduced with transfer of enriched phase into a new medium in the same way as classical counter-current distribution operates [238]. Alternatively, if the partitioning can be adopted in a chromatographic procedure, separation of components of even quite similar partition coefficients should theoretically be possible. Liquid–liquid chromatography does not at first sight seem to be possible, but, in fact, it is possible with one of the liquid phases immobilized. The method has been successfully demonstrated for separations of nucleic acids [247], and recently successful application to proteins has been reported [248]. By preequilibration of a suitable support medium, such as agarose beads, with one of the phases (e.g., the dextran-rich phase), the other phase is the moving buffer in the column (Figure 8.23). Surface tension prevents mixing of the two components; the dextran-rich phase remains within and immediately around the surface of the beads. Simple partitioning of proteins on such a column is possible, but affinity partitioning is even more effective. The system becomes very analogous to conventional affinity chromatography in this case, the only difference being that the bound ligand is not immobilized covalently to a solid support, but trapped within the beads in a soluble form.

Some of the problems associated with the method include, as mentioned above, handling problems with viscous solutions, and the cost of the polymers, in particular, dextran. Consequently, there has been an effort put into finding alternative cheap polymers that can be used on a large scale commercially, such as starch derivatives to replace the dex-

Figure 8.23. Liquid-liquid chromatography; one liquid phase is trapped by surface tension to the beads in the column, and the other phase is mobile.

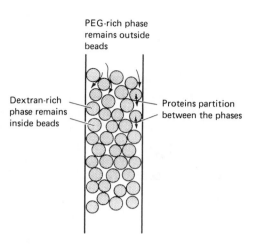

PEG-rich phase remains outside beads

Dextran-rich phase remains inside beads

Proteins partition between the phases

tran. Novel partitioning systems also include the use of Triton X-114 (Section 2.4), which separates into detergent-rich and -poor phases on warming [249], particularly suitable for membrane proteins, and co-polymers of ethylene and propylene oxide, which also have temperature-dependent phase properties [250].

Phase partitioning, because it does not involve adsorption, is gentler than adsorption chromatographic methods, and so may be especially useful for labile enzymes. It is rapid, has high capacity, and requires little in the way of equipment. These many pluses will ensure that the technique has an important future. A recent review can be found in [251].

8.4 Ultrafiltration

A final method to note in this chapter is the use of ultrafiltration, using gas pressure to force liquid through a membrane (see Figure 1.13b). If the membrane pores are such that small protein molecules can pass through with the "ultrafiltrate," a separation is achieved. Larger molecules are retained and concentrated relative to the starting solution. Because of the range of pore sizes in an ultrafiltration membrane, there is no absolute cutoff point. A proportion of molecules of sizes close to the stated cutoff size will pass through, the remainder staying behind. But if the protein of interest has a size much smaller or larger (say, 30–50% at least) than the quoted cutoff size, virtually all of it will appear in either the ultrafiltrate or the "retentate."

The method is, obviously, less discriminating than gel filtration (though based on the same principle, size of molecules), as it only provides two fractions, "bigger molecules" and "smaller molecules." Nevertheless, there are advantages in certain applications, especially when one has a

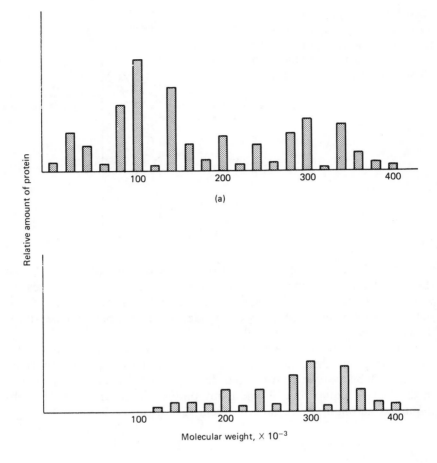

Figure 8.24. A possible distribution of proteins by size and amount before (a) and after (b) ultrafiltration through a membrane with pores of nominal cutoff at 200,000 daltons (proteins all assumed spherical).

large volume of dilute protein solution such as may be obtained from an adsorption column. Before gel filtration this would need concentrating, probably by ultrafiltration, so if a membrane is chosen that just allows retention of the protein required, a separation of the smaller proteins is achieved without any extra effort. This is of most benefit when purifying proteins of larger size. For instance, suppose the protein we are interested in has a molecular weight of 300,000, and other proteins are distributed in size and amount as illustrated in Figure 8.24a. Using a membrane with nominal cut of 200,000, most of the proteins of size less than 150,000 pass through the membrane, and the resulting distribution is shown in

Figure 8.24b. In this example, more than half the protein has been re-moved, which allows the retentate to be concentrated to a smaller volume than if a conventional 10,000–20,000 Da cutoff membrane had been used. Ultrafiltration slows up as the protein concentration increases, so redilution with buffer once or twice may be useful before finally collecting the retentate. This is not so desirable if the protein wanted is in the ultra-filtrate, since it is already dilute, and extra buffer passing through makes it more so. A second ultrafiltration using a membrane with smaller pores can be used to concentrate the first ultrafiltrate.

Ultrafiltration is mainly used as a rapid method for concentrating pro-teins, without expecting significant purification. Membranes with cut-off sizes of 20,000 or 10,000 Da are used, so that few proteins are not retained.

Chapter 9
Purification of Special Types of Proteins

9.1 Recombinant Proteins

In the first edition of this book, the concept of expressing proteins in recombinant form was only just being realized; there were few examples and no general technology to describe. Ten years later, purification of proteins from transformed recombinant host cells has become almost as common as purification from the natural source. The whole process can be divided in two: the molecular biology of gene isolation and expression, and the purification from the recombinant cells. But the latter can be strongly influenced by the former; indeed, the form in which the protein is expressed can include the means of purification, so that the task of protein purification becomes relatively trivial.

Before detailing methods that are specific to recombinant proteins, I shall first summarize the molecular biology steps. One either starts with the protein, goes through the molecular biology, then returns to isolate the protein again, or one has the gene first. Many genes are now being identified and characterized without the protein having first been isolated, and in fact many encode unknown proteins. It is now fully possible to express a gene which encodes a protein of completely unknown physiological role, and purify that protein. But, usually, a protein of known physiological significance is the target.

So genetic techniques may be able to isolate DNA containing the gene, which can then be expressed, and the protein purified. But more commonly, the protein is the starting point; it is purified in sufficient quantity from its natural source so that some amino acid sequence can be determined. If enough suitable sequence is obtained, the molecular biology commences, using a degenerate oligonucleotide to fish for the gene from a "library" of recombinant clones. The demand for information from the minutest amounts of protein has led to ever-increasingly sensitive technologies for amino acid sequencing, such that a single spot off an electrophoretic gel can now provide enough protein. Alternatively, if the protein has been isolated in sufficient quantities already, antibodies (usually polyclonal) can be raised, and the screening of gene libraries can be for expression of protein. This has the advantage over oligonucleotide screening in that only gene fragments that are actually expressed (which is what is ultimately wanted) will be detected. On the other hand, antibodies are not guaranteed to succeed, as expression may not occur in the particular host/vector system chosen; in theory at least, the oligonucleotide should eventually work. A third method of screening an expression library is to detect actual bioactivity (usually enzyme activity) by a specific staining technique. If this succeeds (and it is even less generally applicable than antibody screening), then one probably has the complete gene, and it has not been necessary ever to have isolated the protein. It is perfectly feasible to screen a genomic or cDNA library made from a species that has never been studied biochemically before, for an enzyme which can be detected by a staining technique, isolate its gene, and then express and purify the enzyme.

Terminology of Recombinant Proteins

Recombinant: Describes the technology of transplanting a gene from its natural environment to a host organism. Recombinant protein refers to the foreign protein being expressed in a living organism: also known as heterologous expression.

Host cell: The organism in which the recombinant protein is being expressed.

Vector: The means by which the foreign DNA is transported to and reproduced inside the host cell, usually a plasmid or virus particle.

Expression: Detectable protein being produced from the foreign DNA.

cDNA (copy DNA): DNA artificially produced using reverse transcription from messenger RNA isolated from the donor organism. The technology used does not allow cDNA to be produced from prokaryotes.

Library: A collection of recombinant cells containing different fragments of the donor DNA in the vector. An expression library is one in which the host/vector system is designed to result in expression of protein from these DNA fragments.

A summary of some of the possible procedures is shown in Figure 9.1.

Having obtained the gene, and developed a suitable expression system, the protein is to be purified. In many cases there are still problems to be overcome; these include the question of correct glycosylation (or lack

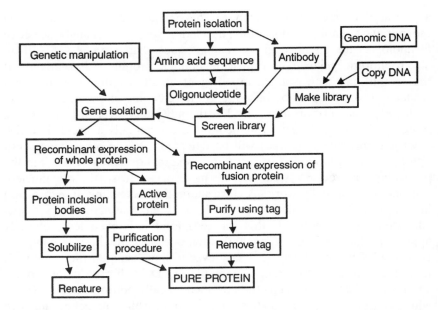

Figure 9.1. Possible steps in the use of recombinant technolgy to purify a protein.

of any) when dealing with extracellular eukaryotic proteins; whether the expressed protein folds correctly in the host organism, especially in regards to disulfide bond formation; and whether the expressed protein remains soluble, or aggregates into "inclusion bodies." The latter consist of insoluble and incorrectly folded recombinant protein in a cellular compartment which may make up much of the cell. Inclusion bodies occur mainly in prokaryotic expression systems, which nine times out of ten means *Escherichia coli*. Their formation can be regarded as a problem; however, advantages are that they indicate that a large proportion of the cellular protein has been expressed as the desired foreign protein, and it is possible to isolate the inclusion bodies relatively cleanly from other proteins by differential centrifugation. The big disadvantage is that the protein is not in a native state, and some solubilization/naturation step is needed. (The word "naturation," meaning to create the native state, is used here instead of *re*naturation, because, strictly speaking, the protein has never yet been in a native state.) Although, in general, inclusion bodies occur more frequently when expressing eukaryotic proteins in *E. coli* than when expressing prokaryotic proteins, there are many exceptions, and there can be no general rule of what to expect; the physical properties of the protein being expressed are all-important.

One of the chief reasons for purifying a protein from a recombinant expression system is quantitative; the natural source contains so little of the protein that purification is too laborious and unfruitful. So it is obviously important that the recombinant expression level is high. Around 1% of the host protein is considered to be minimally satisfactory, but higher levels are preferable; 5–10% is commonly achieved, and makes purification simple. To increase the expression level, many different host/vector systems are available and new ones are continually being marketed. High plasmid number, strong promoters, and inducible expression (e.g., by temperature) may all be considered, but ultimately, both the messenger RNA and the protein itself must have compatible properties with the host system if the protein is to be expressed in high amounts. If the protein is in a bioactive form, its activity must not upset the host cell so much that growth is prevented; if that is the case, then an inducible expression vector must be used, to switch on after growth. Apart from the numerous *E. coli* expression systems, other bacteria, yeasts, mammalian, plant, and insect cells are all available as well-characterized systems, but each requires experience and appropriate equipment.

If the expression level is low, or the purification process not suitable, then fusion proteins may be considered. With cDNA libraries, fusion proteins are the norm, as the cDNA may be incomplete and lack a ribosome binding site needed for translation of the message. It is now common

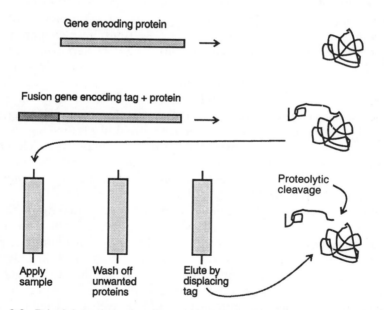

Figure 9.2. Principles of tagging for purification of recombinant proteins. The gene encoding the protein is combined with DNA sequence encoding the tag, which may be a short stretch of amino acids, or a whole protein. The expressed fusion protein is purified using an affinity column specific for the tag. Cleavage of the fused tag using a protease releases the original protein. A necessary feature is that the required protein folds correctly in the fusion form, or that it can fold from a denatured state after removal of the tag.

Table 9.1. Some of the Tags used for Fusion Proteins, to Simplify Purification. Cleavage is usually by proteolytic enzyme, and a specific site other than those indicated below may be engineered at the fusion site

Tag	Ref.	Purification	Cleavage
Polyarginine (C-term.)	[253]	cation exchange	carboxypeptidase B
Polyhistidine	[254]	IMAC	not available
HDHDH (His-Asp)$_n$	[255]	IMAC	renin
Protein A	[256]	IgG column	chemical
GST-transferase	[257]	GSH column	thrombin
Maltose-binding	[258]	amylose column	factor Xa

for fusion proteins to be made in which the fused part is a "tag" added for ease of purification (Figure 9.2) [252]. The tag may range from a few extra amino acids to a complete protein much bigger than the target protein to which it is fused. Some of the tags used currently are listed in Table 9.1; except where indicated, they are fused to the N-terminal (5'-) end.

The expressed fusion protein is passed through the appropriate affinity column, to which it binds and the host proteins do not. Elution is by displacement with salt (cation exchanger), imidazole buffer (IMAC), or by affinity elution (glutathione for the GST-transferase column; maltose for the maltose-binding protein column). The final step is to remove the fused portion to liberate the target protein. This is commonly done with a specific protease, whose cleavage site has been introduced in the fusion. There may be a few amino acids left on the N-terminal of the protein which are not naturally present. In the case of the polyarginine tag, if it is placed at the C-terminus (which is much more difficult than tagging at the N-terminus), it can be removed with carboxypeptidase B, which removes basic residues form the C-terminus of proteins. Of course, the native protein must not have an essential arginine or lysine at its C-terminus, as that would be removed too.

Proteins expressed intracellularly must be separated from the host proteins. Excretion vectors which position the recombinant protein either in the periplasm, or totally excrete into the culture medium, can greatly simplify the purification procedure. E. coli vectors which include leader sequences on the fusion protein that direct the product to the periplasm are available, but total excretion from E. coli is unusual. After harvesting the cells, the periplasmic proteins may be selectively extracted by osmotic shock, sometimes with a little lysozyme to assist the breakup of the cell wall and release of the proteins without rupture of the cell membrane. The quantity of recombinant protein produced in this way is limited to what can be accommodated in the periplasm without destroying the viability of the cell. There is also the problem of proteolytic degradation. Although host strains have been developed to be deficient in proteases, they do exist both in the cytoplasm where their function is, in general, protein turnover, and in the cell membranes and periplasm, where they function in polypeptide transport and processing. In addition, periplasmic proteases function to degrade extrinsic polypeptides to provide an amino acid source. Some of the problems of proteolysis, and possible solutions have been summarized [259].

Total excretion of recombinant protein is more readily achieved using yeasts, insect or mammalian cells as hosts, or using *Bacillus subtilis*, a gram-positive bacterium that naturally excretes large quantities of enzymes into the medium [260]. Although molecular biology manipulations with *B. subtilis* are not as easy as with *E. coli*, the genetics of the organism is now well understood, and suitable host/vector combinations are available. As the cell envelope of gram-positive organisms is much simpler than that of gram-negative, with no outer membrane, proteins directed through the (inner) membrane are immediately freed into the external environment. Large quantities of recombinant protein can be produced without being restricted to how much can be accommodated within the cells. Problems remain, however, and again these concern proteolysis. *B. subtilis* strains naturally produce large amounts of extracellular protease; at least seven different ones have been identified, and it has proved difficult to develop strains lacking all of them. The best strain available has only about 0.5% of the protease of the wild type [261].

The most popular expression system for eukaryotic proteins has been yeast [262]. Next to *E. coli*, the genetics of the yeast *Saccharomyces cerevisiae* (baker's/brewer's yeast) is probably the most understood of any organism. There are many vectors, including *E. coli* compatible plasmids, that can be used for transformation, and glycosylation will usually occur with excreted proteins (but not the same glycosylation as on mammalian proteins). Nevertheless, despite many years of work with yeasts, there are still unsatisfactory aspects, including low expression levels of foreign genes. The genetics and life cycles of yeasts are considerably more complex than *E. coli*, so most workers prefer to start with the bacterial system.

In mammalian cell culture, which is theoretically the best for mammalian proteins, suitable excretion vectors also exist; the culture medium can be collected and replaced if the cells remain viable. Most animal tissue/cell culture media require proteins in the form of fetal calf serum, or artificial media containing serum albumin and other protein factors, so the purification process involves separation from these proteins. Unless expression leads to an external concentration of the recombinant of at least $10 \mu g \, ml^{-1}$, the purification may be tedious.

Large-scale production of mammalian recombinant proteins in whole animals has progressed to the stage at which production is constrained principally by regulatory and ethical issues. One of the most successful procedures is to create transgenic mammals which express the foreign protein in their milk. Cows and sheep producing milk containing foreign proteins in levels of over $10\,\mathrm{mg\,ml^{-1}}$ have been reported. Purification of the valuable product from milk is a relatively simple process.

Most recombinant proteins that have been targeted are those that have potentially high commercial value, such as growth factors, viral proteins for vaccines, therapeutic enzymes, and other proteins for clinical and veterinary use. It happens that most of these proteins are naturally extracellular, and as a result, many have disulfide bonds between cysteines, since extracellular environments (such as blood) are oxidative. Cysteine residues are unstable in oxidative environments unless linked together. The disulfides often confer stability to the proteins as well as constraining them to their correct biologically active conformation. (Conversely, intracellular proteins do *not* have disulfide bonds; any cysteines are kept in the reduced state by cellular glutathione.) The expression of extracellular proteins inside host cells such as *E. coli* leads to problems concerning folding into the correct conformation, since in the reducing environment of the *E. coli* cell, the disulfide bonds cannot form readily. This is one of the main reasons why so often, if a eukaryotic protein is expressed at all in *E. coli*, it forms inclusion bodies of incorrectly folded, insoluble protein. If an expression system such as a mammalian cell producing the protein is not suitable, the final product must be recovered and "natured." This can be done at least partially by dissolution in a denaturant such as urea or guanidine salt, and gently removing the denaturant by dialysis. To allow the disulfides to form correctly, the appropriate reducing environment must be provided, this will consist of a mixture of reduced and oxidized thiols such as cysteine-cystine, glutathione, or dithiothreitol. In addition, the enzyme called protein disulfide isomerase, whose natural function is to assist with the formation of disulfide bonds during protein folding, may help the process. Finally, the presence of chaperonin proteins such as the *E. coli* GroEL and GroES may also help correct protein folding. The progress of "naturation" can only realistically be followed by the development of bioactivity, but it may be possible to recognize the bioactive form among a mixture of peaks in some analytical separation procedure. Some incorrect, degraded, or unfolded protein remaining will need to be separated in final polishing processes of conventional chromatography.

The final comments on recombinant proteins concern the vexing question of glycosylation. As noted above, most proteins of interest in the new biotechnology are extracellular. As well as having disulfide bonds which must be correctly linked up, many such proteins are also glycosylated during their natural excretion process. This glycosylation, which may be quite species-specific, can be the determining factor when choosing an expression system. If the natural protein is glycosylated, then the final product has a requirement either (1) to be glycosylated in exactly the same way, (2) to have some sort of glycosylation, not necessarily the same as the natural form, or (3) it does not need to be glycosylated for its intended use. If (1), then it is probable that the host cells will need to be similar, e.g., mammalian cells for mammalian proteins. If (2), then a larger range of possibilities are available, as yeast and insect hosts are capable of glycosylation during excretion of the recombinant protein. Glycosylation may have one or more of several functions, for instance it may stabilize the protein in an absolute way, or against proteolytic attack. The sugar moieties may have a direct biological role, in which case they are essential to get right. And the glycosyl groups may be antigenic, which may limit the final product's usefulness if it is to be used for *in vivo* therapeutics.

In summary, purification of recombinant proteins can differ from conventional purifications in several ways, since it is possible to modify the level of the protein in the starting material, or one can create a fusion protein that enables simpler purification. Sometimes it will require a naturation step to create bioactive protein. On the other hand, the techniques used in separation from unwanted proteins are essentially the same: affinity chromatography, ion exchange, reverse phase chromatography, and all other techniqes as described for proteins from their natural environment.

9.2   Membrane Proteins

The extraction of proteins from membranes and similar insoluble cell fractions has been described in Chapter 2. The key feature of most processes is the use of a detergent, first to solubilize the membranous structure, and second to stabilize the proteins once extracted. Nevertheless, many peripheral membrane proteins are only loosely associated, and once released do not require any detergent or other solubilizing agent to remain in the buffers. For these proteins, once they have been extracted and any detergent used has been removed, they can be treated like any other soluble protein for purification. But integral membrane proteins are likely to need the presence of detergent in all subsequent procedures, to mimic the lipids of their natural environment. For this reason, purification of integral membrane proteins causes numerous problems and difficulites to overcome.

The detergents used to extract the proteins will normally be chosen according to efficiency of extracting the desired component, but some consideration of the next step should be made. In general, the less detergent present the better, i.e., as low a concentration as possible, in as small a volume as possible. This may involve a decision between using say 10 vol of 0.1% detergent, or 2 vol of 0.5% detergent, which both extract the same amount of the desired protein. The latter is nearly always best, because it means handling smaller volumes (with the possiblility of higher speed centrifugation to clarify), and if desired, it can be diluted to the former condition after extraction by just adding extra buffer. Detergents are quite small molecules, but cluster into micelles which may be of similar size to proteins (Section 2.4). And with lipids present, the sizes of micelles may be very varied. The micelles tend to behave rather like hydrophobic proteins; for example, they aggregate with ammonium sulfate, and on centrifuging this will tend to form a floating layer, and carry many of the hydrophobic membrane proteins with it. Little genuine fractionation of proteins is achieved, nor much separation from excess detergent. On gel filtration, micelles plus associated proteins may spread over a wide size range, including the desired protein, so again little separation may be achieved.

Ion-exchange columns do generally result in some purification and separation from excess micelles, since most detergents used are nonionic and so have no reason to be adsorbed on an ion exchanger. The proteins being purified must be able to be adsorbed by being pulled away from the detergent onto the adsorbent, and this may need a higher charge density than in the absence of detergents. But if the *column buffer* does not contain any detergent, there is a high probabilty that once the washing process commences, the proteins will denature or otherwise become insoluble on the column, and not be recovered. Integral membrane proteins are, by definition, of low solubility in water, and in the absence of detergents will precipitate from solution. Thus, a minimal amount of detergent, if possible less than the critical micelle concentration, should be retained in all buffers being used. The critical micelle concentration for commonly used detergents such as Triton X-100 is quite low (ca. 0.02% w/v), and remembering that it may require typically a 1:1 ratio of detergent to protein by weight to keep it soluble, this level is a constraint on the maximum protein concentration in the process. So whereas normal proteins may be eluted at concentrations of several milligrams per milliliter from ion exchangers, membrane proteins in the presence of limiting amounts of detergent are likely to be eluted at less than $1\,\mathrm{mg\,ml}^{-1}$, and will require much longer salt gradients in which to recover each protein. On the other hand, if a higher detergent concentration is chosen, the micelles forming may complicate the separation and solubility of the protein fractions.

Use of detergents with a high critical micelle concentration, such as octyl glucoside, CHAPS, Na cholate (the last being ionic) may improve the performance of the fractionation procedures, but at a financial cost. Not only do these detergents have critical micelle concentrations in the 0.2–1% w/v range, but the micelles themselves are smaller, in the molecular weight range 4,000–8,000. Thus, many of the problems associated with micelle size can be avoided. The disadvantage is that they are less efficient as detergents, so a higher concentration is needed to maintain protein solubility. The *most* efficient detergents are the strongly ionic sulfates such as dodecyl (lauryl) sulfate, more commonly known as SDS (sodium dodecyl sulfate), and cationic detergents such as cetyl trimethylammonium bromide (CTAB), but these are also denaturing, so most proteins will be solubilized but unfolded. Purification in the presence of these detergents is limited mainly to size separation, in particular gel electrophoresis (see Section 8.2). Refolding to a bioactive protein is often successful for small proteins by dialysis of excess SDS, in exchange for a benign, non-ionic detergent such as Triton X-100.

For carrying out ion-exchange chromatography on a detergent-solubilized membrane protein fraction, buffers and general procedures can be similar to those used for other proteins, except for the requirement of a detergent being present. Remember that an ionic detergent will contribute to the ionic strength: 1% Na cholate is $23\,\mathrm{m}M$. Do not expect to elute large amounts of protein in a small volume unless a high concentration of a detergent with high critical micelle concentration is being used.

Most membrane proteins will be purified in the presence of detergent, but at the end it may be required to remove excess detergent for characterization of the protein. In many cases, removal of the detergent will cause inactivation of the protein, but that is not any concern for, e.g., amino acid sequencing. Special hydrophobic beads for adsorbing detergents are available commercially (also see [263,264]).

Finally, there is the possibility that intrinsic membrane proteins which are very hydrophobic can actually be solubilized in organic solvents, and manipulated in nonaqueous systems. There are reports of such methods, including chromatography in 100% organic solvent [33,34].

9.3   Purification of Antibodies

Antibodies are of major importance in diagnostics, and are finding increasing uses in therapeutics. Purification of the antibodies, especially for therapeutic use, is a major task in protein purification, and justifies a short description in its own right.

Antibodies may be produced in at least three ways:

1. The traditional method of injecting an animal and bleeding it to collect serum containing the antibody;
2. Production of monoclonal antibodies using mouse hybridomas; the antibodies are either collected from the ascites fluid or, more commonly, from the culture supernatant after *in vitro* culturing of the hybridoma cells;
3. Antibodies produced by molecular biology techniques, with expression in bacterial or other heterologous host cells [182].

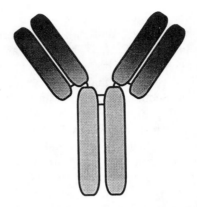

Figure 9.3. Diagrammatic representation of an antibody molecule (IgG immuno-globulin). The separate polypeptide chains are held together with disulfide bonds, and the heavy (long) chains are hinged near the center to create two distinct domains. The antigen binding site is formed at the ends of the adjacent heavy and light chains, indicated by the darker shading.

The antibodies are globulin-type molecules, the γ-globulins of serum, or immunoglobulins. There are several different sorts of γ-globulins; the major ones produced as a response to antigen are of the IgG class, and to a lesser extent, IgM. The structure of all immunoglobulins is based on the familiar Y-shape, consisting of two heavy chains of 53 kDa each, and two light chains of 23 kDa each (Figure 9.3). Thus, the molecular size of IgG is 152 kDa; but IgM consists of five such molecules held together with another short polypeptide, giving it a size of about 800 kDa. All γ-globulins have similar properties, which are quite distinct from most of the other proteins in serum; as a consequence, purification is not difficult, and there are several simple alternative procedures. However, if there is a mixture of different antibodies in the γ-globulin fraction, such as will be present in whole serum from an animal, separation of them is very difficult except by using the specific binding of the antibody you are interested in to its antigen.

The purification method to be adopted will depend on the source material (serum, cell culture, bacterial expression culture, etc.) and the purpose of the purification (research, diagnostic investigation, commercial production). The major methods are are follows:

1. *Ammonium sulfate precipitation.* The γ-globulins precipitate at a lower concentration than most other proteins, and a concentration of 33% saturation is sufficent. Either dissolve in 200 g ammonium sulfate per liter of serum, or add 0.5 vol of saturated ammonium sulfate. Stir for 30 min, then collect the γ-globulin fraction by centrifugation, redissolve in an appropriate buffer, and remove excess ammonium sulfate by dialysis or gel filtration.

2. *Polyethylene glycol precipitation.* The low solubility of γ-globulins can also be exploited using PEG. Add 0.1 vol of a 50% solution of PEG 6,000 to the serum, stir for 30 min, and collect the γ-globulins by centrifugation. Redissolve the precipitate in an appropriate buffer, and remove excess PEG by gel filtration on a column that fractionates in a range with a minimum around 6,000 Da.

3. *Isoelectric precipitation.* This is particularly suited for IgM molecules, and the precise conditions will depend on the exact properties of the antibody being produced. For example, large-scale purification of a hybridoma culture supernatant IgM by isoelectric precipitation at pH 5.0 was carried out [265], but only after ultrafiltration to increase the protein concentration.

4. *Ion-exchange chromatography.* Whereas most serum proteins have low isoelectric points, γ-globulins are isoelectric around neutrality, depending on the exact properties of the antibody being produced. Adsorption to cation exchangers in a buffer of around 6 has been used successfully, with elution with a salt gradient, or even standard saline solution to allow immediate therapeutic use.

5. *Hydrophobic chromatography.* The low solubility of γ-globulins reflects their relative hydrophobic character. In the presence of sodium or ammonium sulfate, they bind to many hydrophobic adsorbents, but the most specific, highly selective adsorbent for the γ-globulins is the "T-gel" described by Porath [140]. This consists of β-mercaptoethanol coupled to divinyl sulfone-activated agarose (Section 6.6), which was originally intended to be a "control" when investigating other similar hydrophobics. But the control itself acted as a very selective adsorbent for γ-globulins in the presence of $0.5\,M\,Na_2SO_4$. Further development of similar salt-promoted hydrophobic adsorbents for γ-globulins is described in [145]. After adsorption, the γ-globulins are eluted by a buffer with low salt content. The emerging fraction may include a little Na_2SO_4 which may need to be removed.

6. *Affinity adsorbents.* *Staphylococcus aureus* has a bad reputation in medical circles, but it has provided protein chemists with a very useful tool. An outer coat protein, known as Protein A, is isolated from the bacterial cells, and it interacts very specifically and strongly with the invariant region (F_c) of immunoglobulins [171]. Protein A has been cloned, and is available in many different forms, but the most useful is as an affinity column: Protein A coupled to agarose. A mixture containing immunoglobulins is passed through the column, and only the immunoglobulins adsorb. Elution is carried out by lowering the pH; different types of IgG elute at different pHs, and so some trials will be needed each time. The differences in the immunoglobulins in this case are not due so much to the antibody specificity, but due to different types of F_c region. Each animal species produces several forms of heavy chain varying in the F_c region; for instance, mouse immuno-

globulins include subclasses IgG_1, IgG_{2a}, IgG_{2b}, and IgG_3, all of which behave differently on elution from Protein A [266].

Some γ-globulins do not bind well to Protein A. An alternative, Protein G from a *Streptococcus* sp., can be used. This is more satisfactory with immunoglobulins from farm animls such as sheep, goats, and cattle, as well as with certain subclasses of mouse and rabbit IgGs.

A recent report describes the use of a synthetic affinity ligand that mimics the bacterial antibody receptor for purifying IgG immunoglobulins [267]. A single step purification from serum produced immunoglobulin at 95% purity.

The most specific affinity adsorbent is the antigen itself. The process of purifying an antibody on an antigen adsorbent is essentially the same as purifying the antigen on an antibody adsorbent. The antigen is coupled to the activated matrix, and the antibody-containing sample applied. Elution requires one of the processes described for weakening the antibody–antigen complex (Section 7.2). This is particularly useful for purifying a specific antibody from a polyclonal mixture.

Chapter 10
Small-Scale and Large-Scale Procedures

10.1 Small-Scale Procedures—Proteins for Sequencing

There are occasions when a purification reaches a stage where there is so little protein left that it is difficult even to handle the samples without severe losses. Small-scale (or microscale) operations might be defined as those in which the total amount of protein being processed is of the order of a milligram or less, and/or the volume is less than 1 ml (Figure 10.1). This small amount may be due to the fact that the protein being isolated makes up a very small proportion of the starting material, so even starting with many kilograms, the amount of the desired protein may be only a milligram *before* accounting for losses during the isolation procedure. Alternatively, it may be that the raw material is not available in any considerable amount, so even an abundant protein in, for example, a small insect, cannot be handled in any great quantity. Two developments over the past decade and more have resulted in small-scale operations becoming commonplace. These are, first, the refinement of high-performance chromatographic equipment, with recent progress in capillary systems that can handle very small volumes. Second, the fact that an end result of only a few micrograms of protein is useful is that it is sufficient for amino acid sequencing. With an appropriate sequence, the gene may be isolated using degenerate oligonucleotides to hybridize with a recombinant library (see Section 9.1). Current technology enables a reliable protein sequence to be determined on about 10 pmol of polypeptide. For a protein of 30 kDa, this is $0.3\,\mu$g. Future technologies will use mass spectrometry to obtain sequences on femtomole amounts; the limitation will then become contamination of samples with extraneous proteins and peptides.

The technologies used in HPLC have been described in essence in Chapter 5. This is the main technique used in small-scale isolations, and

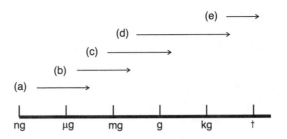

Figure 10.1. Illustration of the scale of protein purifications. (a) Microscale using capillary techniques; (b) HPLC technology; (c) laboratory to pilot scale; (d) commercial production; (e) very large-scale, low-cost proteins.

it can be scaled down to handle submicrogram quantities in capillary columns; these systems are currently under development by a number of companies. For instance, the Pharmacia SMART^{TM} system is designed for processing submicrogram amounts of protein, and has all the design features necessary to overcome the problems of dealing with small volumes. Columns have diameters of 1.6 or 3.2 mm (total volume as little as 0.1 ml), and sample collection down to 5 μl fractions is possible. Detectors have to be very sensitive at this level to monitor the separation. The basic types of protein separation in columns (ion exchange, hydrophobic interaction, reverse phase, gel filtration) are all available.

Alternative processes for obtaining pure samples of proteins present in μg amounts are mainly electrophoretic. Direct elution of protein bands from gels has never been fully successful, but electrophoretic transfer to nitrocellulose, nylon, or other "blotting" papers is now an established technology in biochemical laboratories. Modern protein sequencing equipment can take the section of paper with the adsorbed protein directly into the apparatus without further treatment. Generally the electrophoresis is in two dimensions (see p. 300) so as to separate the protein spots as well as possible. The maximum quantities that can be dealt with are small, but if complete separation is achieved, then this can be an ideal procedure, as it might produce up to a 100-fold purification, and the spot is then immediately available for sequencing.

Preparative gel electrophoretic systems have been developed (also see Section 8.2), the most sophisticated being that developed by Applied Biosystems, Foster City, CA (High Performance Electrophoretic Chromatography) specifically for very small samples. But this system, as with most other preparative electrophoretic equipment, has not always achieved resolutions to the claimed specifications.

Capillary electrophoresis has been introduced as an analytical technique, using only nanoliter volumes of sample; already preparative applications are reported, recovering up to 10 pmol of peptide for sequencing [268]. But in order to have enough final product, it is necessary to have a concentrated sample (tens of micrograms per microliter) in a considerably larger volume than the 50 nl injected into the apparatus; at least 100 pmol are required at this penultimate step to provide 10 pmol for sequencing. Nevertheless, when more sensitive sequencers become routine, preparative capillary electrophoresis (yielding 1 pmol or less of protein) may become commonplace.

Outside of the equipment, handling and transfer of μl volumes remains an important consideration. One of the problems in handling small volumes concerns losses by adsorption on the walls of apparatus, including the amount retained in liquid layers and drops left behind when transferring from one container to another. If the protein concentration is kept high to minimize adsorption losses, then, by definition, any lost drop represents a larger percentage of the total. Alternatively, to recover lost drops by flushing out inevitably results in dilution. Essential to any technology of handling small quantities is the minimization of dead spaces between the active separation unit and the collection of fractions. This is where the commercial developments in HPLC apparatus have achieved the ability both to get highly refined separations with narrow peak widths, and to collect those peaks without significant dilution.

The final step in any purification procedure is the one on the smallest scale, and it should result in a completely pure protein. This final step is usually a chromatography step, either adsorption or gel filtration. In the latter case, dilution of the sample from the original applied mixture is an inevitable result of the process, so the applied sample must be in a minimal volume (which also assists in optimizing the resolution). In the case of adsorption chromatography, the volume of the applied sample is not so important, provided that the proteins are stable and do not get lost by *irreversible* adsorption in the apparatus or on the adsorbent in the column at low concentration. Gradient elution should result in the desired protein coming off in a small volume, often smaller than that applied to the column originally. HPLC equipment is designed primarily for small-scale work, to purify less than 1 mg of protein. For although even small (1 ml) columns can successfully resolve up to 30 mg of protein, optimum performance is at low loading. The most suitable adsorbent materials have the smallest plate height, using microbeads of no more than 10–15 μm diameter. Most monodisperse beads designed for protein HPLC operate at close to ideal behavior, and should result in peaks contained within a total volume of less than half the column volume. With a conventional 1-ml HPLC column, this means in less than 0.5 ml. Collecting the precise peak is a problem for equipment manufacturers, as the drops entering a fraction collector are slightly delayed from the flow cell recording the peak, and the size of the drops becomes a significant concern at low total volume. These problems have been dealt with (see above).

Most small-scale procedures in biochemistry and molecular biology involve handling the sample in 1- or 2-ml Eppendorf microfuge tubes of polypropylene. These are sterile, disposable, nonwetting, and have minimal adsorptive properties. They may be frozen or boiled without harm. Transfer of samples may be done using Pasteur pipettes or even by capillary action, but in either case, care must be taken to ensure that as little as possible is left behind in the transferring equipment.

The above comments have stressed the importance of minimizing volumes, which of course allows one to deal with smaller amounts of protein. But it is not always necessary to keep the volumes small if the protein can be handled as a very dilute solution. For $10\,\mu g$ in $0.1\,ml$ is more difficult to keep track of, transfer without loss, and measure, than the same amount of protein in, say, $5\,ml$. The problem often is that at high dilution, the protein is more likely to inactivate, or generally disappear by adsorbing on surfaces. This applies generally, but in particular to:

1. *Enzymes which are multimeric*: Although subunits of enzymes (and other polymeric proteins not held together by disulfides) have a high affinity for each other, there comes a dilution at which, after a spontaneous dissociation of subunits, the chance of them quickly finding each other again becomes low. If a two-subunit protein is represented by S_1S_2, then the dissociation process can be written:

$$S_1S_2 \underset{k_{ass}}{\overset{k_{diss}}{\rightleftharpoons}} S_1 + S_2$$

   The dissociation rate constant, k_{diss}, is first order; the *proportion* of the protein dissociating in a given time is independent of the concentration. But the association rate constant, k_{ass}, is second order, and so the actual proportion reassociating in a given time depends strongly on the concentration of the two isolated subunits.

   The isolated subunits are nearly always less stable than the natural polymeric state, and their denaturation can occur rapidly. High dilution leads to a larger proportion of the protein being in the dissociated state, hence a faster denaturation overall. (Note that this proportion might be only 1 part in 1,000, even at high dilution; but if the isolated subunits denature a million times faster than the native oligomer, it will soon be lost!)

2. *Hydrophobic proteins*: This includes many hormonal proteins and similar factors which act by binding to receptors, which are often hydrophobic in character. Adsorption of hydrophobic proteins to surfaces is a major problem at high dilution. There are silicone treatments to minimize adsorption, and it is even possible to pretreat with a neutral protein to saturate possible binding sites on surfaces. To give an example of this effect, assume we have $5\,ml$ containing $10\,\mu g$ of protein $(2\,\mu g\,ml^{-1})$, and the container's surface adsorbs $2\,\mu g$ of protein; 20% is lost. But if the protein concentration had been $100\,\mu g\,ml^{-1}$ in the first place, then $2\,\mu g$ out of $500\,\mu g$ would not be missed.

3. *Activity-labile proteins and enzymes*: Loss of bioactivity may be due to the presence of trace amounts of a contaminant in the buffer, to oxidation, or some other reactive process. The larger the proportion of buffer to protein, the greater this loss will be, because there are more reactive molecules per molecule of protein.

Each of these problems is aggravated by high dilution; minimizing volumes is desirable in any process, but particularly so when the amount of protein present is very low.

Having stressed problems of dealing with small amounts of proteins, some of the advantages should be mentioned. The chief of these is speed; most operations are much speeded up on a small scale, and to some extent this can ameliorate the problems referred to above. Whereas a column process may take several hours to complete when using grams of protein, scaled down to milligrams, the same basic process might take only tens of minutes, and on a submilligram scale, the actual separation stage may take only a few minutes. So exposure to conditions that result in bioactivity or quantity loss can be minimized when working on a very small scale. The time taken to prepare the sample, to recover it from the fraction collector, and to get it ready for the next step or for storage may be much greater than the time spent in actual separation.

Another advantage is the fact that such small quantities are being processed that if you *do* have a reasonable large amount of sample, it can be divided among many trials to find the optimum conditions for separation; this can be carried out relatively quickly. Then, having found the optimum conditions, you can scale up to larger columns and volumes. Some of the parameters to consider when scaling up are discussed in the next section.

10.2   Large-Scale Procedures

This section will be divided into two parts: first, some advice on scaling up procedures, but still on a laboratory scale, and second, an outline of really large-scale technologies used in commercial protein production.

Scaling Up in the Laboratory

A most problematical stage in developing a protein purification procedure is when one attempts to scale up a small-scale method so as to get more end product. Each of the various techniques (precipitation, adsorption chromatography, in-solution methods) presents some problems which are likely to require some modifications to the original method. One of the main changes when scaling up is the time factor; handling larger quantities of materials at each step will take longer, and the extra time may cause unexpected changes. Otherwise, the modifications involve mainly equipment; larger-scale equipment has different operating parameters and capabilities.

Starting with precipitation, let us assume that an ammonium sulfate fractionation has been used on 200 ml of extract in the trials, and now we wish to scale up tenfold, using 2 liters of extract. At this level there should not be many problems; however, the whole process involves dissolving the salt at a certain rate, stirring to allow equilibration, then centrifugation. Up to the centrifugation, it should be possible to carry out things in the same way and with the same timing as on the smaller scale. But can the centrifuging be done the same way? The answer is generally no. Probably your small-scale sample was centrifuged at, say 20,000 g for 20 min, the supernatant discarded, and the precipitate dissolved in a certain amount of buffer. Two liters (perhaps nearer 3 liters after adding the salt) can not easily be centrifuged so hard in one batch. Even if the standard workhorse laboratory centrifuge does have a rotor that can take as much as 3 liters, it certainly will not be able to achieve 20,000 g. But you will probably need to split your sample into 2 or more batches, and still not get quite so much g force. In the case of salt or organic solvent fractionation, the concerns are that some of the sample is going to be left for much longer before centrifuging, and the precipitate will not be packed down so hard. Will this affect the total precipitate formed? How about the temperature regime that the samples go through during this time, remembering that solubility is affected by temperature? Is the protein susceptible to inactivation during this period? And if the precipitate is less hard packed, more of the supernatant will be trapped, and so carried through to the next step. These are some of the questions that should be considered when scaling up a precipitation step.

Next, let us consider the numerous problems involved with scaling up column procedures. The major consideration is the dimensions of the column required on scaling. First, was the loading on the original, small-scale column appropriate, as discussed in Chapters 5 and 6? It may have been an underload, in which case the size may not need to be much larger on scaling up. In any case, if the ultimate intention is to routinely purify this protein on a large scale, the economics begin to become important, and you should try to find what is the *maximum* loading you can use before resolution is compromised (this can be done on the small column first). Having optimized the small column, then, theoretically, scale-up should simply involve increasing both the cross-sectional area of the column, and the volumetric flow rate, in proportion to the extra protein being applied. In practice this can be difficult, as manufacturers are still reticent about making short columns of large cross-section for laboratory use, and those that are available tend to be rather costly. Generally, a compromise is made, in which both length and cross section are increased (Figure 10.2).

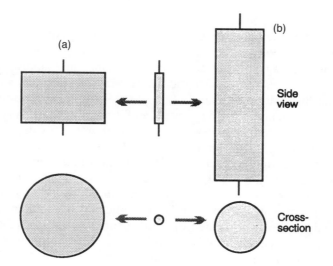

Figure 10.2. Scaling up 100-fold. (a) Ideally, the cross section is increased without changing the column length, but this may be inconvenient in practice. (b) A compromise with an increase in length as well as cross section is illustrated; operation of a column longer than its width is generally easier, although it will be slower.

For example, with a tenfold increase in sample applied, a column of twice the diameter, but $2\frac{1}{2}$ times longer could be used, although a still fatter column would be better. If the column is $2\frac{1}{2}$ times longer, it will require that much extra pressure to attain the same linear flow rate, and the elution will take longer as each fraction has more column to pass through. Also, resolution may be *better* with a longer column, depending on loading and eluted fraction volumes. Extra pressure may not be an option (the beads might start to compress), so a slower flow rate will make things take even longer (but in theory might improve resolution even more). Whatever is done, with a longer column the results will not be identical, and it will take a longer time. So be prepared to modify the procedure at this step when you scale up. There may be a case for changing the grade of adsorbent, especially the bead size. This is discussed in more detail in Section 5.4.

Scaling up in-solution methods, such as gel filtration and affinity partition, do not present many problems other than those associated with having the right equipment. Scaling of electrophoretic methods is not really a consideration, as the equipment is mainly of a fixed design for small-scale work. Gel-filtration separations are not highly dependent on column shape, provided even flow is attainable (see Figure 8.9). So going from a column of 100-ml volume to one of 1 liter can be just a matter of choosing from columns that are available in the laboratory. Typically, a 100-ml column used for gel filtration would have been fairly long and thin, say 16 mm diameter and 50 cm long. A suitable 1 liter column would be 40 mm diameter and about 90 cm long. This is not so far from the ideal of scaling columns in which only the diameter changes. With the equivalent loading amount and sample volume, and appropriately increased flow rate, the results should be very similar, but with better resolution than with the small column, if the same gel filtration beads are used. This is because the efficiency of gel filtration depends very much on the bead size in comparison with the total column. In Chapter 8 it was suggested that a gel filtration column should have half a billion (5×10^8) beads in it for optimum resolution. The 100-ml column referred to above, if packed with $100\,\mu$m beads, would be below this figure, but scaling up tenfold using the same beads, optimum resolution would be obtained.

Phase partitioning is not significantly affected by equipment dimensions, and can be easily scaled up. But manipulations are likely to take longer on the larger scale, so any problems of proteolysis or other types of inactivation will have to be carefully monitored.

Scaling up also means handling much larger volumes of liquid. If temperature changes are involved in the process, then large volumes take much longer to heat up or cool down. On the other hand, a large volume holds its temperature much better without specific control. This is an important consideration in any step that might involve heat-denaturation, since extra time taken warming and cooling would cause more denaturation than on the quicker small scale. Heat transfer is much better through glass than through plastic.

If a large volume is to be applied to a column, and the application time is expected to be many hours, it is often convenient to arrange for this application to be overnight. But, obviously, the system must stop when all has run on. This can be controlled by complex monitoring devices, but the simplest way, if the flow rate is adequate under hydrostatic pressure, is to allow it to run on under gravity, with tubing arranged so that when the last bit has gone on, the flow stops (Figure 10.3).

An alternative method of scaling up at a particular step is not to change any dimension, but to repeat the process several times over until all the sample has been through. To do this really requires some form of automatic control, so that the cycles can continue nonstop without attention. The last sample will have been waiting around much longer—and the purified fraction from the first sample similarly, which might be a problem, but otherwise this is a quite respectable procedure. It is here that fast processes really are an advantage, because if you have 10 subfractions to cycle through, it would not be reasonable if the cycle time was several hours. But a chromatographic procedure in which the separation time is only 10–20 min can be used, since with washing and pre-equilibration steps included, the cycle time can be as little as one hour. With very clean samples, a minimal washing with a pulse of high salt (in ion exchange) can cut this down further, allowing 20 or more subfractions

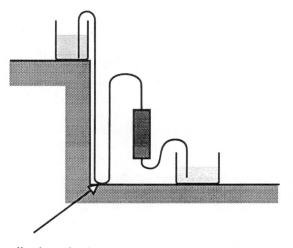

Figure 10.3. Application of a large amount of sample to a column unattended. When the last bit of sample flows down the tubing, the pressure head is lost, and the flow stops. The tubing must be organized so that it dips well below the outlet from the column; it may coil around the floor. It is best to use a wide tubing so that the flow is not unnecessarily restricted by frictional forces.

to be cycled through the system overnight. The apparatus used must have complex control systems and appropriate methods for monitoring, recognizing, and collecting the desired fraction each time.

Scaling up in the laboratory means dealing with a few liters rather than a few hundreds of milliliters. Really large-scale protein production means scaling up to several orders of magnitude greater, and a brief outline of the technologies concerned is presented below.

Commercial-Scale Protein Production

The production of enzymes and other proteins for commercial sale has a number of parameters that one would not normally consider so important in the laboratory. These include the expense of production, the long-term stability of the end product, quality control and consistency, and absence of specific impurities which impair performance in the applications the protein is to be used for. If the protein is for therapeutic use, then the demands on quality are rigorous, and the whole procedure of purification must be approved by regulatory authorities. Enzymes for industry have less stringent purity requirements, though of course they must not have impurities, such as other enzymes, that would affect their performance. Enzymes used for processing in the food industry must be from sources that are approved for human consumption; with microorganisms, these are classified as GRAS, standing for "generally regarded as safe." The same applies to other proteins that may be used as consumer products, such as detergent enzymes and a number of proteins used in the cosmetics industry.

The actual purification processes use essentially the same basic methods as described for small-scale work, but they are chosen as much on the basis of cost as of separation efficiency. Thus, although there have been scale-ups of high peformance chromatography, using the same packing material which requires high pressures, the construction of large columns that can safely be operated at high pressure is expensive and clumsy, and the cost of the adsorbent material used in these pressure columns of a few liters in size has run to millions of dollars. A very valuable end product may justify such practices, but better understanding of the principles of chromatography on scaling up would demonstrate that resolution as good as obtained on a 1-ml column in HPLC, can be obtained with large columns with much larger (and cheaper) bead particles. As has been mentioned before, in column chromatography the resolution is, in theory, approximately constant with the number of beads. A 1-ml column filled with $10 \mu m$ beads has approximately the same number of beads as a 1 liter column filled with $100 \mu m$ beads, which should give as good a resolution using 1,000 times as much protein sample—but the pressure required would be quite low, and not need thick reinforced steel columns.

One of the problems with large columns is that the weight of the adsorbent material causes distortion of the packing, even in quite short, squat configurations. This is because in the sort of column we are familiar with in the laboratory, the interactions of the beads with the wall have a considerable supportive effect over a distance of a couple of centimeters. But wide columns lose that support in the middle, and so there is a tendency for the beads to become distorted and squashed in the central portion. The column will actually acquire a saucer-shaped top surface, with unfortunate consequences for sample application, and flow will be restricted through the center portion. Other than purposely introducing vertical supporting structures within the column, the simplest solution is to keep the column height down to a maximum of about 50 cm. Consequently, production columns may typically have a diameter of 1 meter, but a height of only 30 cm! This requires very careful design to ensure even flow throughout the cross section of the column.

The example above demonstrates that large-scale protein production requires not only biochemists but, even more importantly, chemical engineers to design the equipment, liquid flow schemes, and computerized control and monitoring apparatus. Further discussion of such processes is out of place in this book, but can be found in articles such as [80,83,269].

Chapter 11
Analysis for Purity

11.1 Electrophoretic Analysis

Resolution of separate components of a protein mixture is achieved most clearly using an electrophoretic method. As discussed in Section 8.2, *preparative* electrophoresis has problems in the design and use of apparatus, so despite its obvious potential, it is not often used. But in an *analytical* mode, electrophoresis is a most widely used method; indeed, it is almost obligatory to characterize a purified protein preparation by an electrophoretic technique. Analytical electrophoresis in a gel system requires only 5–25 μg protein (or less with sensitive silver-staining techniques); this is rarely a significant proportion of what is available. Before gel systems were developed, electrophoretic analysis was carried out in the Tiselius free-boundary apparatus, requiring tens of milligrams of protein. This did not resolve closely similar proteins, and analyzing a single sample required a great deal of effort and attention. Paper and other cellulose-based supports were introduced for zone analytical electrophoresis, which eliminated two of the disadvantages of the Tiselius apparatus; only small amounts of protein were needed, and the technique was easy, requiring only simple equipment. But resolution was scarcely any better even on the superior cellulose acetate strips, because separation is based only on a rough charge/size ratio; many proteins move together as a single peak (see Section 8.2).

When Smithies [223] used starch gel as a support medium, an immediate improvement in resolution was achieved. As a result of the small pore sizes in the gel (compared with paper, agar gels, and other materials that had been used up till then), large molecules were retarded. This is a result of the apparent increase of viscosity; effective viscosity becomes dependent on molecular size. Moreover, diffusion is lessened,

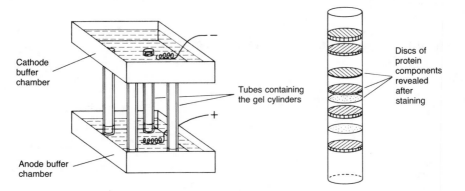

Figure 11.1. Disc gel electrophoresis system. Each protein component resolves as a disc in the tube of polyacrylamide gel.

so very sharp zones could be formed even after overnight electrophoresis. A few years later, Ornstein [224] introduced a synthetic gel medium, cross-linked polyacrylamide, which is more controllable than starch and has some other desirable advantages. Polyacrylamide gel electrophoresis was originally associated with "disc gel electrophoresis," because the system was promoted in which samples are run in individual tubes of gel, resolving into "discs" of protein zones (Figure 11.1). Although still used, disk gel electrophoresis has largely been replaced by slab gels, on which several samples can be run simultaneously with direct comparison of mobility. Thin-slab gel electrophoresis, using polyacrylamide, has been developed extensively over the past few years, and the technical problems associated with pouring and with sample application have been solved, so that the system is as simple as, but more powerful than, disc gels. A variety of commercial apparatuses are available, using gel thicknesses down to less than 1 mm (Figure 11.2) One advantage of a thin gel is that less heat is produced per square centimeter of gel surface for a given applied voltage. Also, during staining and destaining of the protein bands, diffusion of dye is more rapid into the thin gel slab.

There are many distinct electrophoretic procedures using polyacrylamide gel as the medium, sumarized below.

Simple (Native) Gel Electrophoresis

This involves running the sample in a buffer at a pH where the proteins remain stable and in their native form. This method was the original procedure, making use both of differences in charges between proteins, and their different sizes. The buffer chosen depends somewhat on the nature of the proteins, but generally is slightly alkaline, in the pH range of 8–9, where most proteins are negatively charged and so move toward the anode. The anode is normally at the bottom of a vertical gel. It should be noted that there is no provision for proteins that move in the other direction in these systems (Figures 11.1, 11.2); basic proteins disappear into the cathode buffer. If the bulk of the proteins being observed are known to be basic, then a buffer of somewhat lower pH can be employed, and the system operated with the cathode at the bottom of the gel.

The concentration of the polyacrylamide can be adjusted over a wide range according to the size of molecules being separated. Two variations are possible: variation in total acrylamide content, and variation in cross-linker percentage (N,N'-methylene bisacrylamide). Each achieves roughly the same objective; an increase in either reduces the pore size and so slows up larger molecules more. For very large proteins, up to 1,000 kDa, an open, rather difficult-to-handle gel of about 3–4% acrylamide and 0.1% bisacrylamide can be made. For very small proteins, around 10 kDa, 20% acrylamide and up to 1% bisacrylamide may be used. A high percentage of cross-linker tends to make the gels more opaque; 1 part of cross-linker to between 20 and 50 of monomer is the normal range. Usually, gels of 7–10% acrylamide are used. Polymerization is initiated with freshly dissolved ammonium persulfate (1.5–2 mM) together with a free radical scavenger, TEMED (0.05–0.1% vol/vol, N,N,N',N'-tetramethylethylenediamine). Alternatively, photopolymerization using riboflavin may be used; it leaves less harmful residue in the gel. Gelation should

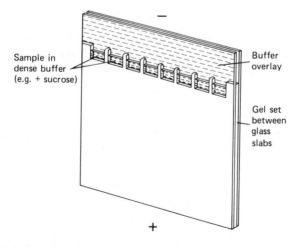

Figure 11.2. Modern thin-gel slab apparatus. Samples are applied in the wells set in the gel using a "comb." Side-by-side comparison of samples is possible in slab systems.

occur within 30 min at room temperature. Further details of suitable buffers and other parts of the system are available from apparatus manufacturers.

Simple electrophoresis can also be carried out in starch gel as the medium. The quality of the starch and its behavior in forming gels can be rather variable, but before polyacrylamide systems were widely used, starch was very popular. It is still particularly useful when detecting enzymic activity after electrophoresis; a starch-gel slab can easily be sliced and stained on one-half for protein, the other half for enzyme activity (see Figure 11.8). Because starch gels are opaque, protein staining is a surface stain, so a larger protein amount is needed than with transparent polyacrylamide gels in which the whole depth of the protein zone can be seen.

Simple electrophoresis in agarose gels does not have the resolving power of smaller-pore gels, so any analysis of purity based on homogeneity in such an electrophoretic system is less convincing.

Urea Gels

The second method is particularly useful for proteins that are insoluble at the low ionic strength needed in electrophoresis. This involves electrophoresis in the denatured state in the presence of between 6 and 8 M urea. The sample is dissolved in urea to completely denature the components first. Generally, a little mercaptoethanol is included to disrupt any disulfide bonds present, and to prevent their formation between thiols on separate polypeptide chains. Each protein is separated according to charge and to *subunit* size. The limitations of these gels are similar to those run in the absence of urea.

Urea-containing starch gels are also satisfactory and can be easier to handle than polyacrylamide-urea gels. Although transparent in the presence of the urea, during staining and washing the urea is lost and starch gel becomes opaque, so again only the surface-stained protein is visible.

SDS Gels

This third method is the one biochemists and protein purifiers are most familiar with. It is probable that many people who use it are not aware that any other method exists. It involves running the electrophoresis after denaturing the proteins with the detergent sodium dodecyl sulfate (Figure 11.3). Commonly known as PAGE-SDS (*p*olyacrylamide *g*el *e*lectrophoresis in *s*odium *d*odecyl *s*ulfate), this high-resolution method has won most protein chemists over. There are two major advantages compared with simple electrophoresis. One is that aggregates and insoluble particles often cause bad results with native gels by blocking the pores; when fully denatured, these aggregates are solubilized and con-

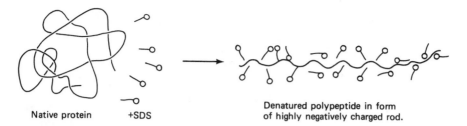

Native protein   +SDS   Denatured polypeptide in form
                                     of highly negatively charged rod.

Figure 11.3. The action of dodecyl sulfate in denaturing proteins.

verted to single polypetides. The other advantage is that the mobility is related to the polypeptide size (see below), so an immediate indication of molecular weight for each component is provided. And although separation is only on the basis of size (whereas native gel electrophoresis also depends on charge), resolution of closely similar components is excellent.

Dodecyl sulfate binds strongly to proteins [270], so that only 0.1% dodecyl sulfate is sufficient to saturate the polypeptide chains, with approximately 1 detergent molecule per 2 amino acid residues. Any oligomeric protein with polypeptides that are not covalently linked is dispersed as individual subunits. In order to disrupt any disulfides, β-mercaptoethanol (ca. 1% v/v) is added, and to ensure complete denaturation, the mixture is boiled for a few minutes. Each dodecyl sulfate carries a negative charge, so a typical polypeptide of molecular weight 40,000 acquires about 180 negative charges—far in excess of any net charge that might exist (at neutral pH) on the polypeptide chain originally. Consequently, the charge/size ratio is virtually identical for all proteins, and separation can occur only as a result of the molecular sieving through the pores of the gel. Despite the fact that the potential of separation of proteins of identical size is not possible in this system, it nevertheless appears to give the sharpest overall resolution and cleanest zones of any method. By making a comparison with a mixture of standard polypeptides of known molecular weight, the whole gel can be calibrated in terms of mobility against size. It is found that a linear plot over a substantial range can be obtained if mobility is plotted against log (molecular weight) (Figure 11.4) [227].

For analyzing the purity of a protein preparation, one hopes that there will be only one component; but if there is more than one (and the desired protein itself can be identified), then, from the subunit size, the nature of the impurities may be guessed if you know a lot about the composition of the material you are working with. Otherwise, the sizes of the impurities may suggest a possible step for removing them, e.g., by gel filtration. Clearly, an impurity with exactly the same subunit size as the desired product will not be detected in this system; however, it can resolve components differing by as little as 1% in molecular weight in op-

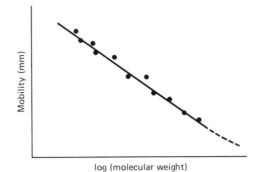

Figure 11.4. Plot of mobility in polyacrylamide-sodium dodecyl sulfate gel electrophoresis versus log of the molecular weight for a range of purified proteins. (From Weber and Osborn [227].)

timum conditions. In fact, the mobility/molecular weight ratio is not followed to within 1% ($\pm$3% is more like the true variation). So strictly speaking, one should speak of resolving components with mobility differences corresponding to 1% of molecular weight; their actual molecular weights could be identical. It should be noted that many proteins have more than one subunit type, so multiple bands are not necessarily an indication of the presence of impurities. If this is suspected, then there should be a rational relationship between the intensities of the multiple components. For example, a protein of subunit structure $\alpha_2\beta_2$, where α is 50 kDa and β is 30 kDa, should show two bands, with the 30 kDa band staining with only 60% of the intensity of the 50 kDa one (as it has only 60% as much protein in it). Some comments on estimating relative intensities are given later.

Numerous buffer systems for use in PAGE-SDS have been evaluated [271]. The most commonly used systems involve a Kohlrausch discontinuity (see Section 8.2) in the buffer between fast (e.g., chloride) ions in the gel, and slow (e.g., glycinate) ions in the cathodic buffer. Small proteins (below about 20 kDa) are not well resolved using this buffer, but an alternative Tris-Tricine buffer system has been found to provide excellent resolution down to 5 kDa [272].

Polyacrylamide gel electrophoresis is a routine procedure in all protein laboratories. The equipment and reagents are all available from various companies, with details of operating procedures provided. SDS gradient gels (see below) are now commonly used.

Gradient Gels

The fourth method to be described is also widely used. It was originally used with native proteins, being designed to get separations according to size only, not charge differences [273]. A slab of polyacrylamide is poured in which the acrylamide concentration varies from a high value at the bottom (usually about 30%) to only 3% at the top. The buffer is a high-pH one, so that most proteins migrate into the gel toward the anode at the bottom. Electrophoresis is continued until all proteins have reached a thickness of gel which prevents them from moving any further; the small molecules may reach 25% acrylamide, large ones remain near the top. As with the dodecyl sulfate system, molecular size (this time of the native protein, not subunits) can be determined by comparison with a standard mixture. The gradient gel system has a similarity with the next, fifth, method; it is run until no further movement occurs, and a "focusing" or concentration of diffuse protein zones can occur, resulting in very fine resolution. Gradient-SDS gels are also used, but in this case the migration does not stop; the gradient simply has the effect of sharpening the smaller-sized polypeptide bands as they run into the thicker gel, compared with running a nongradient SDS system, and allows for a greater range of size separation within the one gel.

Isoelectric Focusing

The principles of isoelectric focusing were described in Section 8.2. Analytical isoelectric focusing in slab gels of polyacrylamide or agarose is a popular very-high-resolution procedure. Components are separated according to their isoelectric points, and bands are sharpened by the "focusing" effect. The gel in this case serves only as a stabilizing medium and should preferably not slow down large molecules. Since molecules eventually have to reach their isoelectric point, viscous barriers to mobility should be avoided. With polyacrylamide, an open-pore gel is indicated; however, these can be difficult to handle, and sometimes polymerization is uneven, giving a nonhomogeneous product. If the proteins of concern are all fairly small (e.g., <100 kDa), then a 5–6% polyacrylamide gel will be satisfactory. Otherwise agarose, which sets to a firm gel at only 2% wt/vol, with large pores, is more appropriate.

Several samples can be applied to the one gel, and it does not matter where they are applied, since each component will eventually find its way to its appropriate spot in the pH gradient. An ampholyte concentration of between 2 and 4% is usually advised. Preset gels containing ampholyte can be purchased; these give consistent results with predictable pH gradients when equilibrium has been reached. Because the conductivity of ampholytes tends to zero as the focusing progresses, it is possible to apply high potentials, $100–200\,\mathrm{V\,cm^{-1}}$ without excessive heating. A high-potential gradient ensures rapid focusing; modern systems can be run to completion in a couple of hours. Immobilized ampholytes for high resolution within a narrow pH range are also available.

One problem with isoelectric focusing is that, on occasion, proteins may associate with ampholyte components, acquiring a combined isoelectric point different from that of the free protein. This can cause multiple minor bands, since ampholytes themselves are not a continuum of species but occur as discrete components. Thus, although a single band on isoelectric focusing is a very good indication of homogeneity, multiple components may not reflect true heterogeneity of the protein sample.

Two-Dimensional Systems

Two-dimensional gel electrophoretic analysis has become a widely used technique. Commonly, a sample strip containing proteins that have been separated by isoelectric focusing is applied along the edge of a gel, and the proteins electrophoresed perpendicular to this strip into a gel containing SDS (Figure 11.5). The final pattern depends on separation according to isoelectric point in one dimension, and according to subunit size in the other direction. Crude cell extracts can give up to 1,000 individually identifiable protein components in two-dimensional systems. However, for analysis of purified enzyme, it should only be necessary to carry out one-dimensional electrophoresis in each of the two systems. Thus, two-dimensional electrophoresis is not a great advantage for analyzing purified preparations.

Capillary Electrophoresis

The most recent development in analytical electrophoresis is the capillary electrophoresis equipment [274]. This extends the dimension of the electrophoresis channel in length, but greatly reduces the cross-sectional area. Consequently, the surface area to volume ratio of the electrophoretic channel is greatly increased, and temperature control can be accurately

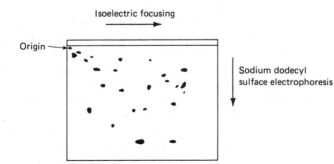

Figure 11.5. Two-dimensional gel electrophoretic analysis.

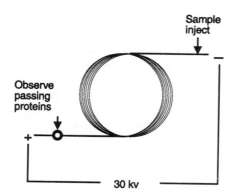

Figure 11.6. Capillary electrophoresis. Capillaries are typically between 25 and 50 μm diameter, and up to 100 cm long, coiled and contained in a capsule for easy fitting and storage. The electric field applied is in the 10s of kvolt range.

maintained even at high current densities. Samples are run either in free solution, or in gels, and isotachophoretic operation is viable (see Section 8.2). Sample size is minute (nanoliters), but should be fairly concentrated for good sensitivity. The detection involves observing protein bands as they pass through a section of the capillary (Figure 11.6). Although only one sample can be analyzed at a time (compared with 10 or more in typical slab gels), the whole procedure is very rapid, does not involve staining, and is quantitated and stored as a record automatically. The equipment can be used for electrophoretic separations other than proteins, and so is a versatile and valuable addition to an analytical laboratory.

Each of the above electrophoretic techniques has some limitations, and a combination of more than one (which two-dimensional electrophoresis is) may be needed to confirm homogeneity. The commonly used PAGE-SDS is excellent for separating subunits differing in size, but sometimes an impurity or an isoform of the protein of identical size is present. And PAGE-SDS cannot distinguish between different conformations (e.g., differing disulfide bonding) of the one protein. Proteins often undergo posttranslational modification, both physiologically significant and accidental. Thus, there may be glycosylation variants, phospho-/dephospho-forms, and deamidated polypeptides. Unless there is a large size variation in glycosylation, none of these will be detected on PAGE-SDS. Native gels can easily show up phosphorylated variants; deamidation results in extra negative charges, and some glycosylation also includes charged sugars. Isoelectric focusing will also resolve these forms. But remember that native gels require that the protein is indeed in a native conformation. A sample that has been purified to apparent homogeneity but in a denatured state will not give a good result on a native gel (but try a urea gel to resolve denatured charge isoforms).

Staining and Detection of Proteins after Electrophoresis

It is not intended to give full details of methodology in this section, but only to outline usual procedures. After electrophoresis, protein zones must be visualized, and this is carried out by a staining procedure, normally involving an organic dye which binds tightly to proteins. Many different dyes can be used, and the main objectives are: (1) sensitivity of detecting small amounts of protein; and (2) proportional staining with all types of protein. Most dyes used tend to be attracted by positively charged groups (lysines, arginines, and histidines) on the proteins, consequently, proteins with higher proportions of these—generally more basic proteins—tend to stain more strongly [275]. Indeed, some acidic polypeptides have escaped detection because they bind so little dye.

The first step is to fix the proteins by denaturation (if not already denatured) in acid. The usual fixatives are either a methanol/acetic acid/water mixture of proportions $3:1:6$, or a protein precipitant, trichloracetic acid, or perchloric acid at $5-10\%$ concentration. Picric acid can also be used for fixing and staining [276]; it has some advantages, in particular the low background staining even without washing. But use of picric acid is not advised because of its chemical instability (explosive!). Except in the cases of sodium dodecyl sulfate gels or isoelectric focusing, the dye may be included in the fixative. For SDS gels the dodecyl sulfate should be washed out first, or the proteins will not bind dye. But with a sufficiently large volume of dye-fixative this can be done in one go. Ampholytes also bind dye, and it is usually best to allow them to diffuse out during the fixing process, before staining.

Staining is done using the dye dissolved in the fixative. There are many different published methods for overcoming various problems, including the speed of the staining process (e.g., [227,277,278,279]). Dyes used include Amido Black, Nigrosine, Coomassie Blue, G-250, and Coomassie Blue R-250. The latter, R-250, is the most widely used dye for staining proteins. Although it does not stain some acidic proteins as well as, say, Amido Black, Coomassie Blue generally has the greater sensitivity. Some methods for staining SDS gels (e.g., [280]) depend not on dyes, but on precipitation of potassium or sodium dodecyl sulfate in the cold gel, or formation of a copper complex. Prelabeling of proteins with fluorescamine [281] has also been used successfully. In this case, the proteins can be observed during the electrophoresis by their fluorescence in near-UV light. Marker proteins that are colored with bound dyes are popular for instant identification of the progress of electrophoresis and calibration of the gel without staining it.

Washing out excess stain can be time-consuming. It is necessary to stain long enough for the dye to penetrate to the center of the gel; then excess dye must diffuse out again. The time depends very much on the gel thickness, hence the modern trend toward very thin gels of 1 mm or less. It may be noted here that starch gels, depending only on surface protein, can be stained in a few minutes and washed quickly after that.

The most sensitive staining method is to use a silver stain, which involves chemistry similar to that used in photography. After fixing the proteins, the gel is treated with silver nitrate. The silver binds to proteins (nonstoichiometrically), and after a reductive and enhancement step, the process is completed in a few hours [282,283,284]. This procedure is particularly useful when only very limited amounts of sample are available. It also enables a gel to be run with less distortion of bands due to overloading. Silver staining is about 10–20 times more sensitive than using Coomassie Blue, depending on the method used and the protein itself. There is much more variation between proteins than with dyes. Very small amounts of impurities can be detected in a sample without overloading the main component excessively.

The ultimate in sensitivity is to label the sample radioactively using one of several possible techniques. After fixing (or, more commonly, the gel is frozen to fix it), autoradiography over any desired period can detect very minute amounts. But at sensitivities down to nanogram amounts, contamination of reagents with polypeptides begins to become a limiting factor. This is not a commonly used method.

Detection of Specific Proteins

One way of determining which of the components found on electrophoresis is the desired protein is to use a specific technique that can detect this protein and no other. There are two main ways: one is to use antibodies to the protein, the other, for enzymes, is to use a specific activity detection method.

Antibodies can be raised against a previously purified sample of the protein, perhaps from a different but related species. In the case of monoclonal antibodies, the protein does not necessarily have to have been completely purified before (Section 7.2). It is most important that the antibody preparation does not contain any extra antibodies against components that might be present in the sample being analyzed. This is a potential problem with polyclonal antibodies, even after affinity purification (Section 9.3). Because antibodies are large proteins, they do not diffuse into gels at a significant rate, and so cannot interact with their antigen when a solution is simply placed on top of the gel. The technique of transferring proteins perpendicularly to the gel, onto an adsorbent paper has been developed. Analogous to the original "Southern blotting" as described by Southern for DNA transfer [285], protein transfer is called "Western blotting" [286,287]. Proteins are transferred by diffusion or electrophoretic migration to a membrane to which they adsorb irreversibly (Figure 11.7). The membrane is usually made of nitrocellulose,

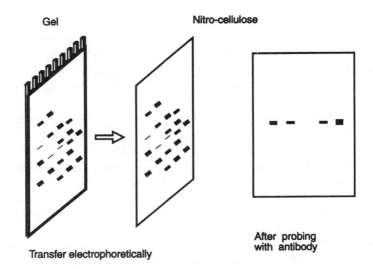

Figure 11.7. Principle of Western transfer. Electrophoresis in a gel is carried out normally, then a sheet of a protein-binding material, commonly nitrocellulose, is placed against the gel (with suitable nonadsorbent spacer so that it does not stick), and the proteins are transferred by electrophoresis perpendicular to the gel slab. After transfer, proteins may be probed with an antibody directed against a specific component.

but other materials including nylon are available. Transfer from SDS-gels is commonest, but native gels and isoelectric focusing gels can also be used. Since antibodies generally recognize even denatured protein, there is not usually any problem, but sometimes an antigenic determinant on a native protein may be destroyed on denaturation. Often there is a degree of refolding (renaturation) during the transfer and washing process, which improves the antigen–antibody interaction.

After transfer, the membrane can be probed with the antibody solution. In order to detect the antibody, excess is washed off, and another antibody, directed against the constant F_c fragment of the γ-globulin is used. This second antibody contains a detection method. Originally, labeling with radioactive iodine was used, but now enzymic methods are routine; the enzyme is conjugated covalently to the antibody. There are several enzymes used (urease, alkaline phosphatase, peroxidase), and each can be detected in several ways. Color reactions are the most popular, but chemiluminescent procedures are now widely used because of their high sensitivity and rapid results [288,289]. As with so many methodologies today, the beginner is advised to consult the literature provided by the supplying companies.

Western blotting requires the availablility of an antibody. If the protein has never been purified before, then it is unlikely that one will be available.

The second method of identifying a specific protein is, when it is an enzyme, making use of its activity. This is mainly restricted to native gels and isoelectric focusing, since denatured enzyme as in PAGE-SDS is inactive. But there are many examples in which enzyme activity has shown up after SDS-electrophoresis, due to limited renaturation as the SDS washes out. It must be carried out without fixing the protein bands (which is a denaturation process), yet the reagents must be able to diffuse into the enzyme site before the enzyme itself diffuses in the gel. The most satisfactory systems are those in which the gel is sliced through the protein zones (Figure 11.8), exposing the enzyme on the surface and eliminating the need for diffusion of reagents. Thick slabs of polyacrylamide gels can be used, although this is more commonly done with starch gels.

There are so many individual methods for staining for enzyme activity that only a brief outline of different principles will be given. The gel can be treated by immersion in the reagents, and it is best if the final product that is visualized (a dye or precipitate) is insoluble, and so stays at the place where it is formed. A soluble color formed at the site of the enzyme will quickly disperse into the bulk liquid. Dehydrogenases are detected by precipitation of *insoluble* reduced tetrazolium dyes, reduced by NAD(P)H being produced at the site of the dehydrogenase (Figure 11.9).

The reagents may be partially immobilized by gelation (e.g., setting in an agar gel on the surface of the electrophoresis gel) or by soaking into filter paper. In these cases, a complex chain of reactions initiated by the presence of the enzyme being detected can be localized; the visible prod-

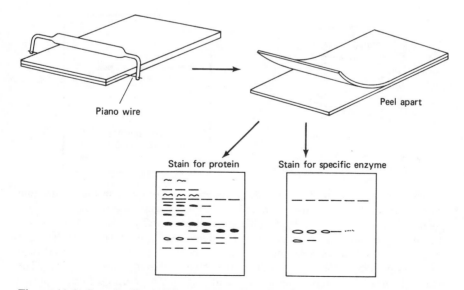

Figure 11.8. Longitudinal slicing of gel slabs to expose protein bands for surface staining.

uct may be soluble, but cannot diffuse away rapidly. Processes akin to coupled enzyme assays (cf. Section 3.4) can be made to occur; production or oxidation of NAD(P)H can also be observed under near-ultraviolet light (blue-green fluorescence). Other reaction systems not practical in enzyme assays may be used on gels for qualitative detection. For instance, phosphates liberated by various phosphatases can be trapped as a precipitate of calcium phosphate [290]. If the calcium phosphate precipitate is not sufficiently clear after washing excess reagents away, the calcium phosphate can be converted to lead phosphate, giving black bands at the site of the original enzyme activity. An extensive summary of enzymic staining methods can be found in [291].

By detecting the particular protein with Western blotting or enzymic staining, you do not get an estimate of purity. But you can find out if any of the bands found by conventional protein staining are alternative forms or degradation products of the main protein.

It is most important when presenting results of purity from gel electrophoretic analysis that some quantitative estimate of any impurity bands is given. It is not satisfactory to say "gave a single band on electrophoresis," if so little protein was applied that only one faint band showed up. It is also desirable to give some probability that the band being referred to is indeed the stated protein and not another protein which happens to be present in even greater amounts. There are many examples in the literature where a protein has been purified, but not to homogeneity, and the major band on the electrophoretic pattern has been assessed to be the

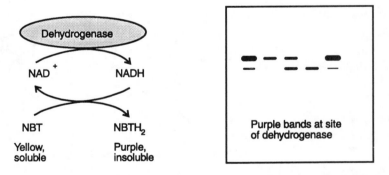

Figure 11.9. Detection of specific dehydrogenase enzymes using the tetrazolium staining method. A solution containing buffer, specific substrate, NAD(P)$^+$, phenazine methosulfate and nitroblue tetrazolium (NBT) is allowed to contact the separated proteins. At the site of the dehydrogenase activity, NAD(P)H is produced. This reduces NBT through the mediation of phenazine. Alternatively, the enzyme diaphorase can mediate the NAD(P)H-reduction of NBT. The oxidized form of NBT is water-soluble and yellow. The reduced form is purple and water-insoluble, so the purple coloration is precipitated at the dehydrogenase site.

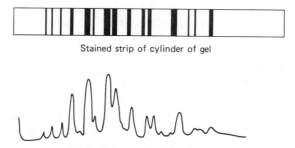

Stained strip of cylinder of gel

Absorption scan of same strip

Figure 11.10. Scanning of stained gels for quantitation of protein zones. It is also possible to scan gels at 280 nm and detect unstained proteins by their natural absorption at this wavelength.

desired protein without any evidence. When an enzyme has been purified, but has only a very low specific activity, the actual enzyme may be only 1% of the total protein, and not even noticed as a band on the gel! Presentation of photographs can assist, though relative intensities can rarely be reproduced faithfully. A diagram of a desensitometer scan of the stained gel (Figure 11.10) is more convincing. The only reservations then concern the relative specific staining intensities of the various components revealed. Making the assumption that all proteins stain equally, precise quantitation of each component can be made with a gel-scanning system.

In summary, analytical gel electrophoresis is one of the most widely used techniques in the biochemical laboratory (and now just as extensively used for nucleic acids as well as proteins), and characterization of a purified protein by one of the techniques described is virtually obligatory. Improvements in commercial apparatus design for running thin polyacrylamide gels allow the inexperienced the opportunity of obtaining high-resolution results easily and quickly. The methods are by no means exclusive; analysis on two different principles provides more information on the likely purity and characteristics of the final protein preparation.

11.2   Other Analytical Methods

The object of analyzing the final product is to determine whether or not it consists of one protein or more. The technique used should make it possible to detect quite small levels of impurity. Gel electrophoresis can detect a single impurity component at 1% of the level of the main component (provided that the two are well resolved), but there are occasions when electrophoresis is not suited to the sample. This can apply to lipoproteins and glycoproteins, in which the nonprotein portion is variable in amount, giving a smeared result. Even a homogeneous protein of this sort will often not run sharply under electrophoresis because it is not all protein. Other methods can give extra information on the degree of inhomogeneity.

In the 1950s and 1960s, analytical ultracentrifugation was one of the main methods of demonstrating protein homogeneity. Largely disused as a technique in recent years, the method of sedimentation equilibrium is making a comeback with the advent of modern ultracentrifuges which are so much easier to operate than the old models. Quite small amounts of impurity can be detected in an otherwise homogeneous protein preparation, by analyzing the deviation from the theoretical straight line of log(concentration) versus (radius)^2 through the sedimentation cell. However, deviations could have interpretations other than the presence of impurities—for instance, a dimerization of the main component.

N-terminal analysis of proteins can give an indication of homogeneity. Provided that the N-terminal amino acid is not acetylated or otherwise blocked, identification of a single type of amino acid with no traces of other minor components is suggestive of a single protein being present. However, the conclusions must be qualified by statements indicating that other proteins could be present but would not be detected if (1) they had blocked N-terminals or (2) they had an N-terminal amino acid identical to the main component. Case (2) is eliminated if a short sequence of the N-terminus is analyzed.

Amino acid composition can also give an idea of the likely heterogeneity of the preparation if two or more amino acid components can be measured very accurately and if the molecular weight of the enzyme subunits is known. If the molar content of these amino acids cannot be expressed by a simple ratio related to the known size of the polypeptide present, then it is probable that a major impurity is confounding the calculation. For instance, if the molecular weight of the polypeptide is known to be close to 30,000 Da, and tryptophan and tyrosine have been determined spectrophotometrically [292] to be present in amounts of 5.1 and 6.5 residues per 30,000, either these determinations are incorrect or there is an impurity with the different amount of tryptophan and tyrosine present. For obvious reasons, such arguments can only be convincing for small polypeptide sizes, and a clear-cut integer amount does not prove that the preparation *is* homogeneous. These calculations are of more value when the preparation is known from other methods to be homogeneous, but the exact molecular weight is uncertain. Then an exact molar ratio of two or more amino acids enables the molecular weight to be calculated with some confidence.

Use of crystallization as a criterion of homogeneity is described in the next chapter; it is by no means unambiguous. Analytical HPLC is the only technique that approaches the resolution of electrophoretic methods for demonstrating homogeneity. As discussed earlier, reverse phase HPLC may not be suited for proteins larger than about 30 kDa, but high-performance ion-exchange or gel-filtration columns can be used in an analytical mode to demonstrate homogeneity.

It is important to note that a demonstration of homogeneity is not so convincing if the analytical technique is identical to a procedure used in the protein's isolation. Thus, a gel-filtration column showing a single smooth peak for the final preparation may conceal several proteins of the same size, especially if a gel-filtration step was used in the purification procedure.

The final word on analytical methods is reserved for one of the latest and most remarkable methodologies, that of mass spectrometry of proteins [293]. Until relatively recently it was not possible to do mass spectrometry on any but the smallest proteins; to get them volatile and flying through the machine was too difficult. Two solutions to this problem have emerged, and commercial machines dedicated to the determination of the molecular weight of biopolymers have come onto the market. The first of these is the matrix-assisted laser desorption time-of-flight system. In this instrument, a very small sample, on the order of a microgram, is ionized, and the majority of the molecules are singly ionized. So a single peak corresponding to each component is obtained. Although not relevant to analysis for purity, the accuracy of the molecular weight determination is as good as 1 part in 10,000. Small components of very similar size can be identified, as they are resolved even if their mass is less than 100 Da different.

The other technique, which has more general application, is the electrospray method. In this case, proteins are volatilized with many charges (mostly protons, sometimes metal ions), and a large array of peaks results for each protein. These are related in that the apparent mass equals true mass divided by the number of charges. The spectrum can be analyzed by suitable algorithms to give a true mass using all the information, which can be accurate to 1 part in 20,000 or better. If the composition of the original sample was not too complex, each peak can be identified, and the mass of each component determined.

The most important application of mass spectrometry for analyzing proteins is not so much for identifying *impurities*, but for spotting and providing information on the precise identity of subfractions of the main component. Thus, posttranslational modifications due to glycosylation can not only be detected, but the precision of the molecular weight determination allows identification of the likely residues separating the different forms. And if the molecular weight has previously been inferred from the base sequence of the gene, this inferred sequence can be checked, and if there is a clear discrepancy it may be due to an otherwise unsuspected posttranslational modification.

Thus there are many different ways of deducing the purity of your preparation. Some are quick and easy to carry out; others require equipment that is far from generally available. The minimum requirement is a convincing electrophoretic gel for final presentation of your results.

Chapter 12
Optimization of Procedures; Final Steps

This chapter is intended for a later stage in purification efforts, when a method has been established, or a published method has been tried, but there may be some dissatisfaction with the overall procedure. The objectives of a scheme for the purification of an enzyme or other protein may be any or all of the following:

1. High overall recovery of activity;
2. High degree of purity of the end product;
3. Reproducibility
   a. within the laboratory,
   b. if scaled up or scaled down,
   c. in other laboratories;
4. Economical use of reagents and simple apparatus;
5. Convenience in terms of working hours.

Although points 1 and 2 are obviously desirable, neither may be necessary for a particular application. If it is a novel procedure resulting in a useful preparation, the method may be publishable, provided that point 3 is covered. For a publishable procedure, point 3c should always be attained—unfortunately, it is the most difficult to confirm! Also unfortunately, it is very often not attained, and many published procedures do not work very successfully in other laboratories for several reasons. Sometimes the originators did not repeat their methods often enough, and so their system was not as tightly defined as they thought. Occasionally there may be vital information missing from the published method. It is also the case that biological raw materials can never be identical; differences in extractability or composition can result in complete failure.

310

Points 4 and 5 above are technical considerations not related to the efficacy of the scheme itself, but to everyday organization and budgeting. The best procedure may be excellent in a well-equipped, well-funded laboratory, but out of the question elsewhere. Also, whereas research students should be prepared to spend 12 or 16 hours nonstop if necessary, it is neither easy nor reasonable to persuade technical staff to do the same; if the process can be (literally) frozen midway and picked up the following morning, everyone would be happier.

12.1   Speed Versus Resolution: The Time Factor

Some of the comments below concern methods of cutting corners, of saving time or reagents in order to fit the exact requirements of the protein purification process. Mostly this means less than ideal operation, and is not for the purist. But speed can be beneficial, lessening inactivation problems, so even if less-than-ideal fractionation procedures are followed for the sake of speed, the end product could be of a better quality.

Any process involving interaction of macromolecules is slow compared with the kinetics of small molecules. It may be in the order of minutes rather than seconds or milliseconds. Consequently, at each stage of protein fractionation there is a need for equilibration; in some cases perhaps 20–30 min is needed to attain (almost) complete equilibrium. Suppose that a process has a half-time of 2 min; the percent completion is plotted in Figure 12.1. Note that at 15–20 min the result is not much short of completion, but at less than 10 min the process is insufficient to be

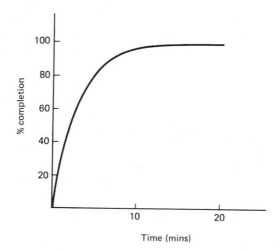

Figure 12.1. Percent completion of a first-order process with a half-time of 2 min.

considered "complete." The optimum time in this example would be around 15 min—longer is unnecessary, unless you need a cup of coffee. Generally, one does not bother to find what the optimum time is, but variation from one preparation to another could be due to cuttings things too short.

Now consider what determines the time factors. First, most processes occurring are diffusion-controlled, whether aggregation of protein molecules in a precipitation process, adsorption to a solid matrix, or movement in free solution. Halving the diffusion coefficient will double the time needed, and since the diffusion coefficient for any particular molecule is dependent on both temperature and viscosity of the medium, these two factors are very important. Traditionally, enzymes have been considered so labile that they must always be kept cold during isolation; many isolations are still done in a cold room at close to 0°C. Yet there are many—probably most—enzymes, and certainly nonenzymic proteins, that are perfectly resistant to inactivation at room temperature, especially in the *short time needed* to complete the fractionation. The 20°C difference between cold-room temperature and average laboratory temperature typically increases chemical reaction rates by three- to fourfold. At 25°C any oxidative, proteolytic, or other detrimental phenomenon other than denaturation can be expected to occur that much faster than at 4–5°C, but by the same token, any fractionation procedures can be expected to take *less* time to reach equilibrium at 25°C. This is of particular importance for column techniques, since the maximum speed of operation of ion-exchange or gel-filtration columns is determined by approach to equilibrium during the chromatography. If the diffusion coefficient for a protein in the solvent is 2 times lower at 5°C than at 25°C, then the flow speed of the column must be 2 times less. The viscosity of water is 1.8 times higher at 5°C than at 25°C; this is the main contribution to diffusion coefficients being smaller at lower temperatures; so a factor of 2 for diffusion is an appropriate estimate. Consequently, operating in a cold room is theoretically more desirable if inactivation phenomena are a real problem, since diffusion is halved, but inactivation may be fourfold less. But consider this point: people working with enzymes will happily leave a solution in the cold overnight, yet would consider leaving it at room temperature for an hour or so as fatal. Now 4–5 h at room temperature is probably equivalent to overnight in the cold, as far as inactivation processes are concerned. (Note that this does *not* apply to thermal denaturation. In the unlikely event that losses at room temperature are due to thermal denaturation, then cooling it will make a very large difference.) So it is not just the immediate processes occurring in the fractionation that need to be considered, but also the breaks between, especially overnight breaks.

Stabilization of enzymes with glycerol or other polyhydroxylated compounds has been found highly beneficial in many cases. Levels of glycerol

Table 12.1. Viscosity of Sucrose and Glycerol
Solutions

	Viscosity (cP)		
	Sucrose		Glycerol
% wt/vol	0°C	25°C	25°C
0	1.78	0.89	0.89
10	2.46	1.18	1.06
20	3.77	1.70	1.54
30	6.66	2.74	2.12
40	14.58	5.16	3.18
50			5.2

from 20 up to 50% v/v have been used. Pure glycerol is nearly 100 times more viscous than water, though the relationship is fortunately not linear for mixtures. Some values for the viscosity of glycerol and sucrose solutions are given in Table 12.1. Note that 25% glycerol is twice as viscous as water, 50% glycerol 5 times as viscous. Thus, processes necessitating glycerol as a stabilizer require extra time in these proportions. A column run in 25% glycerol should be run at about half the speed as in the absence of glycerol. Obviously, the stabilization factor obtained by using glycerol must be higher than this time factor to make it worthwhile.

What are the times needed for almost complete equilibration? No firm figures can be given as each case, obviously, is different (large proteins diffuse more slowly than small ones, which might involve another factor of 2). But in general, the rule-of-thumb figures in Table 12.2 can be used.

Table 12.2. Approximate Times and Flow Rates Needed to Achieve Close to Equilibrium Conditions in Protein Fractionation

Method	5°C	25°C
Ammonium sulfate fractionation	20–30 min	10–15 min
Organic solvent precipitation	15–20 min	(not advised)
Isoelectric precipitation	20–30 min	10–55 min[a]
Ion exchangers, other adsorbents:		
granular types, large beads	5–10 cm h^{-1}	10–25 cm h^{-1}
beads 30–50 μm diameter	15–40 cm h^{-1}	30–80 cm h^{-1}
beads 10–30 μm diameter	40–100 cm h^{-1}	60–200 cm h^{-1}
Immunoadsorbents	Long contact times	
Gel filtration:		
beads 50–150 μm diameter	3–10 cm h^{-1}	5–20 cm h^{-1}
beads 30–50 μm diameter	20–40 cm h^{-1}	30–60 cm h^{-1}

[a] Usually done cold to avoid denaturation at unphysiological pH.

The total time taken for a fractionation step is theoretically independent of volume, but in practice, large volumes inevitably take longer to deal with. Remembering that the faster an isolation procedure is carried out, the less likely the final product is to be degraded, some decisions about compromising on the ideal conditions may need to be taken. For instance, the following might apply: "If I run the column in the cold room at optimum speed, the enzyme should be off at about 8:00 p.m., and the enzyme can be left in the fraction collector tubes until morning. On the other hand, if I run it at room temperature and faster than optimum speed, resolution maybe somewhat worse, but it will be finished by 4:00 p.m., giving me time to concentrate the preparation and freeze it away." In this example, the cold room operation would involve 24 h at 4°C before progressing to the next step. At room temperature, only 5–6 h may pass before storing away the sample in a safe manner. Whereas on one hand, the cold preparation may be more degraded by proteolysis and inactivated by prolonged existence in dilute solution, on the other hand, the room temperature preparation may be rather more contaminated with unwanted proteins. So this is not to recommend one procedure over the other; it is just to point out the alternatives which can be weighed up according to the requirements.

Some proteins are remarkably resistant to any inactivation by all the phenomena described in Section 12.2. In many cases this is just as well, as purification procedures taking weeks to complete have been reported. Extracellular proteins are more resistant because their natural environment is harsher. In these cases, cutting corners for speed should not be necessary once proteolytic problems have been solved. But in other cases, where stability is limited, speed of operation can be much more important than resolution at each step. Processes to be avoided are dialysis and gel filtration on slow media. If it is possible to arrange, one fractionation step should be able to follow from another without extensive preparation. Thus salt fractionation may be followed by a gel filtration step, since the gel filtration will remove the salt as well as fractionate the proteins. But salt fractionation cannot be followed directly by an ion exchange technique, except in exceptional cases where the residual salt does not prevent adsorption. An ion exchange step cannot be followed directly by gel filtration, because the fraction would need concentrating. It is rare that ion exchange elution is obtained in sharp, concentrated peaks (but see discussions of affinity elution, Section 7.4, and chromatofocusing, Section 6.2). Organic solvent fractionation may be followed by ion exchange, although if the enzyme is only weakly adsorbed, endogenous salts co-precipitated by the organic solvent might increase the ionic strength too much to allow adsorption. A summary of some possible combinations of sequential fractionations is given in Table 12.3, together with the processing that may need to be done between fractionations.

Some pairs of sequential fractionation steps which do not involve intermediate treatments are:

Crude extract →   salt precipitation
                           organic solvent precipitation
                           affinity chromatography
                           dye ligand chromatography
                           affinity phase partitioning

Salt precipitation →   gel filtration
                           hydrophobic (salt-promoted)
                               chromatography
                           immobilized metal chromatography

Ion exchange →   affinity chromatography
                           salt fractionation
                           organic solvent fractionation
                           hydrophobic chromatography

Affinity chromatography → ion exchange chromatography
(with affinity elution)   salt precipitation
                           hydrophobic chromatography

This is by no means an exhaustive list.

Remember the capacities of each technique; precipitation methods are particularly suited to the first step, as very large amounts of material can be dealt with. Gel filtration has a moderate capacity unless one is equipped with large-scale columns. True biospecific affinity chromatography has a low capacity, although one may be able to process a lot of protein if very little actually adsorbs.

If only two steps are needed, it is often possible to complete the procedure in one day. But pauses can be tolerated, provided that degradative changes are prevented as far as possible (see Section 12.2). Two main storage processes for overnight or longer are recommended. These are as follows: (1) Storage of an ammonium sulfate pellet in the cold (not frozen)—preferably before dissolving it with buffer. The high salt concentration stabilizes the enzyme and generally inhibits proteases. However, this is not an indefinite storage for impure fractions, and 2–3 days at 4°C is the longest that it should be kept. (2) Frozen storage. If the temperature is low enough, all degradative processes will stop and the sample can be kept indefinitely. A temperature below −50°C is recommended. Normal deep-freeze temperatures are usually suitable for overnight storage, but should not be above −15°C. However, do not try to freeze a dilute protein solution; dilute solutions ($<2\,\mathrm{mg\,ml^{-1}}$) are much more likely to denature on freezing and thawing (dilute solutions also take up more room in a freezer!). Take note of the pH before freezing and make sure

Table 12.3. Possible Sequences of Fractionation Processes, Indicating Which Require Intermediate Processing

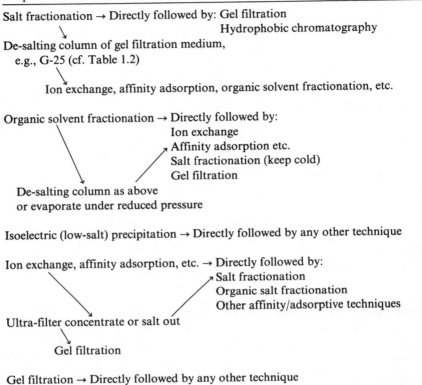

Salt fractionation → Directly followed by: Gel filtration
                                           Hydrophobic chromatography
De-salting column of gel filtration medium,
   e.g., G-25 (cf. Table 1.2)

      Ion exchange, affinity adsorption, organic solvent fractionation, etc.

Organic solvent fractionation → Directly followed by:
                                   Ion exchange
                                   Affinity adsorption etc.
                                   Salt fractionation (keep cold)
                                   Gel filtration
      De-salting column as above
      or evaporate under reduced pressure

Isoelectric (low-salt) precipitation → Directly followed by any other technique

Ion exchange, affinity adsorption, etc. → Directly followed by:
                                             Salt fractionation
                                             Organic salt fractionation
                                             Other affinity/adsorptive techniques
      Ultra-filter concentrate or salt out

            Gel filtration

Gel filtration → Directly followed by any other technique

the solution is well buffered. Freeze in plastic containers (glass containers can break when ice formation increases the volume), and thaw rapidly in warm water. Careful use of a microwave oven may be justified on a large scale. Remember that plastic is a poor heat conductor, and even if the warming water temperature is 50°C, the inside surface will not be higher than 30°C, and with agitation there is little chance of heat denaturation during thawing. Sometimes, freezing is disastrous; there are several examples of enzymes that inactivate in the cold (cold denaturation). This is usually due to dissociation of subunits at low temperatures; the isolated subunits then fall apart.

In summary, the time factor can be very important when dealing with labile enzymes or with extracts containing a lot of proteolytic activity. It may well be better in the long run to rush one step after another and sacrifice some resolution, rather than carry out each process to perfection but take a long time over the isolation. On the other hand, with stable proteins attention to maximum resolution may be best, especially if it eliminates the need for further steps to remove impurities.

12.2 Stabilizing Factors for Enzymes and Other Proteins

Ever since biochemists started purifying enzymes, they have been plagued by the problem of loss of activity. The general reason is that enzymes are often sensitive protein molecules which, when released from their natural protective environment, are subjected to a variety of stresses they may not be able to resist. Extracellular enzymes are an exception simply because they are designed to operate in a more hostile environment. It is for this reason that many of the early successes in purifying proteins and determining structures were with extracellular enzymes, which tend to be small, tough protein molecules, often present as a relatively high percentage of the protein in the fluid concerned. But the natural environment inside a cell is very different. Soluble proteins, whether cytoplasmic or inside organelles, are present in very concentrated soups of protein, usually at least $100\,mg\,ml^{-1}$, and estimated as high as $400\,mg\,ml^{-1}$ in the mitochondrial matrix [58]. There tends to be a relatively low (in some cases zero) oxygen tension, and a variety of reducing compounds are present, such as glutathione, to maintain a high reducing potential. Metabolites are also present in quite substantial amounts, and their presence may stabilize the enzymes. As soon as the tissue is disrupted, the proteins are released from this protective environment, diluted out into an extraction buffer, which is probably fully oxygenated, and proteolytic enzymes which were held in separate lysosomal compartments inside the cell are released also. The enzymes are exposed to three types of threat which may lead to a loss of activity:

1. Denaturation;
2. Inactivation of catalytic site;
3. Proteolysis and other modifications of the polypeptide.

Under type 2 may be included such things as loss of cofactors and dissociable prosthetic groups, as well as chemical changes to the amino acid residues in the active site.

Prevention of Denaturation

Denaturation can be minimized if the principle effectors of denaturation are avoided. These are extremes of pH, temperature, and denaturants such as organic solvents, chaotropic salts, and other solutes that disrupt H-bonds and hydrophobic interactions. The natural pH inside a cell would normally be in the range 6–8, so buffers within this pH range, or as close as possible to the actual pH in the tissue concerned, should protect against pH denaturations. Note that the physiological pH in most animal cells is 7.0–7.5 *at 37°C*. Because of the effects of temperature on the buffering components present (see Section 12.3), the equivalent physiological pH near 0°C for animal tissues is closer to 8.0 than to 7.0. Some problems encountered in controlling the pH when making an extract have been discussed earlier (Section 2.3).

Thermal denaturation has an extremely steep response to temperature (see Section 4.7). As a consequence, losses could increase 20- to 40-fold with an increase of temperature of only 10°C. But most proteins, especially from warm-blooded mammals, show little signs of heat denaturation below 40°C if the pH is near the physiological value—after all, they survived those conditions for perhaps days *in vivo*. Only if dealing with species whose tissues are always cold (e.g., cold-water fish, psychrophilic bacteria) is it likely that room temperature will cause direct heat denaturation.

As mentioned before, some proteins even denature *because of the cold*, due to cold denaturation weakening the hydrophobic forces holding the molecule together. So although it is generally advisable to keep solutions on ice while they are not being subjected to a fractionation process, it is sometimes better to warm them up when carrying out the fractionation. It is often not *necessary* to actually work in a cold room; it is always more comfortable not to do so. As a very general statement, proteins are stable at temperatures up to 10°C above their normal physiological temperature, but exposure to the unnatural environment (especially with extracted membrane proteins) in some cases makes them unstable even at temperatures below physiological.

Avoidance of Catalytic Site Inactivation

The inactivation of enzymes due to a specific effect on the catalytic site is more difficult to avoid. Loss of cofactors can be prevented (if they are known) by adding them back or including them in the buffers used. A preincubation procedure with cofactor before assaying the enzyme may be needed; the apoenzyme is not always less stable than the holoenzyme, so usually most lost activity can be restored. A more serious problem is covalent modifications of the active site due to exposure to the more hostile environment.

The active site of an enzyme contains amino acid residues which are abnormally reactive, and so responsible for the enzyme's catalytic abilities. These residues are also more susceptible to modification. The most troublesome is cysteine; sulfhydryl residues at the active site may be in the ionized form, which is very prone to oxidation. Normally, inside the cell the reducing atmosphere and the presence of other sulfydryl-containing molecules protect these groups. Exposed to a higher oxygen tension, several fates of sulfhydryl are possible. These include:

1. Disulfide bond formations: → -S–S-
2. Partial oxidation to a sulfinic acid: → -SOH
3. Irreversible oxidation to a sulfonic acid: → -SO$_2$H
4. Reaction with metal ions.

Formation of a disulfide bond requires another sulfhydryl to be in the vicinity, and an oxidation process. Disulfide formation is greatly accelerated by the presence of divalent ions which can activate the oxygen molecule and complex with the sulfhydryls. Two protective actions can be taken: (1) removing all traces of (heavy) metal ions using a complexing agent such as EDTA; and (2) including a sulfhydryl-containing reagent in the solution, such as β-mercaptoethanol, or dithiothreitol/dithioerythritol (Cleland's reagents [294]; Figure 12.2).

β-Mercaptoethanol is available as a heavy, odorous liquid. As a protective agent it should be used at a concentration of between 5 and 20 mM; lower concentrations quickly convert to the disulfide, due to oxidation. This is very much pH-dependent; β-mercaptoethanol in solution at lower pH is more resisitant to oxidation. The protective ability can be lost in less than 24 h unless the solution is kept in anaerobic conditions, as any disulfide that has formed can exchange with a still-active sulfhydryl in the protein. Oxidized β-mercaptoethanol can even accelerate the inactivation process (Figure 12.3). And although disulfide bond formation is reversible, it is not a desirable thing to happen.

The dithioanalogs of the reduced sugars threitol and erythritol are white solids which, if pure, have little odor. A concentration of only 0.5–1 mM is sufficient to provide protection for 24 h because the disulfide formed by oxidation is a stable intramolecular disulfide which does not interchange with protein sulfhydryls (Figure 12.4).

Other sulfhydryl protective agents, of similar properties to β-mercaptoethanol, that have been used include 2,3-dimercaptopropanol, thioglycolate, glutathione, and cysteine. Cysteine oxidizes quite rapidly, especially in the presence of heavy metal ions. But its strong metal-ion complexing ability gives it a doubly protective function.

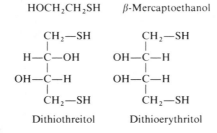

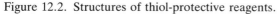

Figure 12.2. Structures of thiol-protective reagents.

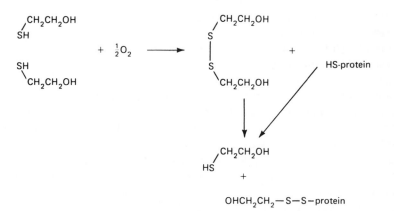

Figure 12.3. Inactivation of an active sulfhydryl in an enzyme by oxidized β-mercaptoethanol.

A routinely successful procedure is to use β-mercaptoethanol diluted 1 part in 1,000 (~12 mM) while running a purification (check if it is required at all), but for longer-term storage use dithiothreitol at 1–5 mM. Preferably, EDTA should be present in all buffers at a concentration of 0.1–0.2 mM. However, there are cases where the presence of this complexing agent is detrimental, since it may remove essential metal ions from enzyme active sites.

It is appropriate here to mention a few words about dilute enzyme solutions. It is well known that very dilute enzyme solutions lose activity quickly, but this loss can be prevented by the inclusion of a relatively high level of another protein, usually bovine serum albumin (BSA). In purification procedures, one would not normally add a contaminating protein, so very dilute solutions should be concentrated as quickly as possible (see Section 1.4). But for enzyme activity measurement, it is likely that the final enzyme concentration during the assay may be as low as $1\,\mu g\,ml^{-1}$, which can lead to rapid inactivation; even diluting the sample just before assay may cause losses. In these cases, the presence of BSA as stabilizer can be justified and should *always* be used. The mode of action of BSA is controversial, but a variety of effects may be involved. Extreme dilution of a protein may lead to dissociation of subunits which either may be

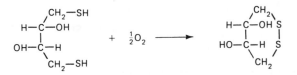

Figure 12.4. Oxidation of dithiothreitol to form the inactive cyclic disulfide.

inactive or, even if fully active, may be unstable and denature rapidly. It is possible that the BSA interacts directly with the dilute protein, either lessening the chances of dissociation or stabilizing the subunits. It may even function rather like a chaperonin in minimizing protein unfolding. Also, a small amount of protein is likely to adsorb to the walls of the container, especially if it is glass. One microgram of protein may be lost on $5\,cm^2$ of container wall, but in the presence of $1\,mg$ BSA, the $1\,\mu g$ that adsorbed would nearly all be BSA, so little of what you want would be lost from the solution. For diluting protein solutions before assay, a suitable buffer containing as much as $10\,mg\,ml^{-1}$ BSA should be used, and the assay mixture itself should have at least $0.1\,mg\,ml^{-1}$ BSA (see also Chapter 3).

Avoidance of Proteolytic Degradation

Living cells contain the elements for their own destruction, namely digestive hydrolytic enzymes such as nucleases, polysaccharide hydrolases, phosphatases, and proteases (proteinases, proteolytic enzymes). In mammalian cells, they are packaged in lysosomes which take up unwanted polypeptides and other polymers for their destruction as required by the physiological circumstances. In plants, digestive enzymes are stored in vacuolar spaces; their action in the cytoplasm may be under the control of a variety of specific inhibitors. In microorganisms, hydrolytic enzymes are often to be found in the periplasmic space in gram-negatives, and in the culture medium with gram-positive bacteria. Their main purpose is to digest extracellular macromolecules into fragments small enough to be taken up into the cell. The process for obtaining an enzyme extract almost inevitably destroys this delicate balance, and results in the mixing of the hydrolytic enzymes with all the other cell contents. The proteolytic enzymes are of particular concern here. There are procedures for minimizing the disruption of lysosomes when extracting soft animal tissues [295], but in many cases these are not applicable when disruption of the cell itself requires vigorous homogenization. Also, on a large scale, the hormonal or enzymic treatments used in these examples to release cell contents gently are not practical. Consequently, we must accept the fact that digestive enzymes will be present and do something about them.

Proteolytic inhibitors are compounds, either chemical or biological, which specifically inhibit the proteolytic enzymes that occur widely in tissues. There are, unfortunately, many different classes of proteolytic enzymes, not all of which can be successfully inhibited. The main categories are [296]:

1. Serine proteases (e.g., trypsin, chymotrypsin, elastase-type activities), EC subclass 3.4.21;

2. Thiol proteases (e.g., papain, yeast proteinase B), EC subclass 3.4.22;
3. Acid proteases (e.g., pepsin, cathepsins D), EC subclass 3.4.23;
4. Metalloproteinases (e.g., collagenases, many microbial neutral proteinases), EC subclass 3.4.24;
5. Carboxypeptidases (remove C-terminal residues), EC subclasses 3.4.16 and 3.4.17;
6. Peptidases (mostly remove N-terminal residues), EC subclasses 3.4.11–15.

The principal attacks on the proteolytic enzyme problem have been directed at the serine proteases, at first using the highly dangerous DFP (di-isopropyl fluorophosphate)—dangerous because it is volatile and attacks another serine hydrolase of vital importance in nerve conduction, human acetyl cholinesterase. The introduction of phenylmethylsulfonyl fluoride (PMSF) [297], which is not such a potent inhibitor of acetyl cholinesterase, has made the enzyme chemists' attack on proteolysis a less dangerous occupation, and PMSF is now widely used. Nevertheless, PMSF should be handled with utmost care. Although it is not very water-soluble, PMSF can be dissolved in acetone or ethanol before dispersing into the extract to a final concentration of $0.5-1 \, mM$. As it hydrolyzes quite rapidly, it should be added directly to the solution containing the serine esterases, not dissolved in the buffer prior to making a homogenate. PMSF also inhibits some thiol proteases and some carboxypeptidases, so it is a very useful reagent; once inhibited, the proteolytic enzymes, not the technician, should be dead, and no more PMSF is needed.

Acid proteases do not act via a serine or cysteine residue, and are unaffected by PMSF. However, fungi of *Streptomyces* spp. [298] produce an extracellular peptide of complex structure which is a potent inhibitor of acid proteases such as pepsin, cathepsin D, and yeast protease A. It is called "pepstatin A" and is effective at concentrations as low as $10^{-7} \, M$. Pepstatin A is available commercially. A variety of similar proteolytic inhibitors have been described, including leupeptin, an inhibitor of some of the more active peptidases. These inhibitors bind tightly to the proteases, but they are reversible, so may need to be replenished during a purification procedure. Pepstatin A, attached to an agarose column as an affinity adsorbent has been used to purify cathepsin D [299]. One could consider passing an extract through a column containing a range of bound proteolytic inhibitors to remove the proteases before commencing the fractionation procedure, but it would not be economical. The class of metalloproteinases, which require divalent metal for activity, can often be inhibited simply by complexing metal ions with EDTA or a stronger complexing agent.

Provided that the enzyme being isolated is not affected by any of the reagents, a suitable treatment for a freshly prepared extract is to include $2-5 \, mM$ EDTA, $0.5-1 \, mM$ PMSF, and $10^{-7} \, M$ pepstatin A.

Other Stabilizing Influences on Proteins

Aqueous 1–2% solutions of proteins are not in a comparable state to the natural environment of the protein. Usually, the nonaqueous components in the cell cytoplasm make up 10–30% wt/vol, and most of these components (mainly proteins) contain a substantial amount of bound or at least loosely associated water. Nuclear magnetic resonance studies on the water of cells have indicated that most of the water is not freely mobile. It is to be expected that proteins have evolved to be stable in these conditions rather than in the artificial solutions of a chemist's laboratory. Consequently, stabilization (against denaturation and other more subtle conformational changes that lead to inactivation) can be accomplished by attempting to mimic the conditions of relatively low water activity. The most widely used method is to include glycerol in buffer solutions. Glycerol, completely water-miscible, forms strong hydrogen bonds with the water, effectively slowing down the motion of the water molecules, and so reducing the water activity. Some analyses of the effect of glycerol suggest that the protein molecules preferentially bind water, and the structure so formed is less able to unfold against the structured glycerol solvent than in water alone [300]. Other interpretations describe the effect of glycerol as a dehydration of the protein.

Glycerol up to 50% wt/vol is used, though the highest concentrations are usually reserved for storage, since the viscosity of such a solution is too high to allow any manipulations other than electrophoresis. But 20–30% wt/vol glycerol mixtures do not have excessive viscosity (see Table 12.1), and most procedures can be carried out in buffers containing this amount of glycerol. Since ion exchangers rely entirely on electrostatic interactions, the presence of glycerol does not affect the behavior to any great extent, except for slowing things down because of the viscosity. But salting-out precipitation with ammonium sulfate may well be affected in terms of percentage saturation required for precipitation, and the precipitates may be impossible to centrifuge if the glycerol concentration is too high. The solvent would then be both too viscous and too dense. The effects of glycerol on hydrophobic, affinity, or other adsorbents cannot be easily predicted. Gel filtration, like ion exchange, should only be slowed, with no effect on the separation achieved.

Stabilization of enzymes with glycerol can have a dramatic effect. Enzymes that otherwise die very rapidly can remain active for weeks, and because the solutions either do not freeze at all or form a thin slurry of ice and concentrated glycerol-solute mixture, storage at very low temperature is rarely detrimental. As an alternative to glycerol, sugar or sugar alcohol solutions (glucose, sucrose, fructose, sorbitol) have been used successfully [301]; the mechanism is presumably similar. Other hydrophilic solutes, e.g., formamide and dioxan, tend to be destructive to enzymes at high concentrations.

The removal of effective water activity seems to be generally beneficial provided that the means of doing it is not a destabilizing influence. High ammonium sulfate concentrations, where the protein might be precipitated, are very stabilizing, and in fact, many enzymes are sold commercially as suspensions in $2-3 M$ ammonium sulfate (see Section 12.5). During protein purifications, it is usually necessary to store fractions overnight or even for longer periods, and consideration must be given to optimize the storage conditions. If an ammonium sulfate fractionation has been carried out, leave it as a centrifuged pellet rather than redissolving it. If storage in the absence of protective agents is unavoidable, consideration should be made of the relative advantages of short-term freezing or storage close to 0°C in the presence of proteolytic inhibitors. Freezing can inactivate enzymes. The susceptibility of different enzymes varies enormously. Remember that during freezing, first pure ice separates, which concentrates the protein and other solutes. Then the least soluble solute will precipitate out; if this is one component of a buffer, the pH will change markedly (see Section 2.1). On freezing sodium/potassium/chloride/phosphate solutions, the pH can change from neutral to 4 or lower; sodium salts aggravate the effect [3,4]. The higher the protein concentration relative to buffer salts, the more it will be capable of acting as a buffer itself, and counteracting radical pH shifts during freezing. Freezing at below the eutectic points of most salts (−25°C), is best, since even if there is a pH shift, true solidification will minimize any denaturation. Freezing at −10 to −15°C is probably no better than not freezing at all, at least for short periods up to 24 h.

12.3   Control of pH: Buffers

Buffer Theory

A key requirement in protein work is to know the pH at all times; to control the pH, a buffer is essential. It is also important to know the characteristics of the buffer being used and to have on hand a range of different buffers so that one most suited to the application is available. Over the decades of protein studies, a great number of buffers have been used, and new ones are still being developed. There are several characteristics we need to know about a buffer:

1. pK_a value;
2. Variation of pK_a with temperature, ionic strength;
3. Anionic, cationic, or multiple charges on buffer species;
4. Interaction with other components, e.g., metal ions;
5. Buffering power;
6. Solubility;

7. Expense;
8. UV absorption.

A buffer maintains pH nearly constant by absorbing protons produced in other reactions, or releasing them if they are consumed. Optimum buffering is at the point where both these capabilities are equal. But for some purposes, e.g., when it is known that protons are to be released, the optimum buffering may be when the basic form is in excess. It depends on how much acidification is expected and how much pH change can be tolerated.

The simplest types of buffer are those in which one form is uncharged:

$$HA = H^+ + A^-$$
$$HB^+ = H^+ + B$$

The dissociation constant is defined in the first case as:

$$K_a = \frac{[H^+][A^-]}{[HA]}$$

or in the second case as:

$$K_a = \frac{[H^+][B]}{[HB]}$$

and is usually expressed in logarithmic form:

$$pK_a = -\log K_a$$

Thus:

$$pK_a = -\log\frac{[H^+][A^-]}{[HA]} = -\log[H^+] - \log\frac{[A^-]}{[HA]} = pH - \log\frac{[A^-]}{[HA]}$$

This is the Henderson-Hasselbalch equation which, when rearranged, indicates the pH obtained with a solution containing both forms of the buffer:

$$pH = pK_a + \log\frac{[\text{basic form}]}{[\text{acidic form}]}$$

This equation applies to both types of buffer, HA and HB^+. A few simple calculations show that $pH = pK_a$ when [basic form] = [acidic form], $pH = pK_a + 1$ when [basic form] = $10 \times$ [acidic form], and $pH = pK_a - 1$ when [basic form] = $0.1 \times$ [acidic form]. Generally, it is desirable to work within about 0.5 unit of the pK_a because then: (1) there is a reasonable amount of each form present; and (2) the pH changes less per proton released or adsorbed the closer one is to pK_a (see Section 6.2; Table 6.4).

Useful parameters to describe buffers include the following:

z = charge on a given species;

$n = 2z - 1$, where z = charge on the acidic buffer form;

I, the ionic strength, $= \frac{1}{2}\Sigma c_i(z_i)^2$, where c_i = concentration of a charged species, and z_i is its charge;

$pK_a^0 = pK_a$ extrapolated to zero ionic srength.

The n value is useful in indicating the effect of ionic strength on buffer pHs (see below). By definition, n is always an odd number.

The concentration of buffer required is really a feature of the system itself rather than the buffer, although if one has to work at the edge of the useful pH range, e.g., around 1 pH unit from pK_a, more buffer is needed to maintain buffering power. It is useful to know what direction the most likely change in pH might be, so that the buffer can titrate in and counteract the change more strongly the more it occurs. For example, suppose we wish to have a solution at pH 6.5 with maximum buffering at a low ionic strength, resisting acidification. An imidazole buffer may be suitable, but as its pK_a is 7.0, acidification from pH 6.5 downward would tend to weaken its power. But if histidine is used (pK_a 6.0), acidification would titrate the buffer and make it stronger as the pK_a was approached.

Many of the buffers which have been used in the past are no longer the best choice because of the ready availability of an alternative. These include maleate, pK_a 6.3 (usually in the form of Tris-maleate which covers a big pH range, though with a weakness around 7.0), because it has UV absorption, and $n = -3$ (see below); barbitone, also known as veronal, pK_a 8.0 (UV adsorption and poisonous); glycylglycine, pK_a 8.2 (Tricine has marginally better characteristics and is cheaper); cacodylate, pK_a 6.2 (poisonous, expensive); and citric acid, pK_2 4.8, pK_3 6.4 (binds to some proteins, complexes metals, and $n = -3$ or -5).

Effect of Temperature, Ionic Strength, and Organic Solvents on pK_a Values

The ionic strength is greater for a given strength of buffer if $|n|$ is greater than 1. Also, n is an important indicator of the effect of ionic strength on the pK_a value (see Appendix C).

The parameter n appears in the simplified Debye-Hückel equation relating pK_a with pK_a^0 (at I = zero):

$$pK_a = pK_a^0 + \frac{0.51\,nI^{1/2}}{1 + 1.6\,I^{1/2}} \quad \text{(at 25°)} \qquad (12.1)$$

Thus values of $|n|$ greater than 1 result in a larger effect of ionic strength on pK_a. This is especially so for citrate (Table 12.4), for which $n = -5$ in the pH range in which citrate is usually employed. Note that for cationic buffers such as Tris, pK_a, rises with ionic strength, whereas for anionic buffers, it falls. The effect of ionic strength should be

Table 12.4. Effect of Ionic Strength of pK_a of Some Characteristic Buffers[a]

	n	pK_a^0	pK_a, $I = 0.01$	pK_a, $I = 0.1$	Concentration of buffering species at the pK_a, if $I = 0.1$, entirely due to buffer
Acetate	-1	4.76	4.72	4.66	0.2
Phosphate	-3	7.20	7.08	6.89	0.05
Citrate	-5	6.40	6.29	5.88	0.022
Tris	$+1$	8.06	8.10	8.16	0.2

[a] Theoretical, calculated from Eq. (12.1) at 25°C.

remembered when making up stock buffers at a strength greater than at which they will be used.

Temperature also affects pK_a values, especially for buffers involving amines. The effect of temperature on the buffering can be judged from the thermodynamic expressions:

$$\frac{d\ln K_a}{dT} = \frac{\Delta H^0}{RT^2}, \quad \frac{dpK_a}{dT} = -\frac{\Delta H^0}{2.3RT^2} \qquad (12.2)$$

and

$$\Delta G^0 = \Delta H^0 - T\Delta S^0 = -RT\ln K_a = 2.3RT\,pK_a \qquad (12.3)$$

Thus, the value of ΔG^0 is large and positive for high pK_a buffers, and depending on the entropy of dissociation, ΔH^0 likewise tends to be large and positive. So the rate of change of pK_a with temperature is likely to be negative and larger the higher the pK_a value [Eq.(12.2)]. On the other hand, low pK_a buffers change less with temperature. The entropy of dissociation is very small for amine buffers which are widely used for pH values above 6, this being a result of no net formation of new charged ions when an $n = +1$ buffer dissociates, e.g., Tris:

$$HTris^+ \rightarrow Tris + H^+ \quad \Delta S^0 = 5.7\,J\,deg^{-1}mol^{-1}$$

The entropy change depends mainly on the degree of participation of water molecules in hydrating the species; the small positive ΔS^0 for Tris (Table 12.5) indicates that the protonated species is probably little more hydrated than the neutral form:

$$HTris^+ \cdot (H_2O)_x \rightarrow Tris \cdot (H_2O)_y + H_3O^+ + (x - y - 1)H_2O$$

when $x \geqslant (y - 1)$.

On the other hand, for an $n = -1$ buffer, two new ions are formed where none existed before, e.g., acetic acid:

$$HAc \rightarrow Ac^- + H^+ \quad \Delta S^0 = -90.5\,J\,deg^{-1}mol^{-1}$$

Assuming that the acetate ion is more hydrated than the neutral acid, several molecules of water become "more ordered," and so there is a large negative entropy change:

$$(x - y + 1)H_2O + HAc \cdot (H_2O)_y \rightarrow Ac^- \cdot (H_2O)_x + H_3O^+$$

where $x > y$.

Similar arguments can be made for $n = -3$ and -5 buffers, where the more charged the ion is, the more hydrated it becomes. Zwitterions which buffer in the higher pH range due to dissociation of a protonated amino group are slightly less affected by temperature than straight amines, for similar reasons: $n = -1$ rather than $+1$. The neutral form is probably well hydrated, as it has both negative and positive charges on it, leading to an intermediate ΔS^0 value, e.g., Tes:

$$HTes \pm Tes^- + H^+ \quad \Delta S^0 = -30\,J\,deg^{-1}\,mol^{-1}$$

A summary of some thermodynamic values for common buffers is given in Table 12.5.

In Figures 12.5 and 12.6, the pK_as of phosphate (pK_2) and Tris are plotted as a function of temperature and ionic strength. Note that the pK_a of Tris changes by 1 pH unit when going between the biochemist's traditional temperatures of 4 and 37°C.

Miscible organic solvents can have a large effect on the pH of a solution. pK_a values are shifted upwards, at least for negative n-value buffers. This is because the dissociation of such a buffer generally involves uptake of water molecules as more hydrated species are produced:

$$HA \quad \rightarrow \quad H^+ \quad + \quad A^-$$
$$\downarrow \qquad\qquad \downarrow$$
$$H^+ x H_2O \quad A^- y H_2O$$

By lowering the water activity when organic solvent is mixed in, the dissociation becomes more difficult, so pK_a rises. On the other hand, the dissociation of an $n = +1$ buffer such as Tris, would not seem to involve a net change in bound water, as ethanol has little effect on the pK_a of Tris (Figure 12.7).

Table 12.5. Thermodynamic Values for a Range of Buffers[a]

	n	pK_a^0	dpK_a/dT (at 25°C)	ΔH^0 (kJ mol^{-1})	ΔG^0 (kJ mol^{-1})	ΔS^0 (J deg^{-1} mol^{-1})
Phosphoric acid pK_1	-1	2.15	$+0.004$	-6.8	12.3	-64
Acetic acid	-1	4.76	-0.0002	0.3	27.3	-90
Citric acid pK_3	-5	6.40	0	0	36.6	-123
Imidazole	$+1$	6.95	-0.020	34.0	39.9	-19
Phosphoric acid pK_2	-3	7.20	-0.0028	4.8	41.3	-122
Tes	-1	7.50	-0.020	34.0	43.0	-30
Tris	$+1$	8.06	-0.028	48.0	46.2	6
Ammonia	$+1$	9.25	-0.031	53.0	53.1	0
Carbonate	-3	10.33	-0.009	15.4	59.3	-147

[a] The values of ΔH^0, ΔG^0, ΔS^0 were calculated from the pK_a and dpK_a/dT, which were obtained from other sources [111, 112].

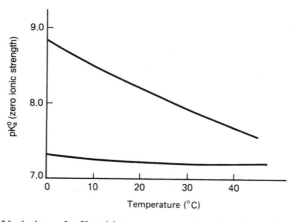

Figure 12.5. Variation of pK_a with temperature for phosphate (lower line) and Tris (upper line). Note that temperature has only a small effect on phosphate, but a large effect on Tris.

A list of useful buffers and their properties is given in Appendix C. More extensive tables can be found elsewhere [110,111]. An ideal series of buffers would consist of two sets, one with $n = -1$, the other with $n = +1$. Suggested buffers for ion exchange chromatography have been listed in Table 6.5. Buffers which are suitable for the pH range 5.0–6.5 are somewhat limited in number.

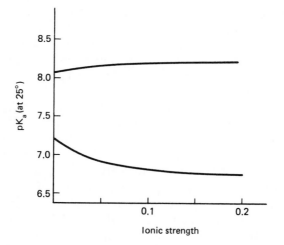

Figure 12.6. Variation of pK_a with ionic strength for phosphate (lower line) and Tris (upper line). The effect on phosphate is greater by a factor of 3 (the ratio of the n values), and is in the opposite direction (since n is negative for phosphate and positive for Tris).

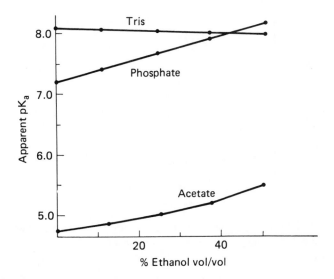

Figure 12.7. Apparent pK_a (measured directly with a glass electrode) of Tris, phosphate and acetate in the presence of ethanol, at 20°C.

Making Up Buffer Solutions

Buffers can be made by weighing the appropriate amount of each form of buffer and making up to correct volume. However, apart from the dangers of making a mistake, it is always best to check the pH with a meter, especially when other additives are needed, e.g., EDTA, mercaptoethanol, magnesium chloride. Some buffer salts cannot be relied on to be completely anhydrous, so precise values of pH, especially at the edge of the useful buffering range, are not reliably obtained without checking.

Ionic strength is not only important in affecting the pK_a values, but may also be a vital variable in other methods such as ion exchange. The ionic strength of a buffer may depend on how it is made up. For instance, "20 mM triethanolamine buffer, pH 7.5" might be 20 mM of triethanolamine adjusted to pH 7.5 with HCl, giving an ionic strength of 0.012 (acid form : basic form = 12 : 8 at pH 7.5); the only ions are H-triethanolamine$^+$ and Cl$^-$. Alternatively, it may be made by adjusting 20 mM triethanolamine chloride with NaOH to pH 7.5, in which case I = 0.020, since 8 mM NaCl would also be present. It might even be that triethanolamine was adjusted with sulfuric acid, in which case I = 0.018. Authors rarely give full details of how they made up their buffers (including temperature); fortunately, it is not *always* critical (see Section 12.4).

So don't waste time weighing out buffer salts to 5 or 6 decimal places. The variable water content alone makes such action irrelevant. Unless otherwise stated, pH values are normally adjusted down with HCl or up with either NaOH or KOH for making up a buffer solution. To maximize buffering power (see Table 6.4) for a given ionic strength, the following should be observed:

1. Choose a buffer with pK_a within 0.5 unit of the desired pH, with n-value of $+1$ or -1;
2. Use the uncharged form of the buffer, and adjust it up ($n = -1$) or down ($n = +1$);
3. For absolute maximum buffering power, choose two buffers, one of $n = -1$ with pK_a 0.5–1 unit higher than the other, of $n = +1$. Mix the two uncharged forms until the appropriate pH is reached.

In example 3, it may be difficult to determine the ionic strength, especially if there are strong interactions between the charged forms of the buffers, but the conductivity is normally very low. Such buffers have been used widely for electrophoresis (see Section 8.2), since their ionic strength is low, e.g., Tris-borate, pH 8–9; imidazole-Mops, pH 6.5–7.5. All ions present are buffers, and the conductivity of the bulky ions is low.

It is useful to have a series of stock buffers 10 times, or even 100 times more concentrated than intended for use. The advantages are: (1) the storage volume is small; (2) bacteriocidal agents added (e.g., azide) are diluted out when the buffer is used; and (3) the buffer can be added to an otherwise buffer-free protein solution without exposing the latter to extremes of pH. But remember that on dilution the pH will change (Figure 12.6, Table 12.4).

There will be frequent occasions when it is necessary to adjust the pH of a protein solution. Consider two situations. The first is a slight adjustment to a pH which is within the buffering range of the buffer salts already present. In this case, a dilute weak acid (e.g., acetic acid, not more than $1\,M$) or base (Tris, $1\,M$) can be used; the solution should be well stirred so that localized acid or alkaline areas around the drops of adjusting solution are quickly dispersed. Only if the pH values required are below 5 or above 8 might some stronger acid or base be needed. If a major adjustment in pH is needed and there is a strong buffer in the solution, more concentrated acid (or base) can be used to lessen the volume increase. So that localized extremes of pH are avoided, the adjusting solution itself can be partly neutralized. Using $5\,M$ acetic acid, pH 4.0 (NaOH) as a concentrated pH-lowering solution, no part of the protein mixture is ever exposed to a pH lower than 4.0. For raising pH values, $5\,M$ ammonia, pH 9.5 (HCl) can be used.

The second situation is where the final pH is outside the buffering range of the salts initially present. One could add suitable buffers before adjusting the pH, but it is generally better to use an adjusting solution that itself will form a buffer when the pH reaches the desired value. For example, suppose a solution is in Tris-chloride, pH 8, and one wants to change the pH to 6.5. By using the acid form of a buffer with pK_a in the range of 6.2–6.8, the pH is lowered and the solution remains buffered. In this example, one would use 1 M Mes or KH_2PO_4 as an acid, or if one wanted the buffer to be of the $n = +1$ type, 1 M histidine-chloride might be appropriate.

The temperature at the time of making up a buffer is very important if precision in pH is necessary. Often in published methods for protein purification, there is a statment that "all procedures were carried out at 4°C," yet the buffers used were almost certainly mixed at room temperature, so the actual pH, at 4°C, would have been different. Most pH meters have a temperature adjustment knob, but this does not mean that by dialing up 4°C, and using the meter at 25°C, the correct compensation has been made. How could the meter know what value of dpK_a/dT is appropriate for the solution? To make up a buffer with precisely known pH at 4°C (the usual cold-room temperature), set the temperature compensation knob at 4°C, place the electrode in standard buffer *at* 4°C, allow it to chill to that temperature, and set the reading to the standard value *at* 4°C. Then adjust the buffer, protein solution, or whatever *at* 4°C, to the desired pH. In some methods based on isoelectric precipitation in the presence of cold organic solvent, the pH must be adjusted at close to 0°C, and the pH value must be correct to ±0.1 unit. The only way to do this reliably is as above. Moreover, as seen in Figure 12.7, the pH changes in the presence of organic solvent.

This section has treated buffers assuming that pH can be measured correctly and accurately each time. Modern pH meters are easy to use and give a sense of confidence of accuracy, especially with a digital readout. Nevertheless, there are pitfalls in using a pH meter, especially with the common combined glass electrode. Even if the instructions are meticulously followed, incorrect readings can be obtained because of poor conductivity, and the development of liquid junction potentials in the ceramic plug between the sample being measured and the KCl inside. This phenomenon has been investigated [302], and it was found that many well-used and some new combined electrodes were giving pH values up to 0.4 unit wrong with buffers that had ionic strengths significantly different from that of the buffer used to standardize the pH meter. It is advisable to check your own meter. Standardize with a 50 mM phosphate buffer, then dilute this buffer tenfold; the pH should read 0.2 unit higher (see Figure 12.6). If not, it would be worth attending to the ceramic plug, *gently* scraping across its surface to clear any clogging. It is best to keep the electrode in a strong KCl solution between measurements.

In conclusion, at each stage you should be sure of the pH. If a method does not state the pH values, then measure them yourself; variable success in applying the method may be due to pH variations in the raw material. If the buffer used in the original method is unavailable, consider whether that buffer might have had some special influence, or whether it could be replaced with another of similar characteristics.

12.4 Following a Published Procedure

This brief section will apply to most people about to embark on a protein purification project. For nine times out of ten, there will already be a lot of information, together with one or more published procedures, about the protein you are after. Let us follow three possiblities. First, the procedure has been carried out in your laboratory many times; perhaps the person who did it has left, and it is your job to produce more of the protein. This should be the easiest of the three possiblities, provided that your predecessor has left adequate notes. Second, the procedure has recently been published, and you have the information in the paper, but no more than this and other earlier publications to go on. The third possiblity is that the procedure you are attempting was published a long time ago (say 20 years or more). In this case, consideration must be made of the use of more modern technologies (equipment, adsorbents, reagents, etc.) which might improve on the published work.

Published work is not always as reproducible as it should be. It may be that the raw material (e.g., growth conditions, age, strain) cannot be duplicated, which sets things off on the wrong track. Otherwise, there may be critical points in the procedure where insufficient information has been presented. This may be due to editorial decisions on minimizing detail in order to reduce length of a published procedure. Or it may be due to the author simply not appreciating that a particular detail is important. That is why an overall background in the principles and important features of protein purification is needed, to overcome any problems that turn up.

If you are simply reproducing a procedure that has been carried out in your laboratory before, there should not be too much of a problem. But if it has been done only once or twice, it may not be documented critically enough, and if you cannot get any extra advice, be prepared to make major modifications.

If yours is the second situation, following a recent publication, then think carefully about each of the individual steps reported, and what is likely to be the most important feature. For example, if they used an ion exchange column of 20 mm diameter, and the only ones you have are 15 or 25 mm diameter, is it really necessary to buy a new column? And if not, what variations in procedure (column length, flow rate) should you make, and what difference to the results is it likely to make, compared with the published work? I hope that readers of this book will be able to answer these questions.

The most important thing is to be able to distinguish trivial information from the vital details, and to criticize steps in the procedure which might benefit from modification. One of the commonest problems is the use of buffers outside their effective range, for instance, "25 mM Tris buffer, pH 6.5" or "10 mM sodium acetate buffer, pH 6.0." Yet by some chance occurrence of undefined conditions in the absence of effective buffering, it might be necessary to follow these instruction exactly to get the desired result! On the other hand, with buffers, just because it says 20 mM that should not imply it would not work at 21 mM or 19 mM, or even 40 mM, provided that ionic strength is not important. Buffers should not have any effect on the system other than to maintain pH and contribute to the ionic strength. But sometimes they are not carefully defined. For instance "50 mM Na acetate pH 4.5" might be made up by taking 50 mM Na acetate and adjusting it to pH 4.5 with acetic acid, or with HCl; in either case the ionic strength would be 0.05. But it might equally have been made up by taking 50 mM acetic acid and adjusting it to pH 5.0 with NaOH. In that case, the ionic strength would be 0.02: a very important difference in cation-exchange chromatography. The other example is the temperature at which the buffer was made up (see Section 12.3). One of the most used buffers, the Tris-glycinate buffer used in the acrylamide gel system described by Laemmli [303], was not precisely defined, because the temperature at which it was made up was not stated. Since the procedure was to take Tris base, and adjust it to the specified pH with HCl, the amount of HCl needed (and so the ionic strength and the conductivity of the buffer system overall) would depend on the temperature at the time, since the pK_a of Tris is highly temperature-dependent (Figure 12.5). Although authors often say that "all operations were carried out at 4°C," it is a fair bet that most times the buffers were made up at an unspecified room temperature.

This brings us to the third situation, in which the method that you are attempting to reproduce was published a long time ago, and may have some steps which would not be done today. The first thing is to check that this method has not in fact been superseded by a more modern method, by using the Science Citation Index. This publication, published by the Institute for Scientific Information, Philadelphia, allows you to work forward in time from a given publication to see if there is anything more up to date that you did not know about. If this gives no help, then you should follow the procedure essentially as published, but consider where modifications might be made. Ion exchangers are very different from 20 years ago, and you should adapt such steps to the modern materials. Old-fashioned buffers such as cacodylate or barbitone can be replaced with their modern counterparts, Mes and Tricine (see Appendix C). And obviously, the make of equipment is not (or should not) be vital to success. This applies equally to recent procedures and to older ones. But do not think that protein purifiers of 20 years ago were not as ingenious, or as careful, as today (possibly the reverse is true). The method should be fully successful without modification if it was properly documented.

Finally, you eventually come to the point where you have to document your work yourself, for someone else to reproduce. This is where you really have to think about what is important in what you have done, because you do not want to get a bad reputation with an irreproducible method. I have left this to the last section in this chapter.

12.5 Final Steps—Storage, Crystallization, and Publication

It is appropriate to conclude this chapter with a section on protein crystallization, as the formation of crystals is often the final stage in purifying and studying proteins, particularly enzymes. Part of this section could have been in Chapter 4, since crystallization, as a precipitation method, is a genuine method that has been used for purifying proteins. But it is more appropriate to discuss it in general at this point. Crystallization is important in at least four areas:

1. As a method of purification;
2. As a confirmation of homogeneity;
3. As a method of stable storage;
4. For determining tertiary structure by diffraction techniques.

Crystals form from supersaturated solutions by aggregation of molecules, occurring in an ordered, repetitive fashion, rather than a random aggregation which leads to amorphous precipitates. This is superficially surprising since crystallization clearly represents an increase in order of the protein molecules, and so a decrease in entropy, of the protein. However, the thermodynamics of the *whole system* is much more complex; the disordering of previously bound solvent molecules more than compensates for the ordering of the protein molecules. Simple organic molecules and inorganic salts can be crystallized from suitable solvents, and the crystals may develop in a few minutes to be several millimeters in size. But protein crystals cannot develop so rapidly because of the much slower diffusion, collision, and rotation speeds of large molecules; to crystallize, molecules must collide at just the right angle to interact with each other in the appropriate way to form a homogeneous crystal. It may take a large number of unproductive collisions before a correctly lined-up collision takes place. So protein crystals grow slowly, especially if they are to attain more than a microscopically visible size. But microcrystals can form quite rapidly in some cases, and can be recognized by the characteristic birefringence occurring when agitating a suspension of asymmetric particles little larger than the wavelength of light. Microcrystals

are elongated and needle-like, and create quite attractive appearances
in suspension when swirled around.

Two crystallographic features which are often (though not necessarily)
related are the morphology of the crystal itself and the geometric nature
of the molecular packing within the crystal. The latter is called the space
group, and need not concern noncrystallographers. The morphology or
shape of the crystals is something which can be observed directly (with
the aid of a microscope), and is frequently characteristic for a particular enzyme. Needles of triangular, square, or hexagonal cross section,
diamond or square plates, chunky cubes and truncated octahedrons,
hexagonal and other bi-pyramids are all observed. Some proteins can be
persuaded to grow in a variety of different morphologies. In general, the
larger crystals must be grown slowly, in very clean conditions that give
few opportunities for seeding multiple-growth areas, and preferably in a
vibration-free building. Large crystals are needed for X-ray diffraction
studies, and are normally grown slowly over a period of weeks or months.
For other applications, rapidity of crystal growth is more important than
size of crystal.

Crystallization for Purification

In enzyme purification, crystallization can be used as a final step; sometimes it does not result in much increase in specific activity, for the
preparation may already have been effectively "pure." To present a final
step of crystallization in a purification table implies a high degree of purity. But crystals can develop in fairly impure protein mixtures. Provided
that the impurities are not also in the supersaturated state, aggregation
will only occur between molecules of the one protein that is supersaturated. If this occurs slowly enough, crystals can develop. The original
method for purifying rabbit aldolase [304] involved making a muscle extract 50% saturated in ammonium sulfate, and removing the precipitate.
One can calculate that the aldolase constitutes about 15% of the protein
left in the supernatant. On increasing the ammonium sulfate concentration
to just 52% saturation, aldolase slowly crystallized out; the presence of
the other 85% of the protein did not prevent crystals from forming. One
can conclude from this that the formation of crystals of the enzyme does
not prove that the solution from which the crystals formed contained only
one pure protein. But the crystals themselves are not likely to contain
much in the way of contaminants, unless there was simultaneous amorphous aggregation of an impurity as the crystals formed. In the later case,
it is not so much that the crystals are impure as that it is not possible to
collect those crystals without also collecting the other precipitated protein.

For purification by crystallization, the penultimate fraction should be
"salted out" with ammonium sulfate, and redissolved in a minimum

amount of a suitable buffer. A brief centrifugation to remove traces of insoluble material should be carried out, then some extra ammonium sulfate, usually as drops of saturated solution, can be added. A turbidity will develop, and it is usually at this point that the solution can be left for hours or even days for crystals to develop from the turbidity. Be careful to ensure that the solution remains turbid if you put it in the cold, since solubilities in ammonium sulfate generally increase in the cold. But sometimes it happens that there is no intermediate amorphous turbidity, but small microcrystals appear almost at once. These are identifiable by a sheen when shaken; more ammonium sulfate can be added gradually as the crystals increase in number (if not in size). It is important to recover as much protein as possible by leaving little in the solution, though there is always the danger of beginning to precipitate the impurities. A second crystallization will enable trace impurities to be left behind. Because their concentration decreases with each cycle, a higher level of ammonium sulfate would be needed each time to cause precipitation of the impurities, whereas no higher level is needed to crystallize what is wanted. The time needed when using ammonium sulfate for crystallizing is not a worry because of the stabilizing influence of the salt. But it is possible to use other precipitants which may expose the sample to the dangers of proteolysis and microbial attack. In this case, crystal formation should occur as rapidly as possible. Insulin prepared from pig pancreas is purified using a crystallization step at low ionic strength in acidic conditions. Rapid formation of microcrystals as a purification step is better than slow formation of large crystals [76,305]. Successive recrystallizations should reach a steady state of no further improvement in specific activity—which is a criterion of homogeneity.

Methods for Crystallization for X-ray Diffraction Studies

Proteins are induced to crystallize by creating solvent conditions that result in the protein solution being supersaturated, leading to protein–protein aggregation. In general it is desirable to avoid any amorphous precipitate, and crystallographers who want large, flawless crystals, attempt to minimize the opportunity for nucleation of crystal formation, while maximizing the degree of supersaturation. Since these are incompatible factors, it is common for them to use a system that results in a gradual decrease in volume as the crystals form, so as to maintain a fairly constant low degree of supersaturation. By removing protein from solution as the crystals form, the remaining solution loses its supersaturation, and it is necessary to reduce its volume to increase the concentration again. Otherwise, considerable amounts of uncrystallized protein will remain, and the crystals already obtained will stop growing. An alternative to reducing the volume is to continually add more precipitant to keep the solution in a supersaturated state until most of the protein has crystallized. The protein concentration at the start is typically around $5\,mg\,ml^{-1}$, rarely less than $2\,mg\,ml^{-1}$, but sometimes much higher starting concentrations have been used.

To get to the supersaturated state, any of the precipitants referred to in Chapter 3 can be used. Although ammonium sulfate is employed most frequently, crystals forming in the salting-in (low ionic strength) range can be very satisfactory for diffraction work, though one must be careful about long-term stability because of possible microbial growth. Organic solvents have occasionally been used successfully, but long-term stability requires low-temperature storage. Polyethylene glycol has been very successful in a number of cases where ammonium sulfate was not successful in growing large crystals, and is becoming as common in protein crystallography as ammonium sulfate. It is particularly successful with proteins that have low intrinsic solubility, such as hydrophobic and large-sized proteins [306,307].

A number of theoretical and practical approaches to the problem of growing large crystals for diffraction have been published [306,308,309, 310,311]. Generally, it requires an extensive study of ranges of conditions, as slight changes in pH, or the presence of small amounts of metal ions can affect the quality of the crystals substantially. The same enzyme from different species may behave quite differently, and the reason why crystallographers often seem to be using enzymes purified from odd sources is that they often find that a particular species' enzyme happens to crystallize much more readily than most others. Hence, sperm whale, lobster, horse, cat, chicken—rather than rat, rabbit, or *E. coli*.

For full, three-dimensional structural studies of any protein, X-ray diffraction must be used on crystals of the pure protein. Although modern methods involving powerful beams of synchroton radiation can get useful information from crystals less than 0.1 mm in size, most crystallographers have access only to conventional sources of X-rays, and need crystals of about 0.3 mm in each dimension. To grow such crystals requires much time and patience, and reproducibility of conditions for crystal growth is often a problem. Crystals must be grown slowly and undisturbed if they are to get large, so it is best to set up a trial just before going on holiday!

If large amounts of protein are available (tens of milligrams), then one can simply put samples of about 0.5 ml in small glass containers, each successive sample having a small increment in precipitant concentration (Figure 12.8). A precipitate immediately forming in about the middle of the range is best, for a priori it is not known whether crystals will develop from a clear solution or from a solution containing an aggregated precipitate. Different sets at various pH values, or containing a substrate, metal ion, etc., can be arranged simultaneously when there is plenty of protein. Protein concentration should be in the range 2–

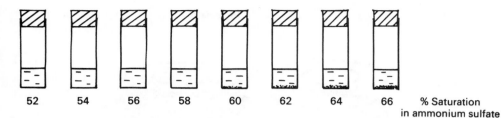

| 52 | 54 | 56 | 58 | 60 | 62 | 64 | 66 | % Saturation in ammonium sulfate |

Figure 12.8. Samples placed in pots at incremental precipitant concentration, and left for crystal development. Some precipitate immediately forms in the higher concentration pots.

$30 \, \text{mg ml}^{-1}$. Glass containers are used so that any crystals can be observed directly with a magnifying lens or a microscope. But it is not common to have such large amounts of protein, and there are better ways of approaching the problem while using much smaller amounts. The two most used involve the gradual change in solvent conditions by increasing the concentration of precipitant over a period of hours to days. These are: (1) dialysis-based systems; and (2) diffusion-based systems. With dialysis, a membrane separates the protein solution, already in a concentration of precipitant close to that required, from a solution of precipitant of higher concentration. Dialysis equilibrates the precipitant concentrations, so that the protein becomes supersaturated. Two methods of achieving this are illustrated in Figures 12.9 and 12.10.

Dialysis equilibration can be carried out in capillary tubes (ca. 2 mm in diameter) containing a small piece of dialysis tubing on one end (the other end should be covered to prevent evaporation). The dialysis tubing can be conveniently held on with a tight ring cut from a piece of silicone or other plastic tubing (Figure 12.9). Fifty microliters of sample occupies 1–2 cm of each tube, and should contain from 0.1 to 0.5 mg of protein. After a few days in the buffer, if no crystals or precipitate form, the tube can be transferred to a buffer of higher precipitant concentration.

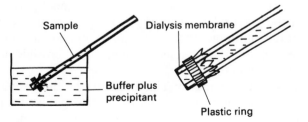

Figure 12.9. Sample for crystallization placed in narrow glass tube with dialysis membrane held on the end. Sample equilibrates with the precipitant buffer, with slow diffusion of precipitant up the sample tube.

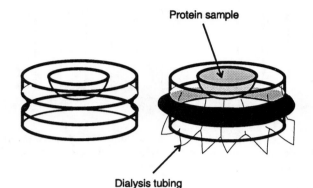

Figure 12.10. Crystallization beads. A grooved cylinder of transparent plastic has a hemispherical depression on one face. The protein sample is placed in the depression, and covered with dialysis tubing, which is stretched around the cylinder, and a rubber ring (black) holds the dialysis tubing taut. The bead is then placed in a container containing liquid with appropriate concentrations of precipitants. There is a rapid equilibration of buffer conditions within a few hours.

Alternatively, the buffer can be opened to the air and so evaporate slowly, until its concentration is sufficient to cause an effect on the protein. If polyethylene glycol (PEG) is the precipitant, osmotic effects will cause the reduction of volume in the tube, concentrating the sample. PEG of MW greater than about 8,000 will not cross the dialysis membrane at all rapidly, and this concentration effect may be so severe that there is little liquid left in the tube. Whenever using PEG, remember that the PEG must be present with the protein from the start, so that its concentration can increase.

When crystals form, they are more dense than the surrounding solution, so they sediment (unless in microgravity, see below). In the simple dialysis method illustrated in Figure 12.9, crystals forming tend to settle down on the dialysis tubing. This is not always a good idea, especially as this is where the precipitant concentration is the highest, and amorphous precipitate may be present. The capillary tube should be kept as near to horizontal as possible to avoid this effect. Occasionally, crystals stick to the walls of the glass tube, which is good for preventing sedimentation, but makes it difficult to get them out. Since crystallography is carried out using crystals mounted in capillary tubes, it is possible to crystallize directly in the tube in this way.

With the dialysis "beads" in Figure 12.10, the dimensions are small, there is not far for any sedimenting crystal to go, and equilibrium is attained rapidly. If no precipitate forms in a few days, the beads can be transferred to a slightly higher precipitant concentration. Alternatively, the outside solution can be uncovered and allowed to concentrate by evaporation. These beads are a useful way of getting a large number of possible conditions surveyed with a small amount of protein, and if successful crystals are obtained, they can be transported easily with minimum damage to the crystals.

The most favored method when sample is limited is the "hanging drop" method (Figure 12.11). In this case, $10-30\,\mu l$ of sample is "hung" from a glass cover slip, or even the plastic lid of a multi-well dish, over a pool of precipitant, and the drop slowly evaporates to equilibrate with the pool below. The chamber must be sealed, and the sample *must* contain some of the precipitant; it could not get there otherwise. If the sample is at, say, 25% saturation with ammonium sulfate, and the pool is 50% saturation, then the sample will halve in volume (and so concentrate the protein) until it, too, is at 50% saturation. Note that a cubic crystal with 0.3 mm sides contains about $15\,\mu g$ of protein, so don't expect many large crystals to form in a drop typically containing only $20-50\,\mu g$ protein. One advantage that the hanging drop method has over the dialysis method is that diffusive equilibration is slower. This gives more chance for single, large crystals to grow, before oversupersaturation results in multiple nucleation sites and/or amorphous precipitation.

Once the approximate conditions have been established for getting crystals, the ideal conditions can be found by narrowing the range of precipitant concentrations (and pH, etc.). To slow the crystal growth may mean making very little difference between the precipitant concentration in the protein sample and the dialysate/diffusate. In the case of the hanging drop method, although it is not physically possible to scale up the

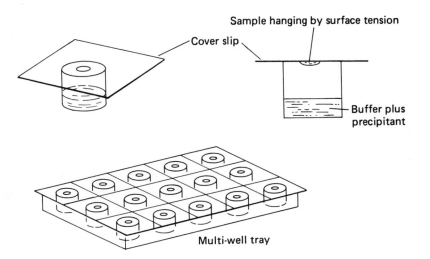

Figure 12.11. "Hanging drop" method for crystallizing from small sample volumes. Slow equilibration by vapor (water) diffusion between the sample and the precipitant solution results in a very slow increase in precipitant concentration in the sample.

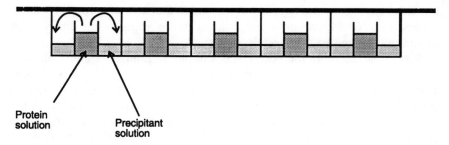

Protein
solution   Precipitant
                         solution

Figure 12.12. Slow crystallization by very gradual increase in precipitant concentration. A relatively large volume of sample, e.g., $200\,\mu l$, is placed in the central well in subprecipitation conditions, and slow diffusion of water vapor into the more highly concentrated precipitant outside takes place. Over a period of days to weeks, the concentration of precipitant in the protein sample reaches protein crystallization point.

size of the drop, an alternative apparatus with the sample in a central well can be used. In this case, the equilibration can be much slower, depending on the relative exposed surface area to volume of the protein sample (Figure 12.12).

The principal variables to try are: protein concentration, pH, presence of metal ions (e.g., Mg^{2+}, Ca^{2+}), substrates, inhibitors, transition state analogs, the addition of small amounts of miscible organic solvent (e.g., dioxan, aliphatic diols), varying temperature, type of precipitant, and, if all else fails, change the source species.

One other variable, available to very few people, is that of gravity. Because of the sedimentation problems referred to above, plus micro-scale convective disturbances, crystals should, in theory, grow better in zero gravity. The experiments carried out so far on the space shuttle generally confirm this (e.g., [312]), but it would have to be a very important project to justify the real expense involved—especially as these experiments have frequently been put forward as one of the main non-military justifications of the whole space shuttle program.

It is highly satisfactory to be able to crystallize a protein after it has been purified. Not only is it demonstrably pure and in a stable condition, but there is the feeling that one can go no further with the purification; the isolation is at an end, and now characterization of the protein can commence.

Conditions for Storage of Purified Proteins

The storage of the purified protein is a most important consideration. If it has been successfully crystallized as described above, then it is likely to be very stable, and this would be ideal.

Crystalline suspensions are stable because the molecules are immobilized and so are unlikely to denature because of thermal agitation. Also, proteolytic enzyme contaminants are most unlikely to cocrystallize (though proteases could absorb to the surfaces of the crystals), so the molecules are protected from proteolytic degradation. Finally, the usual conditions for crystallization, e.g., high concentrations of ammonium sulfate, are bacteriocidal and stabilize protein molecules by dehydration.

Crystallization for the purpose of stabilization and storage does not depend so critically on conditions. Small microscopic crystals are quite suitable. Indeed, large crystals can be a nuisance when trying to get a measured sample from a stock enzyme bottle. Many commercial enzymes are supplied as "a crystalline suspension in $3 M$ ammonium sulfate." Nevertheless, the crystals themselves are quite soft and delicate, and should not be mistreated. Freezing is likely to destroy their structure, but may not do any damage to the bioactivity.

But not all purified proteins can be crystallized, and it is often impracticable to find suitable crystallization conditions at the end of a procedure, even if this is the whole aim of purifying the protein. So long-term storage will need some other approach. The simplest alternative is also in ammonium sulfate, but as an amorphous precipitate. Most of the advantages of crystals apply, except that it cannot be guaranteed that proteolytic enzymes would not be active.

If ammonium sulfate precipitation as a final step is not suitable (for instance, it may be inconvenient to have to remove excess ammonium sulfate before using the protein), then alternative approaches must be sought. The aim for long-term storage is to immobilize the protein, and/or dehydrate it at the molecular level, so that processes that may degrade or otherwise inactivate it are reduced to a minimum. Dehydration at the molecular level implies removal of water molecules around the outside of the protein molecule so that movement of the polypeptide is restricted. High concentrations of structure-forming salts like ammonium sulfate do this, as do polyols such as glycerol. In each case, the water molecules preferentially interact with the solute. So addition of glycerol (or sorbitol, sucrose, etc.) to a concentration of about 50% v/v results in substantial stabilization of proteins in solution. What is more, such a solution can be stored at subzero temperatures without freezing, at least down to $-50°C$ or so. And if the solution does freeze (the freezing point of 50% aqueous glycerol is actually $-23°C$, but supercooling is common), the proteins are protected from damage from ice-crystal formation. Even so, the concentration of glycerol could present some difficulties when the protein is warmed up for whatever use or investigation is planned.

Simple freezing of the protein solution will be quite adequate with many proteins. It is best if the concentration of protein is high ($10 \, mg \, ml^{-1}$ or more) when the freezing takes place; indeed, a high concentration of protein is always best whatever storage method is used, unless one comes up against insolubility problems.

The final alternative is complete dehydration. Generally, drying of proteins will result in inactivation, but freeze-drying (lyophilization) is a safe and gentle process. Freeze-drying must occur with the water always present as ice, and the product is a fluffy powder occupying a large volume for its weight. The characteristics of the powder will depend on the amount of nonprotein material, mainly buffer salts, that are present. Freeze-dried proteins, provided that they are kept in a low humidity, will often keep for years. It is one of the commonest ways of preserving enzymes for commercial use; they can be reconstituted simply by adding water, and there are usually no additives to get rid of. But sometimes bioactivity can be lost during freeze-drying, and additives such as polyols may need to be present. Recent reports have described the remarkable efficacy of the disaccharide trehalose in stabilizing proteins, enabling long-term storage in the dried state at ambient temperatures.

These preservation methods are mostly mutually exclusive. For example, it does not do much good to add glycerol to an ammonium sulfate suspension. Nor can one freeze-dry a glycerol mixture, and freezing ammonium sulfate suspensions can do more harm than good. But temperature should be kept down as far as appropriate for the circumstances: around 0°C for ammonium sulfate suspensions, and as low as freezer space will allow for glycerol mixtures. Freeze-dried samples can be kept at subzero temperatures, but it is not usually necessary—as long as they are dry. Mention must be made again that some proteins are actually destabilized by freezing—by the process of cold denaturation.

Some enzymes and other proteins are susceptible to inactivation by processes that may not be protected by the techniques indicated above, although most things will occur more slowly at reduced temperature and reduced water activity. One of the main processes is oxidative inactivation. Storage can be in anaerobic conditions, but more commonly, a reducing agent such as dithiothreitol is used at a level of 5–10 mM. There may also be a need to protect against proteolytic inactivation by inclusion of appropriate compounds (see Section 12.2).

What is Important for Publication?

This is the converse of Section 12.4. You are not going to follow a published procedure, but publish your own for someone else to follow. It is most important that you have carried out the procedure yourself several times (although occasionally, a single example of a very difficult but important protein purification may qualify for publication). This enables you to know what the variablity is like, and to know the critical features of the process, for it is most likely that, intentionally or otherwise, there will have been some differences each time. Unless you have some feeling for the most important parameters, there can be no certainty that someone else in a different laboratory with different apparatus and source of raw material will be able to reproduce your method. The best procedure, unfortunately rarely possible in the hustle and bustle of modern laboratories, is to get a coworker to use your method without help from you, *after* you have written it in what you hope is the final version. If you can't get someone else to do it, then at least do it yourself using your final protocol. Because if it does not always work as well as you have claimed in your own laboratory, what is the chance of it working somewhere else? Most of the comments in Section 12.4 can be applied to your own work. Ask, "have I defined the buffers fully?"; "are all the numerical values correct?" The latter seems pretty obvious, but still people make 1,000-fold errors, or confuse "mM" with "mmol." If describing the size of a column, is it clear that the cross-section dimension is a diameter, a radius, or a cross-sectional area?

Finally, and this is partly an editorial matter, should a new purification procedure come as part of the Methods section, or as part of the Results? The answer to this really depends on the importance of the purification procedure to the overall paper. If the procedure *is* the main feature, then it should definitely be in the Results section. But if it is peripheral to the main topic, it can be in Methods or, alternatively, in an Appendix. I like to think that the protein purification procedure, if it is significantly novel, should always be part of the Results; someone has spent a fair bit of time getting these "Results," and they should be appropriately acknowledged.

Appendix A
Precipitation Tables

This section contains tables for calculating amounts of precipitant to add to a solution:

Table I: Solid Ammonium Sulfate.
Amount of ammonium sulfate (solid) to be added to a solution already at $S_1\%$ saturation to take it to $S_2\%$ saturation.

For 1 liter of solution:

$$g = \frac{533(S_2 - S_1)}{100 - 0.3S_2}$$

Table II: Organic Solvent, or Saturated Ammonium Sulfate.
Volume of miscible liquid (organic solvent or saturated ammonium sulfate) to be added to a solution already containing $C_1\%$ vol/vol liquid, to take it to $C_2\%$.

For 1 liter of solution:

$$\text{Volume (ml)} = \frac{10(C_2 - C_1)}{100 - C_2}$$

Table I. Ammonium Sulfate, Grams to be Added to 1 Liter

From S_1% \↓ To S_2% →	5	10	15	20	25	30	35	40	45
0	27	55	84	113	144	176	208	242	277
	5	27	56	85	115	146	179	212	246
		10	28	57	86	117	149	182	216
			15	28	58	88	119	151	185
				20	29	59	89	121	154
					25	29	60	91	123
						30	30	61	92
							35	30	62
								40	31
								45	
									50

Table II. Volume of Miscible Solvent, ml to be Added to 1 Liter

From C_1% \↓ To C_2% →	5	10	15	20	25	30	35	40	45
0	52	111	176	250	333	428	538	666	818
	5	55	117	187	266	357	461	583	727
		10	58	125	200	285	384	500	636
			15	62	133	214	307	416	545
				20	66	142	230	333	454
					25	71	153	250	363
						30	76	166	272
							35	83	181
								40	90
								45	
									50

From S₁ %	50	55	60	65	70	75	80	85	90	95	100
0	314	351	390	430	472	516	561	608	657	708	761
5	282	319	357	397	439	481	526	572	621	671	723
10	251	287	325	364	405	447	491	537	584	634	685
15	219	255	292	331	371	413	456	501	548	596	647
20	188	223	260	298	337	378	421	465	511	559	609
25	157	191	227	265	304	344	386	429	475	522	571
30	126	160	195	232	270	309	351	393	438	485	533
35	94	128	163	199	236	275	316	358	402	447	495
40	63	96	130	166	202	241	281	322	365	410	457
45	31	64	97	132	169	206	245	286	329	373	419
50		32	65	99	135	172	210	250	292	335	381
55			33	66	101	138	175	215	256	298	343
60				33	67	103	140	179	219	261	305
65					34	69	105	143	183	224	266
70						34	70	107	146	186	228
75							35	72	110	149	190
80								36	73	112	152
85									37	75	114
90										37	76
95											38

From C₁ %	50	55	60	65	70	75	80	85	90	95
0	1,000	1,222	1,500	1,857	2,333	3,000	4,000	5,666	9,000	19,000
5	900	1,111	1,375	1,714	2,166	2,800	3,750	5,333	8,500	18,000
10	800	1,000	1,250	1,571	2,000	2,600	3,500	5,000	8,000	17,000
15	700	888	1,125	1,428	1,833	2,400	3,250	4,666	7,500	16,000
20	600	777	1,000	1,285	1,666	2,200	3,000	4,333	7,000	15,000
25	500	666	875	1,142	1,500	2,000	2,750	4,000	6,500	14,000
30	400	555	750	1,000	1,333	1,800	2,500	3,666	6,000	13,000
35	300	444	625	857	1,166	1,600	2,250	3,333	5,500	12,000
40	200	333	500	714	1,000	1,400	2,000	3,000	5,000	11,000
45	100	222	375	571	833	1,200	1,750	2,666	4,500	10,000
50		111	250	428	666	1,000	1,500	2,333	4,000	9,000
55			125	285	500	800	1,250	2,000	3,500	8,000
60				142	333	600	1,000	1,666	3,000	7,000
65					166	400	750	1,333	2,500	6,000
70						200	500	1,000	2,000	5,000
75							250	666	1,500	4,000
80								333	1,000	3,000
85									500	2,000
90										1,000

Appendix B
Solutions for Measuring Protein Concentration

(i) Biuret Reaction [268]

Dissolve 1.5 g copper sulfate ($CuSO_4 \cdot 5H_2O$) and 6.0 g sodium potassium tartrate in 500 ml water. Add 300 ml 10% sodium hydroxide, and make up to 1 liter with water. If 1 g potassium iodide is also added, the reagent will keep indefinitely in a plastic container.

Procedure

To 0.5 ml of unknown containing up to 3 mg of protein, add 2.5 ml reagent. Allow to stand for 20–30 min before reading A_{540}.

(ii) Lowry Reaction [229]

Reagent A Dissolve 0.5 g copper sulfate ($CuSO_4 \cdot 5H_2O$) and 1 g sodium citrate (Na_3 citrate) in 100 ml water. This solution will keep indefinitely, and is more stable than the more common copper tartrate mixture [269].

Reagent B Dissolve 20 g sodium carbonate (Na_2CO_3) and 4 g sodium hydroxide in 1 liter water.

Reagent C To 50 ml Reagent B, add 1 ml Reagent A.

Reagent D To 10 ml Folin-Ciocalteau reagent, add 10 ml water.

Procedure

To 0.5 ml of unknown containing up to 0.5 mg protein, add 2.5 ml Reagent C. Mix and let stand 5–10 min. Then add 0.25 ml Reagent D. Mix well and let stand 20–30 min, read color at a convenient wavelength between 600 and 750 nm. Note that variations in the strength of Folin-Ciocalteau reagent may result in unsatisfactory color development. The dilution of this reagent (or the amount added) may be varied to make the pH of the final mixture after adding Reagents C and D between 10.0 and 10.5. Color development is more rapid at higher pH, but it also fades more quickly. For automated systems where the time of reaction is precisely controlled, rapid color development at pH 11.0 may be more suitable, for which the NaOH concentration in Reagent B should be increased by 30–40%.

(iii)   Coomassie Blue Binding [239, 270]

(a) 100 mg Coomassie Brilliant Blue G-250 is dissolved with vigorous agitation in 50 ml 95% ethanol, then mixed with 100 ml 85% phosphoric acid. The mixture is diluted to 1 liter with water, and filtered to remove undissolved dye. This solution is stable for 1–2 weeks, kept in the cold.

(b) 600 mg Coomassie Brilliant Blue G-250 is dissolved in 1 liter 2% perchloric acid, and filtered to remove undissolved material. The absorbance at 465 nm should be between 1.3 and 1.5. This solution is stable indefinitely.

Procedure

(a) To 0.1 ml of sample containing up to 50 μg protein, add 2.5 ml of Coomassie Blue reagent, and read absorbance at 595 nm after 2–30 min.

(b) To 1.5 ml of sample containing up to 50 μg protein, add 1.5 ml Coomassie Blue reagent, and read absorbance at 595 nm after 2–30 min. For dealing with concentrated protein solutions ($>$5 mg ml^{-1}), it is convenient to take 1.5 ml water or dilute sodium chloride, add the sample from a microliter syringe, mix well, then add the dye reagent. Concentrated protein solution will precipitate on contact with the dye reagent. Reagent (a) above gives a higher absorbance reading than reagent (b), but is less stable.

Appendix C
Buffers for Use in Protein Chemistry

Buffers with pK_a values between 4 and 10 have been presented. For pHs significantly below 4, a range of carboxylic acids (e.g., formic acid) is available for $n = -1$ buffers, and glycine and other simple amino acids for pHs significantly above 10, volatile amines such as triethylamine can be used ($n = +1$).

A Selection of Useful Buffers

Name	n	pK_a^0 (25°C)	Comments
Lactic acid	−1	3.86	
Acetic acid	−1	4.76	
Pivalic acid (trimethylacetic acid)	−1	5.03	Not very soluble; unpleasant odor
Pyridine	+1	5.23	Volatile, poisonous
Picolinic acid	−1[a]	5.4	Strong UV absorption
Succinic acid	−3	5.64	
Histidine	+1[a]	6.0	Complexes Me^{2+} strongly
N-Morpholinoethane sulfonic acid (Mes)	−1[a]	6.15	
Bis-(2-hydroxyethyl)imino-tris-(hydroxymethyl)methane (Bis-Tris)	+1	6.5	
N-(2-acetamido)-2-aminoethane sulfonic acid (Aces)	−1[a]	6.9	
Imidazole	+1	6.95	Complexes Me^{2+}
Phosphate	−3	7.20	Stabilizes many enzymes
N-Morpholinopropane sulfonic acid (Mops)	−1[a]	7.2	

(continued)

N-Tris(hydroxymethyl)methyl-2- aminoethane sulfonic acid (Tes)	-1^a	7.5
Carbonic acid	-1	6.3
Triethanolamine	$+1$	7.75
Tris(hydroxymethyl)aminomethane (Tris)	$+1$	8.06
N-Tris(hydroxymethyl)methyl-glycine (Tricine)	-1^a	8.15
Tris(hydroxymethyl)aminopropane sulfonic acid (Taps)	-1^a	8.4
2-Amino-2-methyl-1,3-propanediol	$+1$	8.8
Diethanolamine	$+1$	8.9
Taurine	-1^a	9.1
Ammonia	$+1$	9.25	Volatile
Boric acid	-1	9.23	Complexes with many carbohydrates
Ethanolamine	$+1$	9.5
Glycine	-1^a	9.8
1-Aminopropan-3-ol	$+1$	9.95
Carbonate	-3	10.3

^{a} Indicates zwitterionic buffer in either acidic or basic form.

The following buffers, which used to be quite widely used, are now usually replaced with more suitable buffers, for reasons of cost, toxicity, solubility, or n value.

Old buffer	*Suggested replacement*
Veronal (barbitone, 5:5'-diethyl barbituric acid) pK_a 8.0, $n = -1$	Tricine, pK_a 8.1, n $= -1$ Tris, pK_a 8.1, $n = +1$
Cacodylic acid (dimethyl arsonic acid)	Mes, pK_a 6.2, $n -1$ Histidine, pK_a 6.0, $n = +1$
pK_a 6.2, $n = -1$	Bis-Tris, pK_a 6.5, $n = +1$
Citric acid (third pK_a, 6.4, $n = -5$)	Mes, pK_a 6.2, $n = -1$ Histidine, pK_a 6.0, $n = +1$ Bis-Tris, pK_a 6.5, $n = +1$
Maleic acid (second pK_a, 6.3, $n = -3$) Has UV absorption	as above

Appendix D
Chromatographic Materials

The products of some of the major manufacturers of adsorbents and gel filtration matrices for protein chromatography are listed below. The many makers of reverse phase HPLC adsorbents have not been included, and powdered cellulose supports are not listed. Companies distributing these products vary from country to country, and other manufacturers' products may be more readily available in some places. An extensive listing of scientific suppliers for the American market can be found in [313].

Details given are the manufacturers' information, and are not always comparable. For instance, bead sizes may be given as a range, an average without size distribution, or an average with distribution; a single figure does not necessarily mean a narrow distribution range.

1. Adsorbents

Manufacturer	Trade name	bead size (μm)	adsorbent types	matrix base
Bio Rad	Macro-Prep 50	50	Q-, S-, CM- C1-, C4-	polyacrylic
	Bio-Gel A	80–150	DEAE-, CM-	agarose
	Affi-Gel	80–150	activated affinity heparin Protein A Blue dye, etc.	

(continued)

353

Manufacturer	Trade name	bead size (μm)	adsorbent types	matrix base
Pharmacia	Sephacel	40–160	DEAE-	cellulose
	Sepharose CL-6B	45–165	DEAE-, CM- IMAC-, thiol-, Blue dye, etc. activated affinity	agarose
	Sepharose CL-4B	45–165	C8-, phenyl- Protein A, Protein G, Con A, etc. activated affinity	agarose
	Sepharose Fast Flow	45–165	Q-, S-, DEAE-, CM-	agarose
	Sephadex	*>100	QAE-, SP-, DEAE-, CM-	dextran
	Polybuffer PBE	45–165	polyimine	agarose
	Sepharose High Perf.	†34	CM-, Q-, S-	agarose
	Monobeads	†10 ± 0.2	Q-, S-, polyimine	synthetic polymer
Merck	Fractogel TSK (S)	25–40	SP-, DEAE-, CM-	vinyl
	(M)	45–90	TMAE-, DEAE-, S- CM-	polymer
	LiChrosphere 1,000	5	TMAE-, DEAE-, S-, CM-	silica
Toyo-Soda	Toyopearl (S)	20–50	SP-, QAE-, DEAE, CM-	vinyl
	TSK gel (M)	40–90	C4-, phenyl-, IMAC	polymer
	(C)	60–150	activated affinity, dyes, etc.	
Amicon	Matrex Cellufine	80	DEAE-, CM-	cellulose
	Dye Matrex	80–120	dyes	agarose
	Matrex PAE	10, 20, 50	polyimine	silica
IBF (LKB, Sepracor)	Trisacryl	40–80	DEAE-, SP- reactive affinity	acrylamide polymer
	Zephyr	†20 ± 5	DEAE-, S-	synthetic/ polysaccharide
Alltech	HEMA	†10	SP-, CM-, TMAE-,	acrylate
		20	DEAE-	polymer
		60	activated affinity	
Waters	Protein-Pak	†8	DEAE-, SP-	synthetic
		15		polymer
		40		

* Sephadex beads swell according to buffer ionic strength, especially the ion exchangers, so a definite bead size cannot be given.

† Available in pre-packed columns only.

2. Gel Filtration media

Manufacturer	Trade name	Bead size	sizing ranges (kDa) (globular proteins)	matrix base
Bio Rad	Bio Gel P	20–30 45–90 90–180	1–6, 1.5–20, 2.5–40, 3–60, 5–100	polyacrylamide
	Bio Gel A	40–80 80–150 150–300	10–500, 10–1,500, 10–5,000, 40–15,000, 100–50,000, 1,000–150,000	agarose
Pharmacia	Sephadex	*(20–50) *(40–120)	1.5–30, 1–80, 4–150, 5–300, 5–600	dextran
	Sepharose & Sepharose CL	40–160	10–4,000, 60–20,000 60–200, 70–40,000	agarose
	Sephacryl HR	25–75	1–100, 5–250, 10–1,500 20–8,000, 50–20,000	dextran-acrylamide
	Superdex	†34	3–70, 10–600	agarose-dextran
	Superose	†10 †30	1–300, 5–1,000,	agarose
Merck/ Toyo-Soda	Fractogel TSK HW (S) Toyopearl TSK HW (M) (C)	25–40 32–63 50–100	1–80, 2–800 30–6,000	vinyl polymer
Amicon (Chisso)	Matrex Cellufine	45–105	10–60, 10–120, 10–400, 10–2,000	cellulose
IBF (LKB, Sepracor)	Trisacryl	40–80	1–15, 5–25, 5–65, 10–130	acrylamide polymer

Bio Rad Laboratories, Hercules, CA, USA; **Pharmacia** LKB Biotechnology, Uppsala, Sweden; E. **Merck**, Darmstadt, Germany; **Amicon** Inc., Beverly, MA, USA; **Waters** Chromatography (Millipore Corp.), Milford, MA, USA; **Alltech** Associates Inc., Deerfield, IL, USA; **Toyo-Soda** (Tosoh Co.), Tokyo, Japan; **IBF** Pharmindustrie Reactifs, Ville-la-Garenne, France.

References

1. Saul, A. & Don, M. (1984) A rapid method of concentrating proteins in small volumes with high recovery using Sephadex G-25. Anal. Biochem. 138, 541–543.
2. Helmhorst, E. & Stokes, G.B. (1980) Microcentrifuge desalting: a rapid quantitative method for desalting small amounts of protein. Anal. Biochem. 104, 130–135.
3. Smiley, K.L., Berry, A.J. & Suelter, C.H. (1967) An improved purification, crystallization, and some properties of rabbit muscle 5'-adenylic acid deaminase. J. Biol. Chem. 242, 2502–2506.
4. van de berg, L. & Rose, D. (1959) Effect of freezing on the pH and composition of sodium and potassium phosphate solution: The reciprocal system KH_2PO_4-Na_2PHO_4-H_2O. Arch. Biochem. Biophys. 81, 319–329.
5. Orii, Y. & Morita, M. (1977) Measurement of the pH of frozen buffer solutions by using pH indicators. J. Biochem. 81, 163–168.
6. Tyler, D.D. & Gonze, J. (1967) The preparation of heart mitochondria from laboratory animals. Methods Enzymol. 10, 75–77.
7. Hughes, J., Ramsden, D.K. & Symes, K.C. (1990) The flocculation of bacteria using cationic synthetic flocculants and chitosan. Biotechnol. Tech. 4, 55–60.
8. Harrison, S.T.L., Dennis, J.S. & Chase, H.A. (1991) Combined chemical and mechanical processes for the disruption of bacteria. Bioseparation 2, 95–105.
9. Tsuchihashi, M. (1923) Blood catalase. Biochem. Z. 140, 63–112.
10. Lazarus, N.R., Ramel, A.H., Rustrum, Y.M. & Barnard, E.A. (1966) Yeast hexokinase. I. Preparation of the pure enzyme. Biochemistry 5, 4003–4016.
11. Rutter, W.J. & Hunsley, J.R. (1966) Fructose diphosphate aldolase from yeast. Methods Enzymol. 9, 480–486.
12. Yun, S.L., Aust, A.E. & Suelter, C.H. (1976) A revised preparation of yeast (*Saccharomyces cerevisiae*) pyruvate kinase. J. Biol. Chem. 251, 124–128.

13. Takagahara, I., Suzuki, Y., Fujita, T., Yamauti, J., Fujii, K., Yamashita, J. & Horio, T. (1978) Successive purification of several enzymes having affinities for phosphoric groups of substrates by affinity chromatography on P-cellulose. J. Biochem. 83, 585–597.

14. Ito, Y., Tomasselli, A.G. & Noda, L.H. (1980) ATP:AMP phosphotransferase from baker's yeast. Eur. J. Biochem. 105, 85–92.

15. de la Morena, E., Santos, I. & Grisolia, S. (1968) Homogenous crystalline phosphoglycerate phosphomutase of high activity: A simple method for lysis of yeast. Biochim. Biophys. Acta 151, 526–528.

16. Naganuma, T., Uzuka, Y. & Tanaka, K. (1984) Simple and small-scale breakdown of yeast. Anal. Biochem. 141, 74–78.

17. Andrews, B.A. & Asenjo, J.A. (1987) Enzymatic lysis and disruption of microbial cells. Trends Biotechnol. 5, 273–277.

18. Torner, M.J., Andrews, B.A. & Asenjo, J.A. (1992) Kinetics of protein release from yeast using enzymatic lysis for selective product recovery. Biotechnol. Tech. 6, 371–376.

19. Harrison, S.T.L., Dennis, J.S. & Chase, H.A. (1991) Combined chemical and mechanical processes for the disruption of bacteria. Bioseparation 2, 95–105.

20. Schwinghamer, E.A. (1980) A method for improved lysis of some gram-negative bacteria. FEMS Microbiol. Lett. 7, 157–162.

21. Scopes, R.K., Testolin, V., Stoter, A., Griffiths-Smith, K. & Algar, E.M. (1985) Simultaneous purification and characterization of glucokinase, fructokinase and glucose 6-phosphate dehydrogenase from Zymomonas mobilis. Biochem. J. 228, 627–634.

22. Keilin, D. & Hartree, E.F. (1947) Activity of the cytochrome system in heart muscle. Biochem. J. 41, 500–502.

23. Clarke, F.M. & Masters, C.J. (1976) Interactions between muscle proteins and glycolytic enzymes. Int. J. Biochem. 7, 359–365.

24. Scopes, R.K. (1971) An improved method for the isolation of 3-phosphoglycerate kinase from yeast. Biochem. J. 122, 89–92.

25. Penefsky, H.S. & Tzagoloff, A. (1971) Extraction of water-soluble enzymes and proteins from membranes. Methods Enzymol. 22, 204–219.

26. Timmins, P.A., Leonhard, M., Wettzein, H.U., Wacker, T. & Wette, W. (1988) A physical characterization of some detergents of potential use for membrane protein crystallization. FEBS (Fed. Eur. Biochem. Soc.) Lett. 238, 361–368.

27. Findlay, J.B.C. (1990) Purification of membrane proteins. In: Protein purification applications. (Harris, E.L.V., and Angal, S., eds) pp 59–82, IRL Press, Oxford.

28. Hjelmeland, L.M. (1980) A nondenaturing zwitterionic detergent for membrane biochemistry: Design and synthesis. Proc. Natl. Acad. Sci. 77, 6368–6370.

29. Bordier, C. (1981) Phase separation of integral membrane proteins in Triton X-114. J. Biol. Chem. 256, 1604–1607.

30. Griffin, D.C. & London, M. (1982) Ion-exchange chromatography in the presence of the non-ionic dissociating agent chloral hydrate. Biochem. J. 201, 227–231.

31. Ayala, G., Nascimento, A., Gomez-Puyou, A. & Darszon, A. (1985) Extraction of mitochondrial membrane proteins into organic solvents in a functional state. Biochim. Biophys. Acta 810, 115–122.

32. Luisi, P.L. & Laane, C. (1986) Solubilization of enzymes in apolar solvents via reverse micelles. Trends Biotechnol. 4, 153–160.

33. Chang, N., Hen, S.J. & Klibanov, A.M. (1991) Protein separation and purification in neat dimethyl sulfoxide. Biochem. Biophys. Res. Commun. 176, 1462–1468.

34. Chang, N. & Klibanov, A.M. (1992) Protein chromatography in neat organic solvents. Biotechnol. Bioeng. 39, 575–578.

35. Ayala, G., Nascimento, A., Gomez-Puyou, A. & Darszon, A. (1985) Extraction of mitochondrial membrane proteins into organic solvents in a functional state. Biochim. Biophys. Acta 810, 115–122.

36. Itzhaki, R.F. & Gill, D.M. (1964) A micro-biuret method for estimating protein. Anal. Biochem. 9, 401–410.

37. Lowry, O.H., Rosebrough, N.J., Farr, A.L. & Randall, R.J. (1951) Protein measurement with the Folin phenol reagent. J. Biol. Chem. 193, 265–275.

38. Folin, O. & Ciocalteau, V. (1927) Tyrosine and tryptophan determination in proteins. J. Biol. Chem. 73, 627–650.

39. Peterson, G.L. (1979) Review of the Folin phenol protein quantitation method of Lowry, Rosebrough, Farr, Randall. Anal. Biochem. 100, 201–220.

40. Legler, G., Muller-Platz, C.M., Mentges-Hettkamp, M., Pflieger, G. & Julich, E. (1985) On the chemical basis of the Lowry protein determination. Anal. Biochem. 150, 278–287.

41. Alam, A. (1992) A model for formulation of protein assay. Anal. Biochem. 203, 121–126.

42. Warburg, O. & Christian, W. (1941) Isolierung und kristallisation des Gärungs-ferments Enolase. Biochem. Z. 310, 384–421.

43. Goldfarb, A.R., Saidel, L.J. & Mosovich, E. (1951) The ultraviolet adsorption spectra of proteins. J. Biol. Chem. 193, 397–404.

44. Wetlaufer, D.R. (1962) Ultraviolet spectra of proteins and amino acids. Adv. Prot. Chem. 17, 303–390.

45. Scopes, R.K. (1974) Measurement of protein by spectrophotometry at 205 nm. Anal. Biochem., 59, 277–282.

46. van Iersel, J., Jzn, J.F. & Duine, J.A. (1985) Determination of absorption coefficients of purified proteins by conventional ultraviolet spectrophotometry and chromatography combined with multiwavelength detection. Anal. Biochem. 151, 196–204.

47. Bradford, M.M. (1976) A rapid and sensitive method for the quantitation of microgram quantities of protein utilizing the principle of protein-dye binding. Anal. Biochem. 72, 248–254.

48. Spector, T. (1978) Refinement of the Coomassie blue method of protein quantitation. Anal. Biochem. 86, 142–146.

49. Sedmak, J.J. & Grossberg, S.E. (1977) A rapid, sensitive and versatile assay for protein using Coomassie Brilliant Blue G250. Anal. Biochem. 79, 544–552.

50. Löffler, B.-M. & Kunze, H. (1989) Refinement of the Coomassie Brilliant Blue G assay for quantitative protein determination. Anal. Biochem. 177, 100–102.

51. Smith, P.K., Krohn, R.I., Hermanson, G.T., Mallia, A.K., Gartner, F.H., Provenzano, M.D., Fujimoto, E.K., Goeke, N.M., Olson, B.J. & Klenk, D.C. (1986) Measurement of protein using bicinchonic acid. Anal. Biochem. 150, 76–85.

52. Li, K.W., Gerearts, W.P.M., Van Elk, R. & Joosse, J. (1989) Quantification of proteins in the subnanogram range and nanogram range: Comparison of the Aurodye, Ferridye and Indian Ink staining methods. Anal. Biochem. 182, 44–47.

53. Crabtree, B., Leech, A.R. & Newholme, E.A. (1979) Measurement of enzymic activities in crude extracts of tissues. In: Techniques in the life sciences. Elsevier/North-Holland, Amsterdam, B211, 1–37.

54. Scopes, R.K. (1972) Automated fluorometric analysis of biological compounds. Anal. Biochem. 49, 73–87.

55. Garcia-Carmona, F., Carcia-Canaves, F. & Lozano, J.A. (1981) Optimizing enzyme assays with one or two coupling enzymes. Anal. Biochem. 113, 286–291.

56. Eisenthal, R. & Danson, J. (eds), (1992) Enzyme assays: A practical approach. Oxford University Press, Oxford, pp 336.

57. Folkes, B.F. & Yemm, E.W. (1956) The amino acid content of the proteins of barley grain. Biochem. J. 62, 4–11.

58. Srere, P.A. (1980) The infrastructure of the mitochondrial matrix. Trends Biochem. Sci. 5, 120–121.

59. Torres-Ruiz, J.A. & McFadden, B.A. (1992) Purification and characterization of chaperonin 10 from Chromatium vinosum. Arch. Biochem. Biophys. 295, 172–179.

60. Tanford, C. (1980) The hydrophobic effect: Formation of micelles and biological membranes, 2nd ed. Wiley-Interscience, New York.

61. Arakawa, T. & Timasheff, S.N. (1982) Preferential interactions of proteins with salts in concentrated solutions. Biochemistry 21, 6542–6552.

62. Melander, W. & Horvath, C. (1977) Salt effects on hydrophobic interactions in precipitation and chromatography of proteins: An interpretation of the lyotropic series. Arch. Biochem. Biophys. 183, 200–215.

63. Czok, R. & Bücher, T. (1960) Crystallized enzymes from the myogen of rabbit skeletal muscle. Adv. Prot. Chem. 15, 315–415.

64. Arakawa, T. & Timasheff, S.N. (1984) Mechanism of protein salting in and salting out by divalent cation salt: Balance between hydration and salt binding. Biochemistry 23, 5912–5923.

65. Scopes, R.K. & Stoter, A. (1982) All the glycolytic enzymes from one muscle extract. Methods Enzymol. 90, 479–483.

66. Scopes, R.K. (1984) Techniques for protein purification. In: Techniques in the Life Sciences, vol. BS102 Supplement, 1–46.

67. Curling, J.M. (1980) Methods of plasma protein fractionation. Academic Press, New York.

68. Askonas, B.A. (1951) The use of organic solvents at low temperature for the separation of enzymes: Application to aqueous rabbit muscle extract. Biochem. J. 48, 42–48.

69. Polson, A., Potgieter, J.F., Largier, J.F., Mears, G.E.F. & Joubert, F.J. (1964) The fractionation of protein mixtures by linear polymers of high molecular weight. Biochim. Biophys. Acta 82, 463–475.

70. Thrash, S.L., Otto, J.C. & Deits, T.L. (1991) Effect of divalent ions on protein precipitation with polyethylene glycol: Mechanism of action and applications. Protein Expression Purif. 2, 83–89.

71. Sternberg, M. & Hershberger, D. (1974) Separation of proteins with polyacrylic acids. Biochim. Biophys. Acta 342, 195–206.

72. Larsson, P.-O. & Mosbach, K. (1979) Affinity precipitation of enzymes. FEBS (Fed. Eur. Biochem. Soc.) Lett. 98, 333–338.

73. Flyare, S., Griffin, T., Larsson, P.-O. & Mosbach, K. (1983) Affinity precipitation of dehydrogenases. Anal. Biochem. 133, 409–416.

74. Pearson, J.C., Burton, S.J. & Lowe, C.R. (1986) Affinity precipitation of lactate dehydrogenase with a triazine dye derivative: Selective precipitation of rabbit muscle lactate dehydrogenase with a Procion Blue H-B analog. Anal. Biochem. 158, 382–389.

75. Kuby, S.A., Noda, L. & Lardy, H.A. (1954) Adenosinetriphosphate-creatine transphosphorylase. I. Isolation of the crystalline enzyme from rabbit muscle. J. Biol. Chem. 209, 191–201.

76. Scopes, R.K., Griffith-Smith, K. & Miller, D.G. (1981) Rapid purification of yeast alcohol dehydrogenase. Anal. Biochem. 118, 284–285.

77. Yoshida, A. & Watanabe, S. (1972) Human phosphoglycerate kinase. I. Crystallization and characterization of normal enzyme. J. Biol. Chem. 247, 440–445.

78. Martin, A.J.P. & Synge, R.L.M. (1941) A new form of chromatogram employing two liquid phases: 1. A theory of chromatography. Biochem. J. 35, 1358–1368.

79. van Deemter, J.J., Zuderweg, F.J. & Klinkerberg, A. (1956) Longitudinal diffusion and resistance to mass transfer as a cause of nonideality in chromatography. Chem. Eng. Sci. 5, 271–289.

80. Janson, J.-C. & Hedman, P. (1982) Large-scale chromatography of proteins. Adv. Biochem. Eng. 25, 43–99.

81. Peterson, E.A. & Sober, H.A. (1965) Chromatography of proteins. I. Cellulose ion-exchange adsorbents. J. Amer. Chem. Soc. 78, 751–755.

82. Scopes, R.K. (1981) Quantitative studies of ion-exchange and affinity elution chromatography of enzymes. Anal. Biochem. 114, 8–18.

83. Janson, J.-C. & Ryden, L. (eds) (1989) Protein purification: Principles, high resolution methods, and applications. VCH Publishers, New York.

84. Champluvier, B. & Kula, M.-R. (1992) Sequential membrane-based purification of proteins, applying the concept of multidimensional liquid chromatography (MDLC). Bioseparation 2, 343–351.

85. Nachman, M., Azad, A.R.M. & Bailon, P. (1992) Efficient recovery of recombinant proteins using membrane-based immunoaffinity chromatography. Biotechnol. Bioeng. 40, 564–571.

86. Nakagawa, Y. & Noltmann, E.A. (1965) Isolation of phosphoglucose isomerase from brewer's yeast. J. Biol. Chem. 240, 1877–1881.

87. Giddings, J.C. (1965) Dynamics of chromatography. Part 1: Principles and theory. Marcel Dekker, New York.

88. Done, J.M., Knox, J.H. & Loheac, J. (1974) Applications of high-speed liquid chromatography. John Wiley, London.

89. Muller, W. (1990) New ion exchangers for the chromatography of biopolymers. J. Chromatog. 510, 133–140.

90. Pharmacia Fine Chemicals (1980) Ion-exchange chromatography: Principles and practice. Pharmacia AB Publications, Uppsala.

91. Grossman, L. & Moldave, K. (eds) (1974) Nucleic acid synthesizing systems. Methods Enzymol. 29, 1–230.

92. Chilla, R., Doering, K.M., Domagk, G.F. & Rippa, M. (1973) A simplified procedure for the isolation of a highly active crystalline glucose-6-phosphate dehydrogenase from *Candida utilis*. Arch. Biochem. Biophys. 159, 235–239.

93. Qadri, S.S. & Easterby, J.S. (1980) Purification of skeletal muscle hexokinase by affinity elution chromatography. Anal. Biochem. 105, 299–303.

94. Leaback, D.H. & Robinson, H.K. (1975) Ampholyte displacement chromatography—a new technique for the separation of proteins illustrated by the resolution of N-acetyl-D-hexosaminidase isoenzymes unresolvable by isoelectric focusing or conventional ion-exchange chromatography. Biochem. Biophys. Res. Commun. 67, 248–254.

95. Edmond, P. & Page, M. (1980) Ampholyte displacement chromatography. J. Chromatog. 200, 57–63.

96. Skidmore, G.L. & Chase, H.A. (1990) Two-component protein adsorption to the cation exchanger S Sepharose FF. J. Chromatog. 505, 329–347.

97. Felinger, A. & Guiochon, G. (1992) Comparison of maximum production rates and optimum operating design parameters in overloaded elution and displacement chromatography. Biotechnol. Bioeng. 41, 134–147.

98. Cramer, S.M. (1991) Displacement chromatography. Nature 351, 251–252.

99. Jungbauer, A. (1992) Displacement effects in large-scale chromatography? Biotechnol. Bioeng. 39, 579–587.

100. Goward, C.R., Stevens, G.B., Tattersall, R. & Atkinson, T. (1992) Rapid large-scale preparation of recombinant *Erwinia chrysanthemi* L-asparaginase. Bioseparation 2, 335–341.

101. Peterson, E.A. (1970) Cellulosic ion-exchangers. In: Laboratory techniques in biochemistry and molecular biology, vol. 2. (Work, T.S., Work, E., eds) pp 223–392, North Holland, Amsterdam.

102. Ekstrom, B. & Jacobson, G. (1984) Differences in retention behavior between small and large molecules in ion-exchange chromatography and reversed-phase chromatography. Anal. Biochem. 142, 134–139.

103. LKB Produkter (1981) DEAE/CM Trisacryl M for ion-exchange chromatography. LKB Publications, Bromma, Sweden.

104. Sluyterman, L.A.A. & Elgersma, O. (1978) Chromatofocusing: Isoelectric focusing on ion-exchange columns I. General principles. J. Chromatog. 150, 17–30.

105. Sluyterman, L.A.A. & Elgersma, O. (1978) Chromatofocusing: Isoelectric focusing on ion-exchange columns. II. Experimental verification. J. Chromatog. 150, 31–44.

106. Sluyterman, L.A.A. & Wijdenes, J. (1981) Chromatofocussing. III. The properties of a DEAE-agarose anion exchanger and its suitability for protein separations. J. Chromatog. 206, 429–440.

107. Watts, D.C. & Rabin, B.R. (1981) A study of the "reactive" sulphydryl groups of adenosine 5'-triphosphate-creatine phosphotransferase. Biochem. J. 85, 507–516.

108. Algar, E. & Scopes, R.K. (1979) Yeast phosphoglycerate kinase: Evidence from affinity elution studies for conformational changes on binding of substrates. FEBS (Fed. Eur. Biochem. Soc.) Lett. 106, 239–242.

109. Stewart, A.A. & Scopes, R.K. (1978) Phosphoglycerate kinase from ram testes. Eur. J. Biochem. 85, 89–95.

110. Datta, S.P. & Grzybowski, A.K. (1961) pH and acid–base equilibrium. In: Biochemist's handbook. (Long, C., ed) pp 19–58, Spon, London.

111. Perrin, D.D. & Dempsey, B. (1974) Buffers for pH and metal ion control. Chapman & Hall, London.

112. McKenzie, H.A. & Dawson, R.M.C. (1969) pH and buffers and physiological media. In: Data for biochemical research. (Dawson, R.M.C., Elliott, D.C., Elliot, W.H., and Jones, N.M., eds) pp 475–508, Oxford University Press, London.

113. Barnard, E.A. (1975) Hexokinase from yeast. Methods Enzymol. 42, 6–20.

114. Bernardi, G., Giro, M.G. & Gaillard, C. (1972) Chromatography of polypeptides and proteins on hydroxyapatite columns: Some new developments. Biochim. Biophys. Acta 278, 409–420.

115. Gluekauf, E. & Patterson, L. (1974) The adsorption of some proteins on hydroxyapatite and other adsorbents used for chromatographic separations. Biochim. Biophys. Acta 351, 57–76.

116. Gorbunoff, M.J. (1985) Protein chromatography on hydroxyapatite columns. Methods Enzymol. 117, 370–381.

117. Keilin, D. & Hartree, F.E. (1938) Mechanism of the decomposition of hydrogen peroxide by catalase. Proc. Roy. Soc. 124, 387–405.

118. Bernardi, G. (1971) Chromatography of proteins on hydroxyapatite. Methods Enzymol. 22, 325–339.

119. Er-el, Z., Zaidenzaig, Y. & Shaltiel, S. (1972) Hydrocarbon-coated Sepharoses: Use in the purification of glycogen phosphorylase. Biochem. Biophys. Res. Commun. 49, 383–390.

120. Hofstee, B.H.J. & Otillio, N.F. (1978) Modifying factors in hydrophobic protein binding by substituted agaroses. J. Chromatog. 161, 153–163.

121. Ochoa, J.L. (1978) Hydrophobic (interaction) chromatography. Biochimie 60, 1–15.

122. El Rassi, Z., De Ocampo, L.F. & Bacolod, M.D. (1990) Binary and ternary salt gradients in hydrophobic interaction chromatography of proteins. J. Chromatog. 499, 141–152.

123. Hodgkinson, S. & Lowry, P.J. (1981) Hydrophobic-interaction chromatography and anion-exchange chromatography in the presence of acetonitrile. Biochem. J. 199, 619–627.

124. Hoffman, L.G. (1969) Solubility chromatography of serum proteins. I. Isolation of the first component of complement from guinea pig serum by solubility chromatography at low ionic strength. J. Chromatog. 40, 39–52.

125. Hoffmann, L.G. & McGivern, P.W. (1969) Solubility chromatography of serum proteins. II. Partial purification of the second component of guinea-pig complement by solubility chromatography in concentrated ammonium sulphate solutions. J. Chromatog. 40, 53–61.

126. Fujita, T., Suzuki, Y., Yamauti, J.I., Takagahara, I., Fujii, K., Yamashita, J. & Horio, T. (1980) Chromatography in presence of high concentrations of salts on columns of celluloses with and without ion exchange groups (hydrogen bond chromatography). J. Biochem. 87, 89–100.

127. von der Haar, F. (1976) Purification of proteins by fractional interfacial salting out on unsubstituted agarose gels. Biochem. Biophys. Res. Commun. 70, 1009–1013.

128. von der Haar, F. (1978) The ligand-induced solubility shift in salting out chromatography: A new affinity technique, demonstrated with phenylalanyl- and isoleucyl-tRNA synthetase from baker's yeast. FEBS (Fed. Eur. Biochem. Soc.) Lett. 94, 371–374.

129. Scopes, R.K. & Porath, J. (1990) Differential salt-promoted chromatography for protein purification. Bioseparation 1, 3–7.

130. Porath, J., Carlsson, J., Olsson, I. & Belfrage, G. (1975) Metal chelate affinity chromatography, a new approach to protein fractionation. Nature 258, 598–599.

131. Porath, J. (1992) Immobilized metal ion affinity chromatography. Protein Expression Purif. 3, 263–281.

132. Sulkowski, E. (1985) Purification of proteins by IMAC. Trends Biotechnol. 3, 1–7.

133. Porath, J. & Hansen, P. (1991) Cascade mode multi affinity chromatography. Fractionation of human serum proteins. J. Chromatog. 550, 751–764.

134. Porath, J. & Olin, B. (1983) Immobilized metal ion affinity adsorption and immobilized metal ion chromatography of biomaterials. Serum protein affinities for immobilized iron and nickel ions. Biochemistry 22, 1621–1630.

135. Anderson, L. & Porath, J. (1986) Isolation of phosphoproteins by immobilized metal (Fe^{3+}) affinity chromatography. Anal. Biochem. 154, 250–254.

136. Muszynska, G., Anderson, L. & Porath, J. (1986) Selective adsorption of phosphoproteins on gel-immobilized ferric chelate. Biochemistry 25, 6850–6853.

137. Reed, L.K. & Mukherjee, B.D. (1969) α-Ketoglutarate dehydrogenase complex from *Escherichia coli*. Methods Enzymol. 13, 55–61.

138. Jendrisak, J. (1987) The use of polyethyleneimine in protein purification. In: Protein purification: Micro to macro. (Burgess, R., ed) pp 75–97, A.R. Liss, New York.

139. Burgess, R.R. & Jendrisak, J.J. (1975) A procedure for the rapid, large-scale purification of *Escherichia coli* DNA-dependent RNA polymerase involving Polymin P precipitation and DNA-cellulose chromatography. Biochemistry 14, 4634–4638.

140. Porath, J., Maisano, F. & Belew, M. (1985) Thiophilic adsorption—a new method for protein fractionation. FEBS (Fed. Eur. Biochem. Soc.) Lett. 185, 306–310.

141. Porath, J. (1986) Salt-promoted adsorption: Recent developments. J. Chromatog. 376, 331–341.

142. Lihme, A. & Heegaard, P.M.H. (1991) Thiophilic adsorption chromatography: the separation of serum proteins. Anal. Biochem. 192, 64–69.

143. Oscarsson, S. & Porath, J. (1989) Covalent chromatography and salt-promoted thiophilic adsorption. Anal. Biochem. 176, 330–337.

144. Hutchens, T.W. & Porath, J. (1986) Thiophilic adsorption of immunoglobulins—analysis of conditions optimal for selective immobilization and purification. Anal. Biochem. 159, 217–226.

145. Knudsen, K.L., Hansen, M.B., Henriksen, L.R., Andersen, B.K. & Lihme, A. (1992) Sulfone aromatic ligands for thiophilic adsorption chromatography—purification of human and mouse immunoglobulins. Anal. Biochem. 201, 170–177.

146. Porath, J. & Farnstedt, N. (1970) Group fractionation of plasma proteins on dipolar ion exchangers. J. Chromatog. 51, 479–489.

147. Yon, R.J. (1975) Protein chromatography on adsorbents with hydrophobic and ionic groups: Some properties of N-(3-carboxypropionyl)aminodecyl-Sepharose and its interaction with wheat-germ aspartate transcarbamylase. Biochem. J. 151, 281–290.

148. Yon, R.J. (1977) Biospecific-elution chromatography with "imphilytes" as stationary phases. Biochem. J. 161, 223–237.

149. Maeda, K., Truscott, K., Liu, X.-L. & Scopes, R.K. (1992) A thermostable NADH oxidase from anaerobic extreme thermophiles. Biochem. J. 284, 551–555.

150. Gronman, E.V. & Wilchek, M. (1987) Recent developments in affinity chromatography supports. Trends Biotechnol. 5, 220–224.

151. Turkova, J. (1978) Affinity chromatography. Elsevier Scientific, Amsterdam.

152. Mohr, P. & Pommerening, K. (1985) Affinity chromatography: Practical and theoretical aspects. Dekker, New York.

153. Ostrove, S. (1990) Affinity chromatography: General methods. Methods Enzymol. 182, 357–371.

154. Cuatrecasas, P., Wilchek, M. & Anfinsen, C.D. (1968) Selective enzyme purification by affinity chromatography. Proc. Natl. Acad. Sci. USA 61, 636–643.

155. March, S.C., Parikh, I. & Cuatrecasas, P. (1974) A simplified method for cyanogen bromide actrivation of agarose for affinity chromatography. Anal. Biochem. 60, 149–152.

156. Wilchek, M., Oka, T. & Topper, Y.J. (1975) Structure of a soluble superactive insulin is revealed by the nature of the complex between cyanogen-bromide-activated Sepharose and amines. Proc. Natl. Acad. Sci. USA 72, 1055–1058.

157. Pepper, D.S. (1992) Some alternative coupling chemistries for affinity chromatography. Methods Mol. Biol. 11, 173–196.

158. Sunderberg, L. & Porath, J. (1974) Preparation of adsorbents for biospecific affinity chromatography. I. Attachment of group containing ligands to insoluble polymers by means of bifunctional oxiranes. J. Chromatog. 90, 87–98.

159. Bethell, G.S., Ayers, J.S., Hancock, W.S. & Hearn, M.T.W. (1979) A novel method of activation of cross-linked agaroses with 1,1-carbonyldiimidazole which gives a matrix for affinity chromatography devoid of additional charged groups. J. Biol. Chem. 254, 2572–2574.

160. Nilsson, K. & Mosbach, K. (1980) p-Toluensulfonyl chloride as an activating agent of agarose for the preparation of immobilized affinity ligands and proteins. Eur. J. Biochem. 112, 397–402.

161. Nilsson, K. & Mosbach, K. (1981) Immobilization of enzymes and affinity ligands to various hydroxyl group carrying supports using highly reactive sulphonyl chlorides. Biochem. Biophys. Res. Commun. 102, 449–457.

162. Ersson, B. (1977) A phytohaemagglutinin from Sunn Hemp seeds; purification by a high capacity biospecific affinity adsorbent and its physico-chemical properties. Biochim. Biophys. Acta 494, 51–60.

163. Lee, C.-Y. & Chen, A.F. (1980) Immobilized coenzymes and derivatives. In: Pyridine nuceleotide coenzymes. (Everse, J., Anderson, B.M., and Yon, K.S., eds) Academic Press, New York.

164. O'Carra, P., Barry, S. & Griffin, T. (1974) Spacer arms in affinity chromatography: Use of hydrophilic arms to control or eliminate nonbiospecific adsorption effects. FEBS (Fed. Eur. Biochem. Soc.) Lett. 43, 169–175.

165. Lowe, C.R. (1977) The synthesis of several 8-substituted derivatives of adenosine 5'-monophosphate to study the effect of the nature of the spacer arm in affinity chromatography. Eur. J. Biochem. 73, 265–274.

166. Wright, C.L., Warsy, A.S., Holroyde, M.J. & Trayer, I.P. (1978) Purification of the hexokinases by affinity chromatography on Sepharose-N-aminoacylglucosamine derivatives. Biochem. J. 175, 125–135.

167. Aplin, J.D. & Hall, L.D. (1980) Sepharose 4B as a matrix for affinity chromatography: A spin-labeling investigation using nitroxides as model ligands. Eur. J. Biochem. 110, 295–309.

168. Morgan, M.R.A., George, E. & Dean, P.D.G. (1980) Investigation into the controlling factors of electrophoretic desorption: A widely applicable, nonchaotropic elution technique in affinity chromatography. Anal. Biochem. 105, 1–5.

169. Dunbar, B.S. & Schwoebel, E.D. (1990) Preparation of polyclonal antibodies. Methods Enzymol. 182, 663–670.

170. Secher, D.S. & Burke, D.C. (1980) A monoclonal antibody for large scale purification of human leukocyte interferon. Nature 285, 446–450.

171. Kessler, S.W. (1975) Rapid isolation of antigens from cells with a staphylococcal protein A-antibody absorbent: Parameters of the interaction of antibody-antigen complexes with protein A. J. Immunol. 115, 1617–1624.

172. Spielman, H., Erikson, R.P. & Epstein, C.J. (1974) The production of antibodies against mammalian LDH-1. Anal. Biochem. 59, 462–467.

173. Goding, J.W. (1983) Monoclonal antibodies: Principles and practice. Academic Press, London.

174. Livingstone, D.M. (1974) Immunoaffinity chromatography of proteins. Methods Enzymol. 34, 723–731.

175. van Wezel, A.L. & van der Marel, P. (1982) The application of immunoadsorption on immobilized antibodies for large scale concentration and purification of vaccines. In: Affinity chromatography and related techniques (Gribnau, T.C.J., Visser, J., and Nivard, R.J.F., eds) pp 282–292, Elsevier Scientific, Amsterdam.

176. Vidal, J., Godbillon, G. & Gadal, P. (1980) Recovery of active, highly purified phosphoenolpyruvate carboxylase from specific immunoadsorbent column. FEBS (Fed. Eur. Biochem. Soc.) Lett. 118, 31–34.

177. Biveau, D. & Daussant, J. (1981) Immunoaffinity chromatography of proteins: A gentle and simple desorption procedure. J. Immunol. Methods 41, 387–392.

178. Singh, P., Lewis, S.D. & Schafer, J.A. (1979) A support for affinity chromatography that covalently binds amino groups via a cleavable connector arm. Arch. Biochem. Biophys. 193, 284–293.

179. Eisenbarth, G.S. (1981) Application of monoclonal antibody techniques to biochemical research. Anal. Biochem. 111, 1–16.

180. Secher, D.S. & Burke, D.C. (1980) A monoclonal antibody for large scale purification of human leukocyte interferon. Nature 285, 446–450.

181. van Oss, C.J., Good, R.J. & Chaudhury, M.K. (1986) Nature of the antigen–antibody interaction. Primary and secondary bonds: Optimal conditions for association and dissociation. J. Chromatog. 376, 111–119.

182. Chiswell, D.J. & McCafferty, J. (1992) Phage antibodies: Will new "coliclonal" antibodies replace monoclonal antibodies? Trends Biotechnol. 10, 80–84.

183. Haeckel, R., Hess, B., Lauterborn, W. & Wurster, K. (1968) Purification and allosteric properties of yeast pyruvate kinase. Hoppe-Seyler's Z. Physiol. Chem. 349, 699–714.

184. Kopperschläger, G., Freyer, R., Diezel, W. & Hofmann, E. (1968) Some kinetic and molecular properties of yeast phosphofructokinase. FEBS (Fed. Eur. Biochem. Soc.) Lett. 1, 137–141.

185. Thompson, S.T., Cass, K.H. & Stellwagen, E. (1975) Blue dextran-Sepharose: An affinity column for the dinucleotide fold in proteins. Proc. Natl. Acad. Sci. 72, 669–672.

186. Biellmann, J.F., Samama, J.P., Bränden, C.I. & Eklund, H. (1979) X-ray studies on the binding of Cibacron Blue F3G-A to liver alcohol dehydrogenase. Eur. J. Biochem. 102, 107–110.

187. Watson, D.H., Harvey, M.J. & Dean, P.D.G. (1978) The selective retardation of $NADP^+$-dependent dehydrogenases by immobilized Procion HE-3B. Biochem. J. 173, 591–596.

188. Ashton, A. & Polya, G.M. (1978) The specific interaction of Cibacron and related dyes with cyclic nucleotide phosphodiesterase and lactate dehydrogenase. Biochem. J. 175, 501–506.

189. Clonis, Y.D. & Lowe, C.R. (1980) Triazine dyes, a new class of affinity labels for nucleotide-dependent enzymes. Biochem. J. 191, 247–251.

190. Qadri, F. & Dean, P.D.G. (1980) The use of various immobilized triazine affinity dyes for the purification of 6-phosphogluconate dehydrogenase from *Bacillus stearothermophilus*. Biochem. J. 191, 53–62.

191. Scopes, R.K. (1984) Use of differential dye-ligand chromatography with affinity elution for enzyme purification: 2-Keto 3-deoxy 6-phosphogluconate aldolase from *Zymomonas mobilis*. Anal. Biochem. 136, 525–529.

192. Neale, A.D., Scopes, R.K., Kelly, J.M. & Wettenhall, R.E.H. (1986) The two alcohol dehydrogenases from *Zymomonas mobilis*. Eur. J. Biochem. 154, 119–124.

193. Nagata, Y., Maeda, K. & Scopes, R.K. (1992) NADP-linked alcohol dehydrogenases from extreme thermophiles: Simple affinity purification schemes, and comparative properties of the enzymes from different strains. Bioseparation 2, 353–362.

194. Stead, C. (1987) The chemistry of reactive dyes. In: Reactive dyes in protein and enzyme technology. (Clonis, Y.D., Atkinson, A., Bruton, C.J., and Lowe, C.R., eds) pp 125–160, Macmillan, Stockton Press, London.

195. Clonis, Y.D. (1987) Appendix: The disclosed chemical structures of some triazine dyes. In: Reactive dyes in protein and enzyme technology. (Clonis, Y.D., Atkinson, A., Bruton, C.J., and Lowe, C.R., eds) pp 193–199, Macmillan, Stockton Press, London.

196. Hanggi, D. & Carr, P. (1985) Analytical evaluation of the purity of commercial preparations of Cibacron Blue F-3GA and related dyes. Anal. Biochem. 149, 91–105.

197. Atkinson, T., Hammond, P.M., Hartwell, R.D., Hughes, P., Scawen, M.D., Sherwood, R.F., Small, D.A.P., Bruton, C.J., Harvey, M.J. & Lowe, C.R. (1981) Triazine-dye affinity chromatography. Biochem. Soc. Trans. 9, 290–293.

198. Lowe, C.R., Hans, M., Spibey, N. & Drabble, W.T. (1980) The purification of inosine 5'-monophosphate dehydrogenase from *Escherichia coli* by affinity chromatography on immobilized Procion dyes. Anal. Biochem. 104, 23–28.

199. Hughes, P., Sherwood, R.F. & Lowe, C.R. (1984) Studies on the nature of transition-metal-ion-mediated binding of triazine dyes of enzymes. Eur. J. Biochem. 144, 135–142.

200. Hughes, P. & Sherwood, R.F. (1987) Metal ion-promoted dye-ligand chromatography. In: Reactive dyes in protein and enzyme technology. (Clonis, Y.D., Atkinson, A., Bruton, C.J., and Lowe, C.R., eds) pp 125–160, Macmillan, Stockton Press, London.

201. Hughes, P., Lowe, C.R. & Sherwood, R.F. (1982) Metal ion-promoted binding of proteins to immobilized triazine dye affinity adsorbents. Biochim. Biophys. Acta 700, 90–100.

202. Sherwood, R.F., Melton, R.G., Alwan, S.H. & Hughes, P. (1985) Purification and properties of carboxypeptidase G2 from *Pseudomonas sp* strain RS-16: Use of a novel triazine dye method. Eur. J. Biochem. 148, 447–453.

203. Scopes, R.K. (1986) Strategies for enzyme isolation using dye-ligand chromatography. J. Chromatog. 376, 131–140.

204. Kroviarski, Y., Cochet, S., Vabon, C., Truskolaski, A., Bovin, P. & Bertrand, O. (1988) New strategies for the screening of a large number of immobilized dyes for the purification of enzymes. J. Chromatog. 449, 403–412.

205. Mottl, H. & Keck, W. (1992) Rapid screening of a large number of immobilized textile dyes for the purification of proteins: Use of penicillin-binding protein 4 of *Escherichia coli* as a model enzyme. Protein Expression Purif. 3, 403–409.

206. Gianazza, E. & Arnaud, P. (1982) A general method for fractionation of plasma proteins: Dye-ligand affinity chromatography on immobilized Cibacron Blue F3-GA. Biochem. J. 201, 129–134.

207. Santambien, P., Hulak, I., Girot, P. & Boschetti, E. (1992) Elisa-based quantification of Cibacron Blue F3GA used as ligand in affinity chromatography. Bioseparation 2, 327–334.

208. Stewart, D.J., Purvis, D.R., Pitts, J.M. & Lowe, C.R. (1992) Development of an enzyme-linked immunoadsorbent assay for C.I. Reactive Blue-2 and its application to a comparison of the stability and performance of a perfluorocarbon support with other immobilized C.I. Reactive Blue-2 affinity adsorbents. J. Chromatog. 623, 1–14.

209. Lowe, C.R., Burton, S.J., Burton, N.P., Alderton, W.K., Pitts, J.M. & Thomas, J.A. (1992) Designer dyes: 'Biomimetic' ligands for the purification of pharmaceutical proteins by affinity chromatography. Trends Biotechnol. 10, 442–448.

210. Pogell, B.N. (1962) Enzyme purification by selective elution with substrate from substituted cellulose columns. Biochem. Biophys. Res. Commun. 7, 225–230.

211. Black, W.S., van Tol, A., Fernando, S. & Horecker, B.L. (1972) Isolation of a highly active fructose diphosphatase from rabbit muscle: Its subunit

structure and activation by monovalent cations. Arch. Biochem. Biophys. 151, 576–590.

212. Scopes, R.K. (1977) Purification of glycolytic enzymes by using affinity elution chromatography. Biochem. J. 161, 253–263.
213. Scopes, R.K. (1977) Multiple enzyme purifications from muscle extracts by using affinity elution chromatographic procedures. Biochem. J. 161, 265–277.
214. von der Haar, F. (1974) Affinity elution principles and applications to purification of aminoacyl-t-RNA synthetases. Methods Enzymol. 34, 163–171.
215. von der Haar, F. (1974) Affinity elution as a purification method for aminoacyl-tRNA synthetases. Eur. J. Biochem. 34, 84–90.
216. Reuter, R., Naumann, M. & Kopperschlager, G. (1990) New aspects of dye-ligand affinity chromatography of lactate dehydrogenase applying spacer-mediated beaded cellulose. J. Chromatog. 510, 189–195.
217. O'Carra, P. & Barry, S. (1972) Affinity chromatography of lactate dehydrogenase. FEBS (Fed. Eur. Biochem. Soc.) Lett. 21, 281–285.
218. Kadonaga, J.T. & Tjian, R. (1986) Affinity purification of sequence-specific DNA-binding proteins. Proc. Natl. Acad. Sci. USA 83, 5889–5993.
219. Porath, J. & Flodin, P. (1959) Gel filtration: A method for desalting and group separation. Nature 83, 1657–1659.
220. Whitaker, J.R. (1963) Determination of molecular weights of proteins by gel filtration on Sephadex. Anal. Chem. 35, 1950–1953.
221. Andrews, P. (1965) The gel filtration behavior of proteins related to their molecular weight over a wide range. Biochem. J. 96, 595–606.
222. Belew, M., Porath, J., Fohlman, J. & Janson, J.-C. (1978) Adsorption phenomena on Sephacryl S-200. J. Chromatog. 147, 205–212.
223. Smithies, O. (1955) Zone electrophoresis in starch gels; Group variations in the serum proteins of normal human adults. Biochem. J. 61, 629–641.
224. Ornstein, L. (1964) Disc electrophoresis. I. Background and theory. Ann. N.Y. Acad. Sci. 121, 321–349.
225. Rüchel, R. & Brager, M.D. (1975) Scanning electron microscopic observations of polyacrylamide gels. Anal. Biochem. 68, 415–428.
226. Morris, C.J.O.R. & Morris, P. (1971) Molecular-sieve chromatography and electrophoresis in polyacrylamide gels. Biochem. J. 124, 517–528.
227. Weber, K. & Osborn, M. (1969) The reliability of molecular weight determination by dodecyl sulphate-polyacrylamide gel electrophoresis. J. Biol. Chem. 244, 4406–4412.
228. Goodman, W.F. & Baptist, J.N. (1979) Isoelectric point electrophoresis: A new technique for protein purification. J. Chromatog. 179, 330–332.
229. Furlong, C.E., Cirakoglu, C., Willis, R.C. & Santy, P.A. (1973) A simple preparative polyacrylamide disc gel electrophoresis apparatus: Purification of three branched-chain amino acid binding proteins from Escherichia coli. Anal. Biochem. 51, 297–311.
230. Azuma, J.I., Kashimura, N. & Komano, T. (1977) A new convenient column for gel electrophoresis, isoelectric focusing, and zonal electrophoresis. Anal. Biochem. 81, 454–457.
231. Orr, M.D., Blakeley, R.L. & Panagon, D. (1972) Discontinuous buffer systems for analytical and preparative electrophoresis of enzymes on polyacrylamide gel. Anal. Biochem. 45, 68–85.

232. Righetti, P.G., Barzaghi, B. & Faupel, M. (1988) Large-scale electrophoresis for protein purification: Exploiting isoelectricity. Trends Biotechnol. 6, 121–126.
233. Righetti, P.G., Gianazza, E., Brenna, O. & Galante, E. (1977) Isoelectric focusing as a puzzle. J. Chromatog. 137, 171–181.
234. Nguyen, N.Y., Rodbard, D., Svendsen, P.J. & Chrambach, A. (1977) Cascade stacking and cascade electrofocusing: Their interconversion and fundamental unity. Anal. Biochem. 77, 39–55.
235. Nguyen, N.Y. & Chrambach, A. (1977) Natural pH gradients in buffer mixtures: Formation in the absence of strongly acidic and basic anolyte and catholyte, gradient steepening by sucrose, and stabilization by high buffer concentrations in the electrolyte chambers. Anal. Biochem. 79, 462–469.
236. Prestidge, R.L. & Hearn, M.T.W. (1997) Preparative flatbed electrofocusing in granulated gels with natural pH gradients generated from simple buffers. Anal. Biochem. 97, 95–102.
237. Birkenmeier, G., Kopperschläger, G., Albertsson, P.-A., Johansson, G., Tjerneld, F., Akerlund, H.E., Berner, S. & Wickstroem, H. (1987) Fractionation of proteins from human serum by counter-current distribution. J. Biotechnol. 5, 115–129.
238. Johansson, G., Joelsson, M. & Akerlund, H.-E. (1985) An affinity-ligand gradient technique for purification of enzymes by counter-current distribution. J. Biotechnol. 2, 225–237.
239. Dobry, A. & Boyer-Kawenoki, F. (1947) Phase separation in polymer solution. J. Polymer Sci. 2, 90–100.
240. Albertsson, P.-A. (1971) In: partition of cell particles and macromolecules, 2nd ed. pp 24–25, Wiley, New York.
241. Albertsson, P.-A., Cajarville, A., Brooks, D.E. & Tjerneld, F. (1987) Partition of proteins in aqueous polymer two-phase systems and the effect of molecular weight of the polymer. Biochim. Biophys. Acta 926, 87–93.
242. Hartman, A., Johansson, G. & Albertsson, P.-A. (1974) Partition of proteins in a three-phase system. Eur. J. Biochem. 46, 75–81.
243. Pike, R.N. & Dennison, C. (1989) Protein fractionation by three-phase partitioning (TPP) in aqueous/t-butanol mixtures. Biotechnol. Bioeng. 33, 221–228.
244. Forciniti, D., Hall, C.K. & Kula, M.R. (1991) Protein partitioning at the isolelectric point: Influence of polymer molecular weight and concentration and protein size. Biotechnol. Bioeng. 38, 988–994.
245. Kopperschläger, G., Lorenz, G. & Usbeck, E. (1983) Application of affinity partitioning in an aqueous two-phase system to the investigation of triazine dye enzyme interactions. J. Chromatog. 259, 97–105.
246. Hustedt, H., Kroner, K.H., Menge, U. & Kula, M.-R. (1985) Protein recovery using two-phase systems. Trends Biotechnol. 3, 139–144.
247. Müller, W., Bünemann, H., Schuetz, H.-J. & Eigel, A. (1982) Nucelic acid interacting dyes suitable for affinity chromatography, partitioning, and affinity electrophoresis. In: Affinity chromatography and related techniques. (Gribnau, T.C.J., Visser, J., and Nivard, R.J.F., eds) pp 437–444, Elsevier Scientific, Amsterdam.
248. Walsdorf, A. & Kula, M.R. (1991) Properties of supports for the partition chromatography of proteins. J. Chromatog. 542, 55–64.

249. Ramelmeier, R.A., Terstappen, G.C. & Kula, M.R. (1992) The partitioning of cholesterol oxidase in Triton X-114-based aqueous two-phase systems. Bioseparation 2, 315–324.

250. Alred, P.A., Tjerneld, F., Kozlowski, A. & Harris, J.M. (1992) Synthesis of dye conjugates of ethylene oxide-propylene oxide copolymers and application in temperature-induced phase partitioning. Bioseparation 2, 363–373.

251. Walter, H., Johansson, G. & Brooks, D.E. (1991) Partitioning in aqueous two-phase systems: Recent results. Anal. Biochem. 197, 1–18.

252. Sassenfeld, H.M. (1990) Engineering proteins for purification. Trends Biotechnol. 8, 88–93.

253. Brewer, S.J. & Sassenfeld, H.M. (1985) The purification of recombinant proteins using C-terminal polyarginine fusion. Trends Biotechnol. 3, 119–122.

254. Smith, M.C., Furman, T.C., Ingolia, T.D. & Pidgeon, C. (1988) Chelating peptide-immobilized metal ion affinity chromatography. J. Biol. Chem. 263, 7211–7215.

255. Vosters, A.F., Evans, D.B., Tarpley, W.G. & Sharma, S.K. (1992) On the engineering of rDNA proteins for purification by immobilized metal affinity chromatography: Applications to alternating histidine-containing chimeric proteins from recombinant *Escherichia coli*. Protein Expression Purif. 3, 18–26.

256. Nilsson, B., Holmgren, E., Josephson, S., Gatenbeck, S., Philipson, L. & Uhlen, M. (1985) Efficient secretion and purification of human insulin-like growth factor I with a gene fusion vector in *Staphylococci*. Nucleic Acids Res. 13, 1151–1162.

257. Smith, D.B. & Johnson, K.S. (1988) Single-step purificaiton of polypeptides expressed in *Escherichia coli* as fusions with glutathione S-transferase. Gene 67, 31–40.

258. Moks, T., Abrahmsen, L., Holmgren, E., Bilich, M., Olsson, A., Uhlen, M., Pohl, G., Sterky, C., Hultberg, H., Josephson, S., Holmgren, A., Jornvall, H. & Nilsson, B. (1987) Expression of human insulin-like growth factor I in bacteria: Use of optimized gene fusion vectors to facilitate protein purification. Biochemistry 26, 5239–5244.

259. Enfors, S.-O. (1992) Control of *in vivo* proteolysis in the production of recombinant proteins. Trends Biotechnol. 10, 310–315.

260. Harwood, C.R. (1992) *Bacillus subtilis* and its relatives: Molecular biological and industrial workhorses. Trends Biotechnol. 10, 247–256.

261. Wu, X.-C., Lee, W., Tran, L. & Wong, S.-L. (1991) Engineering a *Bacillus subtilis* expression–secretion system with a strain deficient in six extracellular proteases. J. Bacteriol. 173, 4952–4958.

262. Guthrie, C. & Fink, G.R. (eds) (1991) Guide to yeast genetics and molecular biology. Methods Enzymol. 194.

263. Furth, A.J. (1980) Removing unbound detergent from hydrophobic proteins. Anal. Biochem. 109, 207–215.

264. Robinson, N.C., Wiginton, D. & Talbert, L. (1984) Phenyl-Sepharose mediated detergent-exchange chromatography: Its application to exchange of detergents bound to membrane proteins. Biochemistry 23, 6121–6126.

265. Steindl, F., Jungbauer, A., Wenisch, E., Himmler, G. & Katinger, H. (1987) Isoelectric precipitation and gel chromatography for purification of monoclonal IgM. Enzyme Microb. Technol. 9, 361–364.

266. Pepper, D.S. (1992) Selection of antibodies for immunoaffinity chromatography. Methods Mol. Biol. 11, 135–171.
267. Ngo, T.T. & Khatter, N. (1992) Avid AL, a synthetic ligand affinity gel mimicking immobilized bacterial antibody receptor for purification of immunoglobulin G. J. Chromatog. 597, 101–109.
268. Hathaway, G.M. (1992) Capillary electrophoresis of peptides as a preparative technique for sequencing, mass, and composition analysis. Appl. Biosyst. Reporter 17, 3–11.
269. Wheelwright, S.M. (1991) Protein purification: Design and scale up of downstream processing. Hanser: Munich.
270. Reynolds, J.A. & Tanford, C. (1970) Binding of dodecyl sulphate to proteins at high binding ratios: Possible implications for the state of proteins in biological membranes. Proc. Natl. Acad. Sci. 66, 1002–1007.
271. Burg, A.F. (1981) Evaluation of different buffer systems for PAGE-SDS. J. Chromatog. 213, 491–500.
272. Schagger, H. & Von Jagow, G. (1987) Tricine-sodium dodecyl sulfate-polyacrylamide gel electrophoresis for the separation of proteins in the range from 1 to 100 kDa. Anal. Biochem. 166, 368–379.
273. Arcus, A.C. (1970) Protein analysis by electrophoretic molecular sieving in a gel of graded porosity. Anal. Biochem. 37, 53–63.
274. Tal, M., Silberstein, A. & Nusser, E. (1985) Why does Coomassie Brilliant Blue R interact differently with different proteins? J. Biol. Chem. 260, 9976–9980.
275. Davis, M.T. & Lee, T.D. (1992) Analysis of peptide mixtures by capillary high performance liquid chromatography: A practical guide to small-scale separations. Protein Sci. 1, 935–944.
276. Stephano, J.L., Gould, M. & Rojas-Galacia, L. (1986) Advantages of picrate fixation for staining polypeptides in polyacrylamide gels. Anal. Biochem. 152, 308–313.
277. Diezel, W., Kopperschläger, G. & Hofmann, F. (1972) An improved procedure for protein staining in polyacrylamide gels with a new type of Coomassie Brilliant Blue. Anal. Biochem. 48, 617–620.
278. Blakesly, R.W. & Boezi, J.A. (1977) A new staining technique for proteins in polyacrylamide gels using Coomassie Brilliant Blue G250. Anal. Biochem. 82, 580–582.
279. Wilson, C.M. (1979) Studies and critique of amido black 10B, Coomassie Blue R, and fast green FCF as stains for proteins after polyacrylamide gel electrophoresis. Anal. Biochem. 96, 263–278.
280. Higgins, R.C. & Dahmus, M.E. (1979) Rapid visualization of protein bands in preparative SDS-polyacrylamide gels. Anal. Biochem. 93, 257–260.
281. Pace, J.L., Kemper, D.L. & Ragland, W.L. (1974) The relationship of molecular weight to electrophoretic mobility of fluorescamine-labeled proteins in polyacrylamide gels. Biochem. Biophys. Res. Com. 57, 482–287.
282. Sammons, D.W., Adams, L.D. & Nishizawa, E.E. (1981) Ultrasensitive silver-based color staining of polypeptides in polyacrylamide gels. Electrophoresis 2, 135–140.
283. Lasne, F., Berzerara, O. & Lasne, Y. (1983) A fast staining method for protein detection after isoelectric focusing in agarose gels. Anal. Biochem. 132, 338–341.

284. Porro, M., Viti, S., Antoni, G. & Saletti, M. (1982) Ultrasensitive silverstain method for the detection of protein in polyacrylamide gels and immunoprecipitates on agarose gels. Anal. Biochem. 127, 316–321.

285. Southern, E.M. (1975) Detection of specific sequences among DNA fragments separated by gel electrophoresis. J. Mol. Biol. 98, 503–517.

286. Towbin, H., Staehelin, T. & Gordon, J. (1979) Electrophoretic transfer of proteins from polyacrylamide gels to nitrocellulose sheets: Procedure and some applications. Proc. Natl. Acad. Sci. USA 76, 4350–4354.

287. Matsudaira, P. (1987) Sequence from picomole quantities of proteins electroblotted onto polyvinylidene difluoride membranes. J. Biol. Chem. 262, 10035–10038.

288. Thorpe, G.H.G. & Kricka, L.J. (1986) Enhanced chemiluminescence reactions catalyzed by horseradish peroxidase. Methods Enzymol. 133, 331–353.

289. Arakawa, H., Maeda, M. & Tsuji, A. (1991) Chemiluminescent assay of various enzymes using indoxyl derivatives as substrate and its applications to enzyme immunoassay and DNA probe assay. Anal. Biochem. 199, 238–242.

290. Schäfer, H.J., Scheurich, P. & Rathgeber, G.R. (1978) A simple method of activity staining for phosphate-releasing enzymes in electrophoresis gels. Hoppe-Seyler's Z. Physiol. Chem. 359, 1441–1442.

291. Gabriel, O. & Gersten, D.M. (1992) Staining for enzymatic activity after gel electrophoresis—review. Anal. Biochem. 203, 1–21.

292. Edeloch, H. (1967) Spectroscopic determination of tryptophan and tyrosine in proteins. Biochemistry 6, 1948–1954.

293. Geisow, M.J. (1992) Mass measurment at high molecular weight—new tools for biotechnologists. Trends Biotechnol. 10, 432–441.

294. Cleland, W.W. (1964) Dithiothreitol, a new protective reagent for SH groups. Biochemistry 3, 480–482.

295. de Duve, C. & Baudhuin, P. (1966) Peroxisomes (microbodies and related particles). Physiol. Rev. 46, 323–357.

296. Nomenclature Committee of the International Union of Biochemistry (1992) Enzyme nomenclature. Academic Press, New York.

297. Gold, A.M. (1967) Sulfonylation with sulfonyl halides. Methods Enzymol. 11, 706–711.

298. Umezawa, H., Aoyagi, T., Morishima, H., Matsuzak, M., Hamada, M. & Takeuchi, T. (1970) Pepstatin: A new pepsin inhibitor produced by actinomyces. J. Antibiot. 23, 259–262.

299. Kregar, I., Uhr, I., Smith, R., Umezawa, H. & Turk, V. (1977) In: Intracellular protein catabolism, vol. 2. (Turk, V., Marks, N., eds) pp 250–254. Plenum Press, New York.

300. Gekko, K. & Timasheff, S.N. (1981) Thermodynamic and kinetic examination of protein stabilization by glycerol. Biochemistry 20, 4677–4686.

301. Smith, M.B., Oakenfall, D.G. & Back, J.F. (1978) Thermal stabilization of protein by sugars and polyols. Proc. Aust. Biochem. Soc. 11, 4.

302. Illingworth, J.A. (1981) A common source of error in pH measurements. Biochem. J. 195, 259–262.

303. Laemmli, U.K. (1970) Cleavage of structural proteins during the assembly of the head of bacteriophage T4. Nature 227, 680–685.

304. Taylor, J.F., Green, A.A. & Cori, G.T. (1948) Crystalline aldolase. J. Biol. Chem. 173, 591–604.
305. Scopes, R.K. (1974) A simple procedure for crystallization of marsupial muscle phosphorylase. Proc. Aust. Biochem. Soc. 7, 6.
306. McPherson, A. (1976) Crystallization of proteins from polyethylene glycol. J. Biol. Chem. 251, 6300–6306.
307. Gernert, K.M., Smith, R. & Carter, D.C. (1988) A simple apparatus for controlling nucleation and size in protein crystal growth. Anal. Biochem. 168, 141–147.
308. Gulewicz, K., Adamaik, D. & Sprinzl, M. (1985) A new approach to the crystallization of proteins. FEBS (Fed. Eur. Biochem. Soc.) Lett. 189, 179–182.
309. Pusey, M.L. (1986) An apparatus for protein crystal growth studies. Anal. Biochem. 158, 50–54.
310. Eisenberg, D. & Hill, C.P. (1989) Protein crystallography: More surprises ahead. Trends Biochem. Sci. 14, 260–264.
311. DeLucas, L.J. & Bugg, C.E. (1987) New directions in protein crystal growth. Trends Biotechnol. 5, 188–193.
312. Day, J. & McPherson, A. (1992) Macromolecular crystal growth experiments on international microgravity laboratory. Protein Sci. 1, 1254–1268.
313. Guide to scientific products (1993) Science 259, Supplement Jan 29.

Index